AF750960

Tropical Cyclone Activity over the North Indian Ocean

M. Mohapatra • B.K. Bandyopadhyay
L.S. Rathore
Editors

Tropical Cyclone Activity over the North Indian Ocean

Springer

CP

Editors
M. Mohapatra
India Meteorological Department
New Delhi, India

B.K. Bandyopadhyay
India Meteorological Department
New Delhi, India

L.S. Rathore
India Meteorological Department
New Delhi, India

Co-published by Springer International Publishing, Cham, Switzerland, with Capital Publishing Company, New Delhi, India.

Sold and distributed in North, Central and South America by Springer, 233 Spring Street, New York 10013, USA.

In all other countries, except SAARC countries—Afghanistan, Bangladesh, Bhutan, India, Maldives, Nepal, Pakistan and Sri Lanka—sold and distributed by Springer, Haberstrasse 7, D-69126 Heidelberg, Germany.

In SAARC countries—Afghanistan, Bangladesh, Bhutan, India, Maldives, Nepal, Pakistan and Sri Lanka—printed book sold and distributed by Capital Publishing Company, 7/28, Mahaveer Street, Ansari Road, Daryaganj, New Delhi 110 002, India.

ISBN 978-3-319-40574-2 ISBN 978-3-319-40576-6 (eBook)
DOI 10.1007/978-3-319-40576-6

Library of Congress Control Number: 2016959391

© Capital Publishing Company 2017
This work is subject to copyright. All rights are reserved by the Publisher, whether the whole or part of the material is concerned, specifically the rights of translation, reprinting, reuse of illustrations, recitation, broadcasting, reproduction on microfilms or in any other physical way, and transmission or information storage and retrieval, electronic adaptation, computer software, or by similar or dissimilar methodology now known or hereafter developed.
The use of general descriptive names, registered names, trademarks, service marks, etc. in this publication does not imply, even in the absence of a specific statement, that such names are exempt from the relevant protective laws and regulations and therefore free for general use.
The publisher, the authors and the editors are safe to assume that the advice and information in this book are believed to be true and accurate at the date of publication. Neither the publisher nor the authors or the editors give a warranty, express or implied, with respect to the material contained herein or for any errors or omissions that may have been made.

Printed on acid-free paper

This Springer imprint is published by Springer Nature
The registered company is Springer International Publishing AG Switzerland

Preface

Tropical cyclones (TCs) are one of the most devastating natural disasters. The North Indian Ocean is one of the highly vulnerable regions for TC activity. This may be attributed to various factors including geographical conditions and limitations of observations, prediction systems, understanding of physical processes, early warning systems and disaster management processes, apart from the socio-economic conditions of the region. Understanding the patterns of genesis, intensity and movement of TCs and associated adverse weather like heavy rainfall assumed even more importance in recent years in the scenario of global climate change.

Because of the significance of the TCs to India, the India Meteorological Department (IMD), Ministry of Earth Sciences (MoES), Government of India, organised a National Workshop on Tropical Cyclones during 24–25 July 2014 at New Delhi, in order to shed light on the scientific basis and the complexities inherent in combating hazardous impacts of TCs. The purpose of the workshop was to advance the science of TC monitoring and prediction, particularly for nations surrounding the North Indian Ocean.

Considering the significant findings presented in the workshop by various scientists and the recommendations made in the workshop, it was decided to publish the selected papers presented during the workshop as a book after the peer review of the manuscripts. We requested several scientists who participated in the workshop for their interest in developing a volume dedicated to the science of TCs over the North Indian Ocean region. The response was overwhelming, and these authors have generously contributed to the chapters considered in this volume.

This book is relevant to TC forecasters and researchers, managers, policymakers and graduate and undergraduate students. The papers presented in the book also intend to stimulate thinking and hence further research in the field of TCs, especially over the Indian Ocean region. We have attempted to offer the recent progress on understanding and prediction of tropical cyclogenesis, intensification and movement as well as landfall processes like heavy rainfall based on latest observational and numerical weather prediction (NWP) modelling platforms. Further attempt has been made to include the TC management issues like early warning system, recent high-impact TC events, disaster preparedness, policy decisions, etc. We hope this book will provide high-quality reference materials to all the users as mentioned above and will incite further research and their applications in the management of TCs over the Indian Ocean region.

As editors of this volume, we are highly thankful to all the authors for their efforts and cooperation in bringing out this publication. We are sincerely thankful to all reviewers, viz. Dr. D.R. Pattanaik, Dr. G.C. Debnath, Dr. S. Balachandran, Ms. Suman Goyal, Dr. Kamaljit Ray, Dr. S.D. Kotal, Dr. Geeta Agnihotri, Mr. R.P. Sharma, Dr. B. Geetha and Ms. Monica Sharma of the India Meteorological Department, Dr. Someshwar Das and Dr. R. G. Ashrit of the National Centre for Medium Range Weather Forecasting and Dr. Osuri Krishna of the National Institute of Technology, Rourkela, for their continued efforts in reviewing and adding value to the manuscripts. We are grateful to IMD, MoES and all the members of the Local Organising Committee at New Delhi. We want to place our appreciation in record to the Cyclone Warning Division of the India Meteorological Department for the tireless efforts made for the organisation of the conference and significant contribution in the edition and compilation of the manuscripts and publication of this volume.

New Delhi, India

M. Mohapatra
B.K. Bandyopadhyay
L.S. Rathore

Contents

About the Editors

M. Mohapatra is the Head of Weather Forecasting Services of the India Meteorological Department and WMO-recognised Regional Specialised Meteorological Centre for Tropical Cyclones at IMD, New Delhi. His main research interests include high-impact weather events, especially tropical cyclones. He has 22 years of experience in meteorological services and research and is the author of 60 research papers published in peer-reviewed journals. He has received a number of recognitions including the 25th Biennial Mausam Award and Young Scientist Award of the Ministry of Earth Sciences (MoES), Government of India, for his research contributions in the field of atmospheric sciences.

B.K. Bandyopadhyay got his postgraduate degree in solid state physics in 1976 from the Indian Institute of Technology, Kharagpur. He joined as a Research Scholar the Indian Institute of Tropical Meteorology, Pune, and during the next three years, he was associated with research on microphysical characteristics of clouds. He joined the India Meteorological Department in 1981 and was engaged in operational weather forecasting for the past 30 years which mainly included cyclone and heavy rainfall warning services and allied meteorological research. He has made significant research contributions mainly on tropical cyclones. He has about 40 research publications in the national and international journals.

L.S. Rathore is the Director General of Meteorology, India Meteorological Department, and Permanent Representative of India to the World Meteorological Organization. He is Co-Vice Chairman of the Intergovernmental Board on Climate Services (IBCS) and former Vice President of the Commission for Agricultural Meteorology, WMO, and presently on its management board. He is former President of the Indian Meteorological Society and President of the Association of Agro-meteorologists. He made a significant contribution in setting up an integrated agro-meteorological service in India. He has 35 years of experience in meteorological services and research and has published about 100 research papers and seven books. He is recipient of the Dr. Lakhi Ram Memorial Award, 2011, constituted by the Society for recent development in agriculture. He has been conferred fellowship by the Indian Meteorological Society.

Part I
Tropical Cyclone Impact and Early Warning System

Collaborative Mechanism for Tropical Cyclone Monitoring and Prediction over North Indian Ocean

L.S. Rathore, M. Mohapatra, and B. Geetha

1 Introduction

Tropical Cyclones (TCs) are intense synoptic scale weather systems which originate over warm oceans of the world, develop into massive vortices composed of swirling winds, intense clouds and torrential rains by drawing energy from the ocean. When they move over land, they cause large scale destruction to life and property over the coastal areas of the world. India, with an extensive coastline of about 7500 km is vulnerable to the destructive features associated with landfalling TCs of the North Indian Ocean (NIO) basin comprising of the Bay of Be ngal (BOB) and the Arabian Sea (AS).

Generally, under favourable environmental conditions, a pre-existing low pressure area develops into a cyclonic disturbance (CD) (maximum sustained wind speed (MSW) of 17 knots or more) which intensifies into a TC (MSW: 34 knots or more) or a severe TC (MSW of 48 knots or more) or even a very severe TC (MSW of 64 knots or more). On an average, about 11 CDs develop over the NIO during a year including 9 and 2 over the BOB and AS (Mohapatra et al. 2014). Out of these, about five intensify into TC (4 over BOB and 1 over the AS), 3 into severe TC (2 over the BOB and 1 over the AS) and 1–2 into very severe TC. Low lying coastal belts of West Bengal, Odisha and Andhra Pradesh have borne the brunt of the fury of these very severe TCs (IMD 2002; Mohapatra et al. 2012a; Mohapatra and Sharma 2015). Adverse impact of destructive TCs in the past over Indian coasts emphasized the need for a storm warning service in the country. As an effort towards mitigating the disastrous effects of landfalling TCs, the Government of India appointed a committee in 1865 to formulate a scheme for TC warning and based on the recommendations of the committee, the first storm warning centre in

L.S. Rathore (✉) • M. Mohapatra • B. Geetha
India Meteorological Department, Mausam Bhawan, Lodi Road, 110003 New Delhi, India
e-mail: ls.rathore@imd.gov.in

© Capital Publishing Company 2017

M. Mohapatra et al. (eds.), *Tropical Cyclone Activity over the North Indian Ocean*,
DOI 10.1007/978-3-319-40576-6_1

India was established at Kolkata in 1865 (Mohapatra et al. 2012b). Subsequently, India Meteorological Department (IMD), the National Weather Service of India was established in 1875 and the important mandate of monitoring and forecasting of TCs over the NIO was vested with it. Since then, IMD is involved in detection, tracking and forecasting the movement and intensity of TCs over the NIO, which has been a highly challenging task during the early years with data sparse sea areas of BOB and AS. Further, the following few peculiarities of the NIO basin added to the challenging task:

1. Though TCs are in general seasonal phenomena, with most tropical ocean basins having maximum frequency of formation during the late summer to early autumn period, TC frequency over the NIO shows bimodal character with primary peak during the post-monsoon season (October to December) followed by the secondary peak during the pre-monsoon season (March to May). During the southwest monsoon season of June to September, intense systems usually do not develop due to northward shift of the convergence zone over the land and high vertical wind shear (Rao 1976).
2. NIO basin is a smaller oceanic basin compared to other vast oceanic basins of the globe and, it's further division into two sub-regions, viz., the BOB and the AS with intervening South Asian land mass adds to its peculiarity.
3. The average life period of a TC over the NIO is only about 3–5 days, in contrast to longer life periods over other oceanic basins, which reduces the forecast lead time for landfall forecast.
4. The unique coastal geometry as well as bathymetry and topography of the BOB further pose greater challenges to TC forecasting.

Despite these constraints, IMD has been striving hard towards reliable early warning services.

In the year 1914, an arrangement was made for receiving voluntary weather observations by wireless directly from ships at sea. Also, weather bulletins were transmitted to the ships by wireless communications since the same year. Thus, an exchange of information between the Indian Meteorological Service and Ships at sea began in 1914, and it continues even to this day. This service, which initially took the form of wireless transmission to individual ships, was soon enlarged into a regular twice-a-day broadcast service. These bulletins were broadcast from coastal radio stations to all ships. Later on NAVTEX stations of shipping authorities of Government of India were used for transmission of coastal weather bulletins (NAVTEX is an international automated direct-printing service for delivery of navigational and meteorological warnings and forecasts, as well as urgent marine safety information to ships). At present, coastal weather bulletins are being transmitted up to 465 km in the sea through NAVTEX stations along east and west coast of India as well as Andaman and Nicobar islands and Lakshadweep islands.

Around 1924–1925, IMD introduced an arrangement of issuing on loan, meteorological instruments to individual Masters of ship, so as to assist them in taking meteorological observations. Later a system of recruitment of ships for voluntary meteorological work known as Voluntary Observing Fleet (VOF) was commenced in 1946 under the auspices of International Meteorological Organisation. About 200 VOFs have been registered with IMD for taking meteorological observations over the BOB and AS region. With the advent of weather satellites in 1960s, lack of observations over the sea region has been addressed and subsequent technological advancements led to improve satellite observational systems with high resolution state-of-the-art instrumentation for microwave observations over the oceanic region.

Recently, there has been a paradigm shift in early warning services of TCs over the NIO due to various initiatives of IMD and the Ministry of Earth Sciences (MoES), Government of India including the modernisation programme of IMD (Mohapatra et al. 2013a). Under this programme, there has been upgradation of observational systems, monitoring, analysis and prediction techniques, generation and dissemination of warning products, capacity building, outreach programmes, liaison with disaster management agencies, etc. Collaboration among various institutes in national and international levels in all these aspects has also helped tremendously in improving the early warning service of TCs over the NIO.

In this paper, specific international and national collaborative efforts undertaken by IMD so far and the improved results achieved in TC monitoring and prediction over NIO are presented. Section 2 deals with current status of TC forecasting in India, Sect. 3 on the collaborative mechanisms involved in TC monitoring, forecasting and warning dissemination, Sect. 4 discusses conclusions and future scope.

2 Current Status of TC Forecasting in India

2.1 Early Warning System in TC Forecasting

TC forecasting basically deals with prediction of genesis, location/track and intensity of the TC during the next few days. Also it aims at predicting associated adverse weather like heavy rain, gale wind, high waves, storm surge and coastal inundation. The early warning component of TC disaster management includes skill in monitoring and prediction of TC, effective warning products generation and dissemination, coordination with emergency response units and the public perception about the credibility of the official predictions and warnings. IMD's early warning system is shown schematically in Fig. 1. It is important to continuously upgrade all the components of early warning based on latest technology for effective management of TCs. In this regard, it may be mentioned that, as a part of its modernisation programme, IMD has embarked upon replacement of the existing cyclone detection radars (CDRs) along the coastline by the state-of-the-art Doppler Weather Radar (DWRs) and is also constantly upgrading Numerical Weather Prediction (NWP) capabilities.

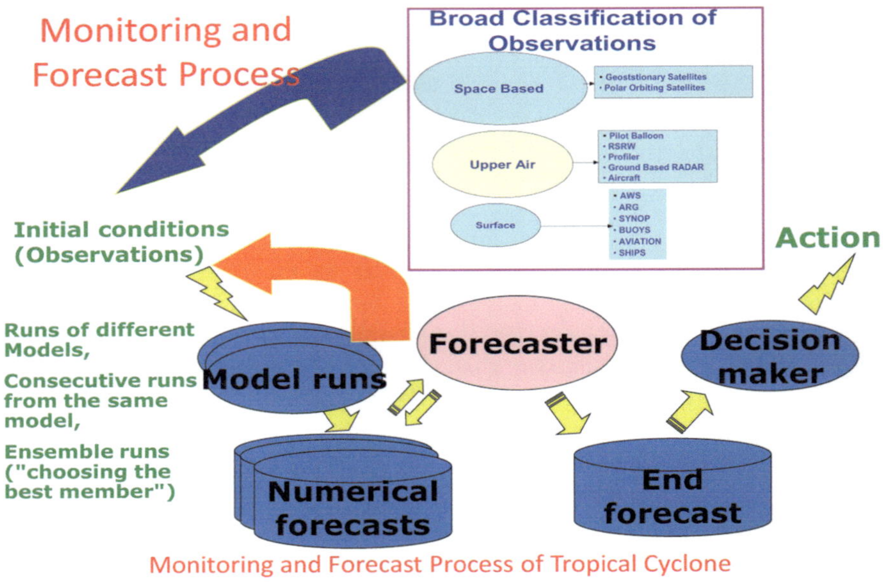

Fig. 1 Monitoring and forecasting process of tropical cyclone (Mohapatra et al. 2013a)

2.2 TC Analytical Procedure

IMD's TC analysis procedure is detailed in Standard Operating Procedure Manual (IMD 2013). The TC analysis, prediction and decision-making process are made by blending scientifically based conceptual models, dynamical and statistical models, meteorological datasets, technology and expertise. For this purpose, a decision support system (DSS) in a digital environment is used to plot and analyse different weather parameters, satellite, Radar and NWP model products.

In this hybrid system, synoptic method could be overlaid on NWP models supported by modern graphical and Geographical Interface System (GIS) applications to produce high quality analyses and forecast products, prepare past and forecast tracks upto 120 h, depict uncertainty in track forecast and to forecast wind in different sectors of TC. Also, additional help is taken from websites to collect and analyse radar data and products from IMD's radar network and neighbouring countries, satellite imageries and products from IMD and international centres and data, analysis and forecast products from various national and international centres. The automation of the process has increased the efficiency of system, visibility of IMD and utility of warning products leading to minimum loss of life (Mohapatra et al. 2013a).

2.3 TC Forecasting

Under the collaborative efforts of the World Meteorological Organisation and the United Nations' Economic and Social Commission for Asia and Pacific (WMO-ESCAP), IMD is also serving as Regional Specialised Meteorological Centre for Tropical Cyclones (RSMC, New Delhi) and extends TC forecasts to the WMO-ESCAP panel member countries in the NIO region. RSMC, New Delhi keeps a continuous watch over the BOB and the AS for monitoring and prediction of cyclogenesis. It issues Tropical Weather Outlook for the NIO region at 0600 UTC every day (based on observations at 0300 UTC of that day) for the benefit of the member countries. The outlook describes the current weather situation over BOB and AS and also provides probability of cyclogenesis during next 72 h based on NWP, synoptic, statistical and dynamical–statistical inputs. This probabilistic forecast is issued in terms of nil, low, fair, moderate and high probability corresponding to 0 %, 1–25 %, 26–50 %, 51–75 % and 76–100 % probability of occurrence.

Although, the synoptic, statistical and satellite/radar guidances help in short range forecast (upto 12/24 h), the NWP guidance is mainly used for 24–120 h track and intensity forecasts. Consensus forecasts that gather all or part of the numerical forecast, synoptic and statistical guidance are utilised to issue official forecast. IMD introduced the objective cyclone track forecast valid for next 72 h in 2009 and up to 120 h in 2013. When the low pressure system (LPS) is a depression (17–27 knot), the track forecast is issued subjectively indicating the expected direction of movement and also the probable area of landfall (if any). This forecast is issued five times a day based on 00, 03, 06, 12 and 18 UTC observations. From the stage of deep depression (28–33 knots), in case it is anticipated to intensify into a TC (34 knots or more), IMD issues TC track and intensity forecasts valid upto 120 h. This forecast is also issued five times a day based on 00, 03, 06, 12 and 18 UTC observations. When the LPS attains the intensity of a TC, TC forecast is issued eight times a day at the interval of 3 h, i.e., based on 00, 03, 06, 09, 12, 15, 18 and 21 UTC observations. The forecasts are issued about 3 h after the above mentioned observation time. The *cone of uncertainty (COU)* in the forecast track of the centre of a TC and the likely error in the forecast track based on predictive skill of past years has been introduced with effect from the TC, 'WARD' during December, 2009 (Mohapatra et al. 2012c). It is helpful to the decision makers as it indicates the standard forecast errors in the forecast. The observed track lies within the forecast COU in about 60–70 % of the cases like other Ocean basins. A typical track forecast along with the COU issued in respect of TC Phailin (08–-14 October 2013) is presented in Fig. 2. Considering the recent improvements in track forecast, the radii of circles used to construct cone of uncertainty has been reduced since 2014. The radii used for the purpose are 20, 35, 55, 85, 110, 150, 185, 220, 250, 280, 305, 335 & 360 km for forecast lead times of 00, 06, 12, 18, 24, 36, 48, 60, 72, 84, 96, 108 and 120 h.

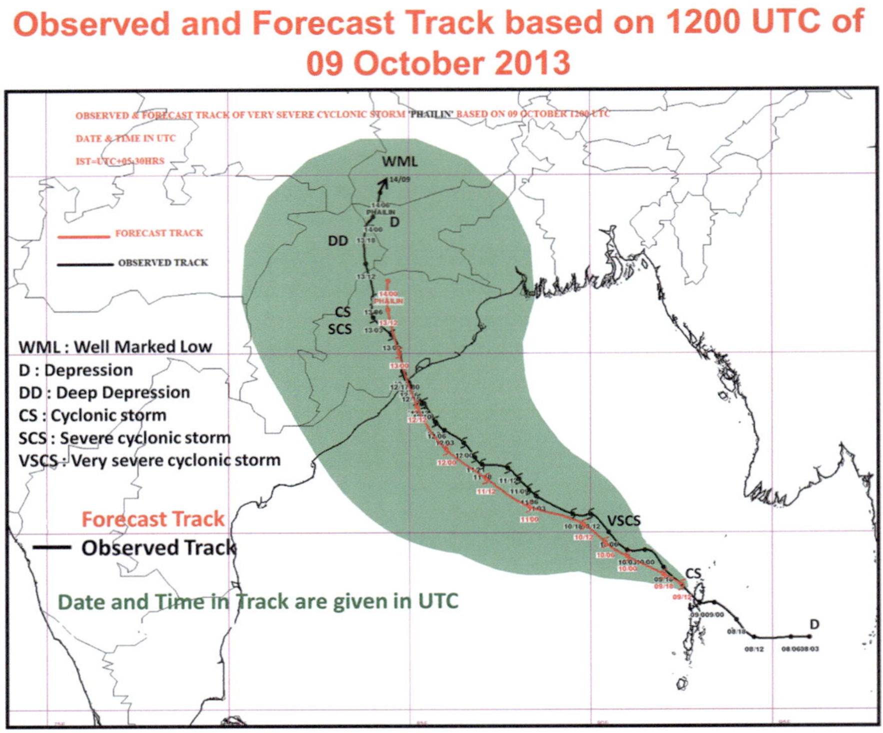

Fig. 2 An example of forecast track along with cone of uncertainty issued on 9th October 2013 in respect of VSCS Phailin (8–14 Oct 2013)

For intensity forecasting, in the satellite method, region of maximum reflectivity and mesoscale vortices are assumed to be associated with higher wind. In radar technique, the direct wind observations are available through radial velocity measurements. The wind estimates from satellite and radar and other observations are extrapolated to forecast the wind. MSW is also available from other sources like Scatterometry wind from Satellite, Buoy, Ships etc., apart from estimate by Dvorak technique. Though the wind forecasts by the NWP models are underestimated, the initial condition of wind from the model can be corrected based on actual observations and accordingly model forecast wind are modified. A statistical–dynamical model (Kotal et al. 2014) has been implemented for real time forecasting of 12 h intensity up to 72 h. For the real-time forecasting, model parameters are derived based on the forecast fields of IMD GFS model.

The cyclone wind radii representing the maximum radial extent of winds reaching 34 knots (kts), 50 (kts) and 64 (kts) in each quadrant (NW, NE, SE and SW) of TC are generated as per requirement of ships. The initial estimation and forecast of the wind radii of TC is rather subjective and strongly dependent on the data availability, climatology and analysis methods. The subjectivity and reliance on climatology is amplified in the absence of aircraft observations. However,

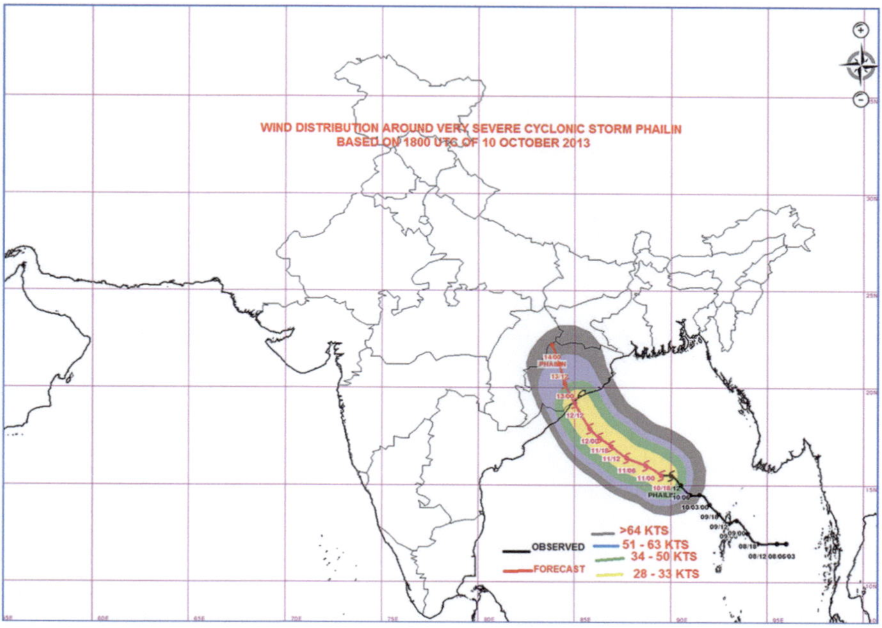

Fig. 3 An example of wind radii forecast issued in respect of VSCS Phailin (08–14 October 2013) based on 1800 UTC of 10th October 2013

recently with the advent of easily accessible remotely sensed surface and near surface winds (e.g., Ocean Sat., Special Sensor Microwave Imager (SSMI), low level atmospheric motion vectors and Advanced Microwave Sounder Unit (AMSU) retrieval methods), multi satellite surface winds, Doppler Weather Radar (DWR), coastal wind observations and advances in real time data analysis capabilities, IMD introduced TC wind radii monitoring and prediction product in October 2010 valid upto 72 h (Mohapatra and Sharma 2015). It has been extended to 120 h in 2013. The consensus forecast issued by IMD is based on numerical forecast from NWP models and synoptic & statistical guidance (IMD 2013). An example of wind radii forecast issued by IMD in respect of TC Phailin (08–14 October 2013) is shown in Fig. 3.

2.3.1 Adverse Weather Forecasting

A TC causes three types of adverse weather, viz., heavy rain, gale wind and storm surge during its landfall. Forecasting procedures for these adverse weather phenomena (IMD 2013) are presented herewith briefly.

The forecast/warning of heavy rainfall includes (i) time of commencement, (ii) duration, (iii) area of occurrence and (iv) intensity of heavy rainfall. The methods for prediction of heavy rainfall include (i) synoptic, (ii) climatological,

(iii) satellite, (iv) Radar and (v) NWP techniques. Although NWP models provide prediction of rainfall for different lead period; satellite and radar provide quantitative precipitation estimates during past 3/12/24 h. The intensity and spatial distribution of rainfall estimated by satellite and radar are extrapolated to issue forecast. In synoptic and climatology method, synoptic climatology of rainfall intensity and spatial distribution are used. The final forecast is the consensus arrived from various methods as mentioned above.

The forecast of gale wind along the coast for landfalling TCs includes (i) time of commencement, (ii) duration, (iii) area of occurrence and (iv) magnitude of gale wind. The methods for prediction of gale wind include (i) synoptic, (ii) climatological, (iii) satellite, (iv) radar, (v) NWP and (vi) dynamical statistical techniques.

Storm surge is the rise of sea water above the astronomical tide due to TC. The storm surge depends on pressure drop at centre, radius of maximum wind, point of landfall, and interaction with sea waves, astronomical tide, rainfall, river run off, bathymetry, coastal geometry, etc. The forecast of storm surge includes (i) time of commencement, (ii) duration, (iii) area of occurrence and (iv) magnitude of storm surge. The methods for prediction of storm surge and coastal inundation include (i) IMD Nomogram (Ghosh 1977), (ii) IIT Delhi Storm surge model (Dube et al. 2009) and (iii) Indian National Centre for Ocean Information Services (INCOIS), Hyderabad Storm surge and coastal inundation model (Rao et al. 2012).

2.4 TC Forecast Accuracy

Mohapatra et al. (2013b) evaluated the TC track forecast issued by IMD during 2003–2011 (9 years) by calculating the Direct Position Error (DPE) and skill in track forecast. Figure 4 shows the 24-, 48- and 72-h TC track forecast errors and skill of IMD verified against the IMD best-track dataset for TCs over the NIO during 2003–2013. The average DPE is about 124, 202 and 268 km and skill is about 36 %, 53 % and 62 %, respectively for 24, 48 and 72 h forecasts over the NIO as a whole during 2009–2013. It indicates that the error has decreased and skill has increased during the recent years. There is also similar improvement in landfall forecast of TCs (Mohapatra and Sharma 2015).

IMD's official TC intensity forecast error and skill during 2005–2013 are presented in Fig. 5. Similar to track forecast skill, TC intensity forecast skill has also improved over the years (Mohapatra et al. 2013c).

Track predictions of very severe cyclonic storm (VSCS), Phailin and Madi over BOB during 2013 are very good examples of the success stories (RSMC, New Delhi 2014). The correct track forecast, especially landfall point and time with 4 days lead period in the former case led to large scale evacuation and hence minimum loss of human lives (only 21 people died). Similarly, the correct track forecast in case of VSCS Madi which had the rarest of rare tracks led to no evacuation of coastal population leading to saving of economy.

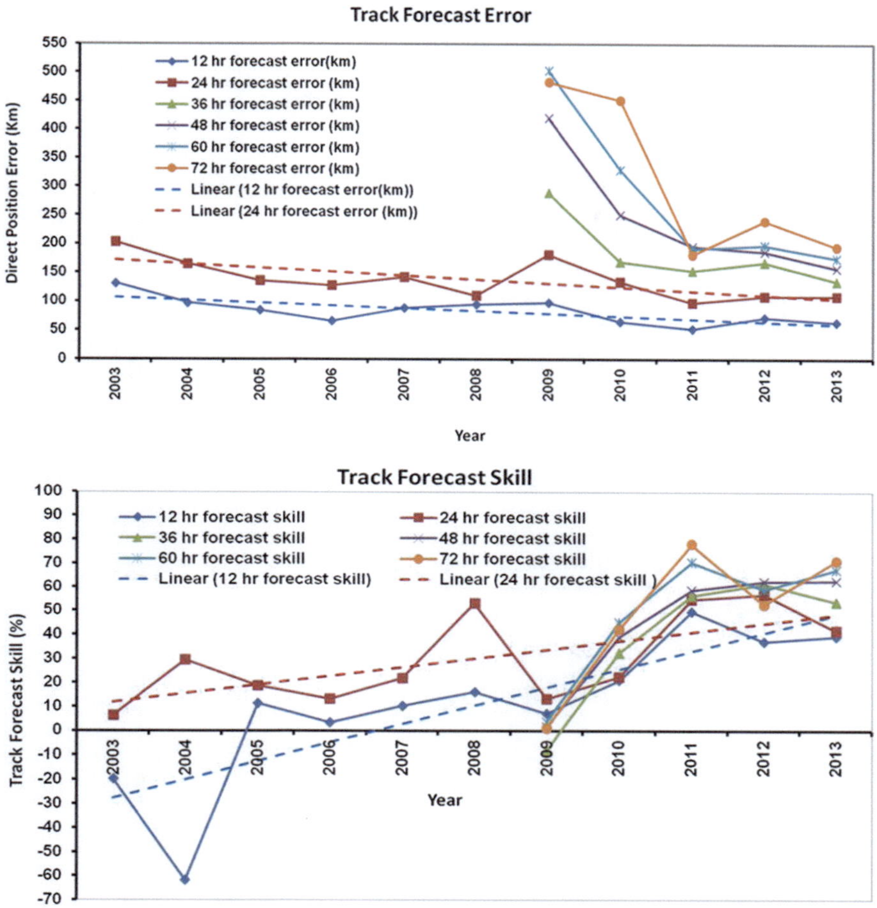

Fig. 4 IMD's Official track forecast error and skill during 2003–2013 over the North Indian Ocean

It should be mentioned here that these improvements in TC forecasting can be attributed to many factors including modernisation programme of IMD and Ministry of Earth Sciences (MoES), Government of India. Collaboration with a host of national and international agencies involved in the highly challenging task of early warning process also contributed significantly to this success. Generation of high resolution data and products, access to numerical model products, real time data exchange and early dissemination of warnings are integral parts of the collaborative mechanism. In the following section, the major collaborations are discussed to bring out the roles of the collaborating agencies towards the progress achieved in TC forecasting and warning.

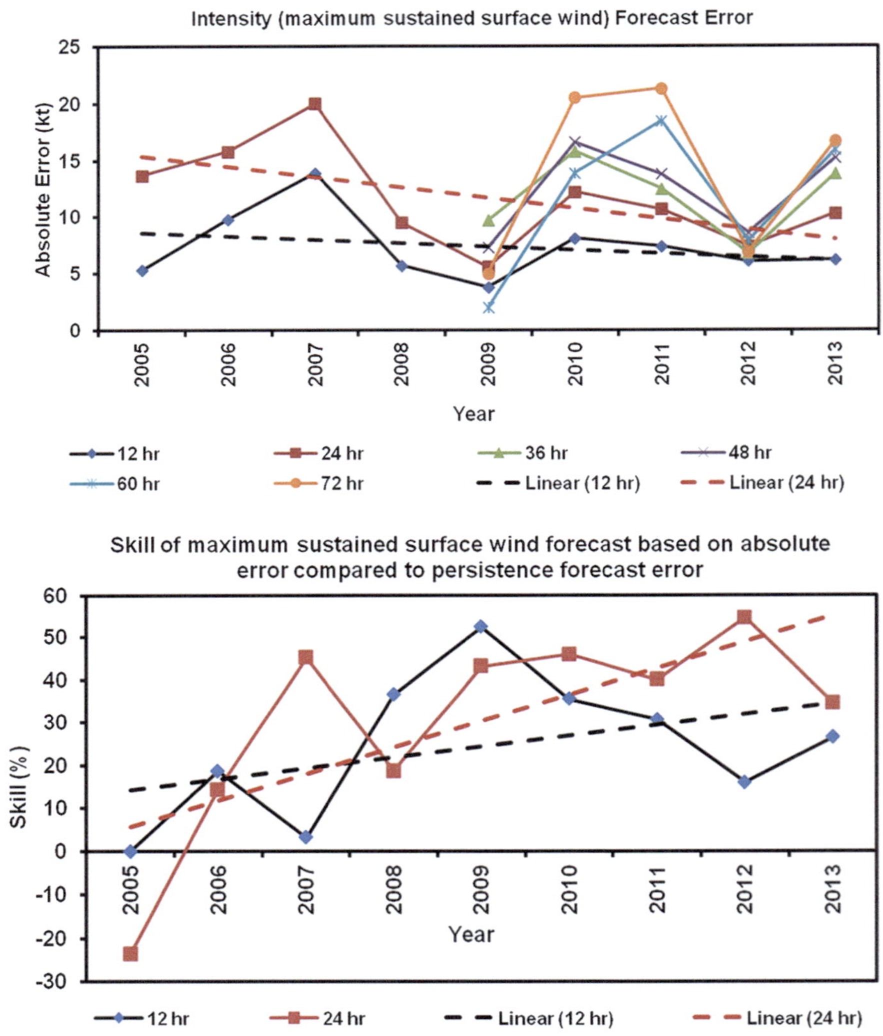

Fig. 5 IMD's Official TC intensity forecast error and skill during 2005–2013 over the North Indian Ocean

3 International and National Collaborations in Early Warning Mechanism

Commissioning of state-of-the-art observing systems throughout the country with their networking and integration, utilising them in high resolution numerical models in high performance computing facility, their visualisation, archival and dissemination to the user community in a skilful manner have resulted from several initiatives undertaken to augment the existing observational set up, telecommunication networks, analytical tools and generation of warning products as detailed below.

3.1 Observational System

Oceanic observations are vital for monitoring and prediction of TC. The NIO being a data sparse region, TC monitoring mainly depends on satellite based observations. International collaborations for satellite-based TC monitoring started with the launch of TIROS (Television and Infra Red Observational Satellite) series of polar orbiting weather satellites by the USA in 1960 (Koteswaram 1971a) when the first Automatic Picture Transmission (APT) facility, donated by the USA, was installed at Mumbai in December 1963 and real time satellite imageries were received for TC analysis and forecasting (Mohapatra et al. 2012b). Subsequently, launch of INSAT series of satellites by Indian Space Research Organisation (ISRO) since 1982 with on-board meteorological pay loads enabled availability of satellite products more frequently for continuous monitoring of TCs over the NIO. High-resolution cyclone-specific satellite imageries and products are generated using meteorological payloads on board INSAT-3A/3D and Kalpana geostationary satellites in collaboration with ISRO. Under international understanding, US National Oceanic & Atmospheric Administration (US-NOAA), Moderate Resolution Imaging Spectro-radiometer (MODIS) and EUROPEAN-Meteorological Operational (METOP) satellite data are received at IMD's three ground receiving and processing stations and the imageries and products generated are uploaded in the web and made available to the TC forecaster on real time. A sample product (NOAA-OceanSAT-2 wind observation) is presented in Fig. 6. In an Memorandum of Understanding (MoU) between the Ministry of Earth Sciences (MoES), India and NOAA, USA signed in 2008, both countries have agreed upon an open data policy to share past and current *in situ* and satellite data related to earth observations and earth sciences (www.moes.gov.in).

US National Aeronautics and Space Administration (NASA)–Japan collaborative project for rainfall measurement, viz., Tropical Rainfall Measuring Mission (TRMM) generates rain rate estimates based on sensors on-board TRMM satellites. Using this data, under IMD–National Centre for Medium Range Weather Prediction (NCMRWF) collaboration, gauge merged rainfall data at $0.5 \times 0.5°$ resolution is generated daily (Mitra et al. 2009). This product is being used for validating NWP-based TC rainfall forecasts.

Satellite-based products generated by Co-operative Institute of Meteorological Satellite Studies (CIMSS), University of Wisconsin, USA and TC specific products of Co-operative Institute for Research in Atmosphere (CIRA) at the Colorado State University, USA, US-NOAA-Atlantic Oceanographic and Meteorological Laboratory's (AOML) oceanographic products and Madden Julian Oscillation (MJO) forecasts from various global NWP centres are regularly used by TC forecasters for analysis.

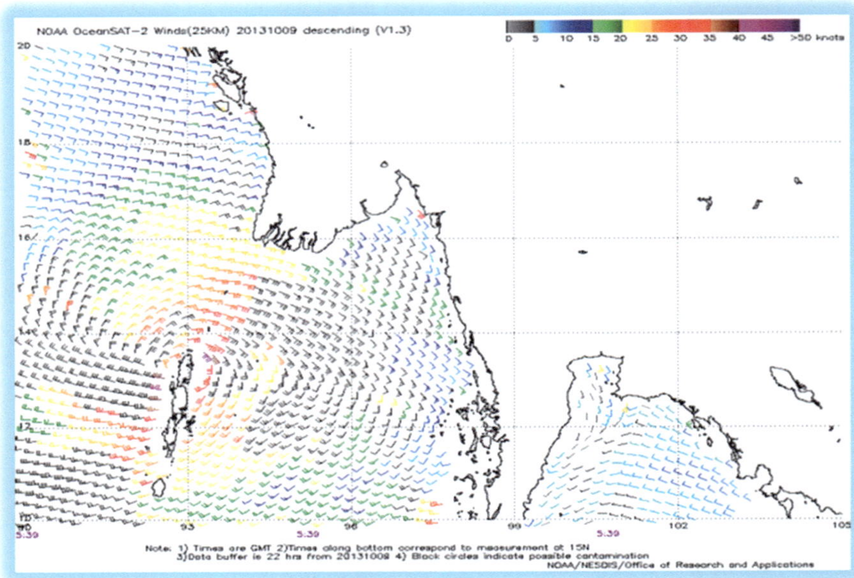

Fig. 6 A sample of NOAA-OceanSAT-2 wind observation

Under the Ocean Observation Network (OON) programme of MoES, the Ocean Observation Systems (OOS) of National Institute of Ocean Technology (NIOT), develops and maintains data buoy observational network (Fig. 7a) since 1996 which provide very vital data of surface pressure, wind, sea surface temperature and wave height over the data sparse sea region. Surface wind and mean sea level pressure data over the BOB on 27 December 2011/0300 UTC in association with the VSCS Thane (25–31 December 2011) is shown in Fig. 7b. INCOIS provides information on wave heights. These observations are crucial to study the ocean parameters as well as to validate remotely sensed observations from satellite, NWP-based products and issue of sea area and coastal weather bulletins (RSMC, New Delhi 2014). Figure 7c presents sea surface temperature and salinity variations observed from the data buoy observation during TC Phailin (08–14 October 2013).

With the introduction of high-resolution satellite-based observations and data buoy network, the services of VOF have reduced considerably during the recent years.

The coastal observations are very important to the structure characteristics of TC as well as to determine the landfall characteristics. The TCs crossing the border areas of two countries can be properly monitored only with the observations from the neighbouring countries. It may be mentioned here that observations of DWR Kolkata contribute significantly in tracking TCs moving towards Bangladesh coast. For example, special hourly observations taken at CDR Paradip and subsequent round-the-clock observations taken at DWR Kolkata provided valuable inputs in monitoring and prediction of TC Rashmi (25–27 October 2008) that crossed

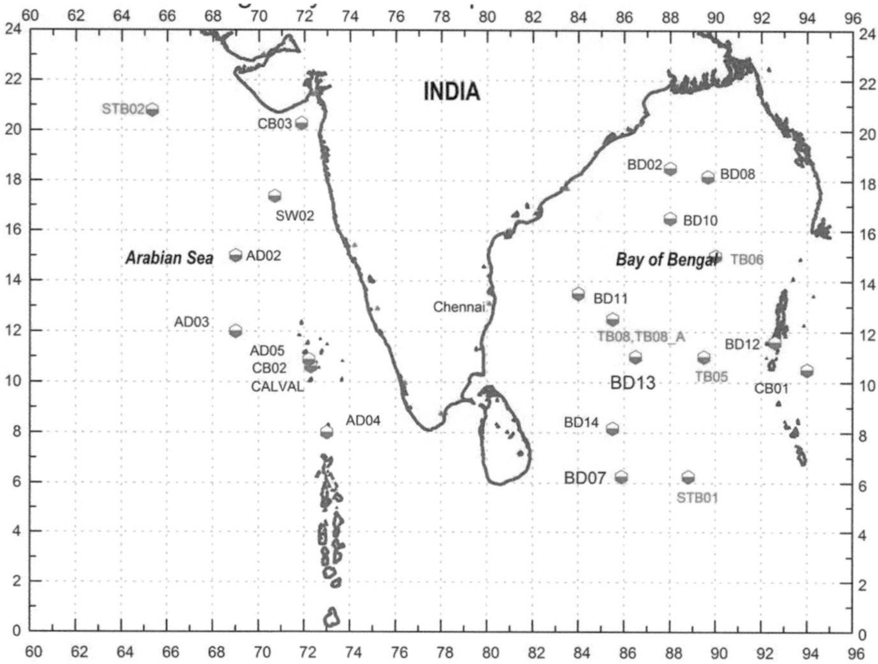

Fig. 7a NIOT's buoy network

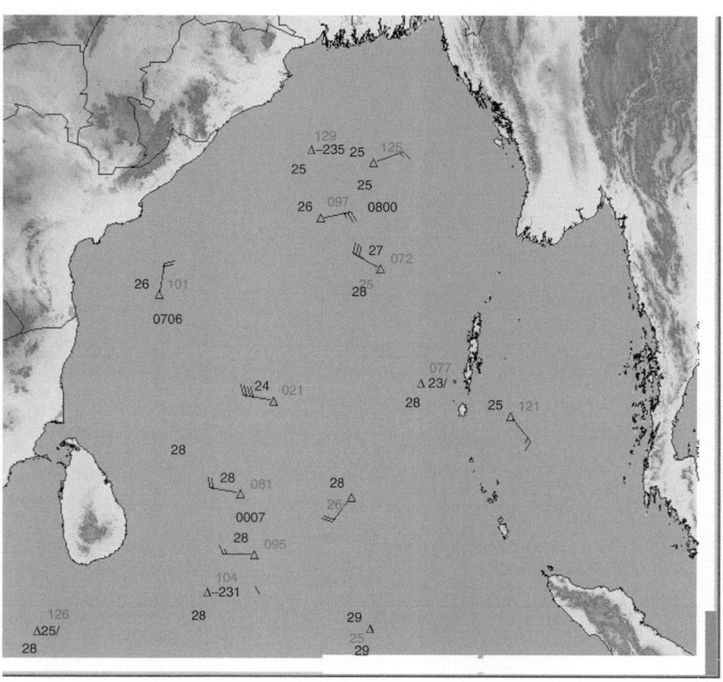

Fig. 7b Plots of ocean buoy observations on 27th December 2011/0300 UTC in association with the VSCS Thane (25–31 Dec 2011) over the BOB

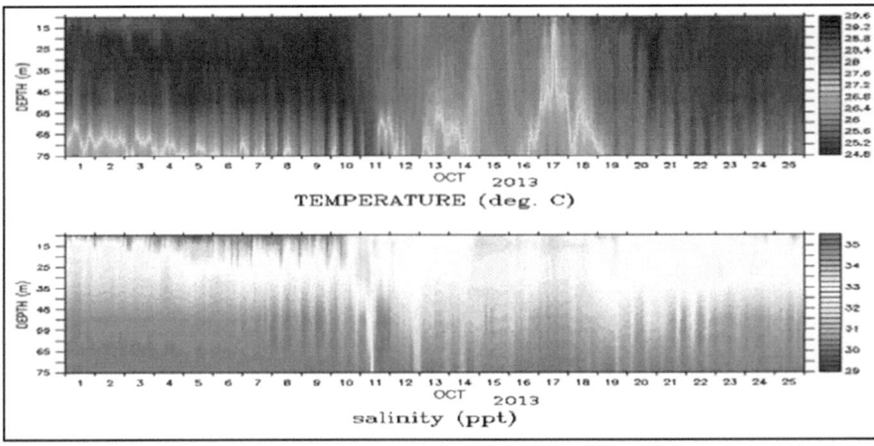

Temperature and Salinity variations during Phailin Cyclone at BD10_16.5N/88E in October 2013

Fig. 7c Temperature and salinity variation studies in respect of TC Phailin (08–14 Oct, 2013) based on buoy data (Source: National Institute of Technology (NIOT), Chennai, India)

Bangladesh coast. In this context, the collaboration achieved through WMO and WMO/ESCAP panel on TCs are noteworthy. A dense network of 675 Automatic Weather Stations (AWS) and 1300 Automatic Raingauge Stations (ARG) were planned and implemented in a phased manner.

The then existing CDRs installed along the coastal region in the 1970s are being replaced by DWRs. DWRs at Chennai, Kolkata, Visakhapatnam and Machilipatnam are of German-make. However, it should be mentioned here that under IMD–ISRO collaboration, an indigenous DWR is functioning at Sriharikota since 2004. Recently, India's BEL has installed at Mumbai and Bhuj.

3.2 *Data Exchange and Automation*

Observational data would transform into meaningful analysis and forecast products only when they are made available to the forecasting and modelling communities in real time. Hence, global data exchange is one of the key objectives of the WMO and Global Telecommunication System (GTS) is one of its chief components for near real time data exchange. World Weather Watch Program (WWWP) is an important project of WMO wherein GTS has long been its key elements. In accordance with the WMO's objectives, IMD set up directorate of telecommunications in the year 1969 to cater to the needs of national meteorological service and to strengthen the meteorological telecommunication in India. Since its inception, IMD maintains an extensive telecommunication network for speedy transmission of meteorological information over the globe. IMD serves as a Regional Telecommunication Hub (RTH) of the WMO. Initially, it was a manual or semi-automatic operational frame work requiring a large number of personnel to carry out the day-to-day operational activities. However, gradually, the communication system, techniques and protocols have undergone significant changes on adopting world-wide application of TCP/IP based services. In its present form, the existing RTH Automatic Message Switching System 'TRANSMET' is a state-of-the-art technology system which has the ability to retrieve message from e-mail and submit that message to GTS in addition to various other facilities. Under the automation project, manual decoding, plotting and analysis of limited volume of surface and upper air observations have been replaced by automatic decoding and analysis. Use of High Performance Computing System (HPCS) for running NWP models has shifted the focus from telecommunications to Information Systems and Services in IMD since 2009. The WMO Information System (WIS) is a project designed to fill gaps in the existing regional and global connectivity. Under this project, in addition to GTS, another component, DAR (Discovery, Access and Retrieval) is also integrated into all WMO programmes. With the upgradation of IMD's telecom network into Information Systems and Services, IMD serves as a Global Information System Centre (GISC) under the WIS project.

3.3 NWP Guidance

The seed for NWP was sown in India in late fifties (Das and Bose 1958), but, its further evolution commenced only in 1970s. An objective technique for TC track prediction was developed in India for the first time based on climatology and persistence (CLIPER) by Sikka and Suryanarayana (1972). Subsequently, a few analogue and simple dynamical models were developed by Indian scientists. A detailed review is available in Sikka (2006).

IMD acquired a mainframe third generation computer system IBM 360/44 with 256 KB memory in 1973 and several versions of the primitive equation model were developed in IMD and the Institute of Tropical Meteorology, Pune, presently known as Indian Institute of Tropical Meteorology (IITM) as a collaborative activity (Datta et al. 2008). Forecasting the movement of TCs over the Indian seas using a non-divergent barotropic model was attempted by Sikka (1975) from IITM. Around the same time, numerical prediction of storm surge associated with TCs over BOB was also attempted under joint efforts of IMD and Indian Institute of Technology (IIT) New Delhi (Das 1972, 1994; Das et al. 1974). In the year 1979, Centre for Atmospheric Sciences (CAS) was set up at IIT New Delhi with IMD as a co-sponsor. Since then, CAS, IIT New Delhi has been actively involved in developing improved models for storm surge prediction in the Indian region under collaboration with Florida State University (FSU), USA (Dube et al. 1994, 2009).

Under INDO-US Science and Technology Initiative in 1985, the then latest NWP models and data assimilation systems capable of ingesting all synoptic and asynoptic data including remote sensing data received on GTS in large numbers, developed in National Meteorological Centre, Washington, FSU and other research institutions of USA were implemented in IMD (Datta et al. 2008). Further, development of a Limited Area Model (LAM) for the Indian region was started at IIT-Delhi in collaboration with the Naval Research Laboratory (NRL), North Carolina State University, USA and the Indian Air Force in mid-eighties (Mohanty et al. 1989; Tyagi et al. 1994). In the mean time, a Centre for Atmospheric and Oceanic Sciences (CAOS) was established at Indian Institute of Science (IISc), Bangalore which was another major development for studies on atmosphere and ocean models in India (Das 2008).

A major boost to NWP in India occurred with the establishment of NCMRWF in 1988 and subsequent procurement of first supercomputer CRAY-XMP-14. Initially, a R40 model of NCEP, USA was installed at NCMRWF with the help of Centre for Ocean Land and Atmosphere (COLA), USA. Simultaneously, the global model T79 of ECMWF was tested for its suitability over the Indian region. However, a major problem of assimilating data from the GTS into the model was faced. Subsequently, an agreement was signed with National Centre for Environmental Prediction (NCEP), USA to obtain its Global Forecasting System (GFS) including T80 model, data decoder and a Global Data Assimilation System (GDAS) based on Spectral Statistical Interpolation (SSI) scheme based on which routine 3-day forecasts were commenced at NCMRWF from June 1994 (Das 2008).

Around the same time, a semi-implicit semi-Lagrangian multi layer primitive equation Limited Area Model (LAM) adopted from Florida State University (Krishnamurti et al. 1990) was installed and used in IMD for TC track prediction up to 48 h since 1994 after installing CYBER-2000U computer system (Tyagi et al. 2008). Here, a synthetic analytical vortex based on satellite observations was used to define the storm circulation. Subsequently, a specialized Tropical Cyclone model known as Quasi-Lagrangian Model (QLM) was adopted for track and intensity prediction of TCs upto 72 h from 1998 after Mathur (1991).

With the advancements in internet technology and USA's open data access policy, the NCEP–GFS model products were made available in the public domain in the early twenty-first century. The model has evolved from the earlier NCEP models of 1980s and 1990s. In its present version, the GFS model (T574) is a spectral model with an approximate horizontal resolution of 13 km for the first 10 days and 27 km from 240 to 384 h (16 days). In the vertical, the model is divided into 64 layers and temporally, it produces forecast output every hour for the first 12 h, 3 hourly through day 10 and 12 hourly through day 16.

Under the modernisation programme of IMD, a High Performance Computing System (HPCS) was commissioned in December 2009 and US-NCEP based Global Forecast System (GFS T574/L64) was made operational using Global Statistical Interpolation (GSI) scheme as the global data assimilation method. The model is run twice a day (based on 00 and 12 UTC initial conditions). NWP based objective forecast products are prepared to support cyclone warning service. Initially, forecasts were generated up to 72-h lead time during the period 2009–2012 and subsequently since 2013, 120-h forecast guidance is provided to operational forecasters. In addition to this, the regional Advanced Weather Research and Forecasting (WRF) Model with 3 dimensional variational (3D VAR) data assimilation is being operated twice a day, at 27, 9, and 3 km horizontal resolutions for the forecast up to 3 days using initial and boundary conditions from the IMD GFS-T574/L64 (horizontal resolution over the tropics ~22 km).

Further, Hurricane WRF (HWRF) (ver 3.2+) is also installed in IMD and commissioned since 2013 in collaboration with NCEP, USA for tracking TCs over the NIO. The model which was run initially at 27×9 km resolution is now being run at 9×3 km resolution over a domain of $80 \times 80^{\circ}$ with a moving nest of $6 \times 6^{\circ}$ centred at the vortex centre. The model has special features such as vortex initialisation, coupled with Ocean model to take into account the changes in SST during the model integration, tracker and diagnostic software to provide the graphic and text information on track and intensity prediction for real-time operational requirement.

Using the NCEP based IMD-GFS model products, a dynamical–statistical genesis potential parameter (GPP), for the NIO basin has been developed (Kotal and Bhattacharya 2013) as the product of four variables, viz., vorticity at lower levels, middle tropospheric relative humidity, middle tropospheric instability, and the vertical wind shear. The GPP is used for predicting cyclogenesis at their early development stages. The grid point analysis and forecast of the genesis parameter up to 7 days are generated on real time for operational forecasting guidance.

NCMRWF's GFS (NGFS) is also based on NCEP-GFS and is presently run at a horizontal resolution of about 23 km (T574) with 64 vertical levels. The NGFS replaces the model vortex with a synthetic vortex developed using the TC vital data provided by IMD (based on satellite and synoptic data) and generates forecast guidance on real time. Under a joint collaboration with United Kingdom Meteorological Office (UKMO) signed in 2008, UKMO's Unified Model (MO) global forecast suite (version 7.4) was implemented in NCMRWF in April 2010 with 4D-VAR data assimilation scheme (www.ncmrwf.gov.in/). The model run (NCUM) generates 168-h deterministic forecasts based on 00 UTC analysis every day and the products are available in the web for the TC forecaster. Presently, the model is upgraded to version 7.6 and is run at resolution T512L70 since June 2011. Further, probabilistic forecasting technique based on NCEP-Global Ensemble Forecast System (GEFS), has also been introduced at NCMRWF (NGEFS). The NGEFS runs at resolution T574L64 and generates forecasts up to 10 days.

In addition, NWP guidance from many global model products such as US-NCEP, European Centre for Medium Range Weather Forecast (ECMWF), Japan Meteorological Agency (JMA) and the cloud-resolving version of National Oceanic and Atmospheric Administration (NOAA)-HWRF are now readily available to forecasters. Aside from the individual model guidances, IMD has also implemented Multi Model Ensemble (MME) prediction using five global models – IMD GFS T574, ECMWF T799, JMA T899, UKMO and NCEP GFS.

Under the joint project of World Weather Research Program (WWRP) and Tropical Cyclone Program (TCP) of the WMO-ESCAP, a guidance of tropical cyclone forecasts based on the THORPEX (THe Observing system Research and Predictability EXperiment, a World Weather Research Program under WMO) Interactive Grand Global Ensemble (TIGGE) in near real-time for the WMO/ESCAP and Typhoon Committee Members and forecasters participating in the Severe Weather Forecasting Demonstration Project (SWFDP) in Southeast Asia was implemented in IMD in 2012 with the support of JMA (http://tparc.mri-jma.go.jp/cyclone/). The products generated include deterministic and ensemble TC track forecasts, Strike Probability Maps, Strike probability for cities within the range of 120 km (RSMC, New Delhi 2014). A sample of deterministic forecast of track of TC Phailin during October 2013 by various models along with the observed track is presented in Fig. 8. A sample probabilistic forecast based on TIGGE-Ensemble Prediction Scheme (EPS) for landfall of TC Phailin is presented in Fig. 9.

IIT-Delhi and INCOIS-Hyderabad run storm surge prediction models have been implemented by RSMC, New Delhi to provide storm surge guidance to disaster managers on real time. Storm surge guidance based on IIT-Delhi model is being provided to WMO/ESCAP panel countries since April 2009. A sample storm surge guidance by IIT-Delhi and INCOIS in association with landfall of TC Phailin (08–14 October 2013) is presented in Fig. 10.

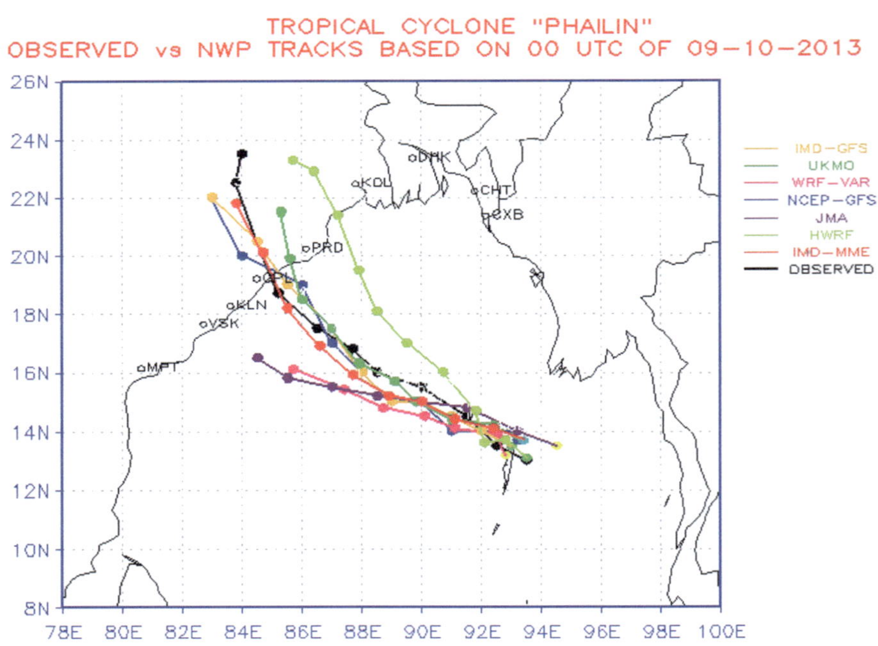

Fig. 8 A sample deterministic track forecast of TC Phailin (8–14 October 2013) by NWP models based on 09 October 2013/0000 UTC initial conditions

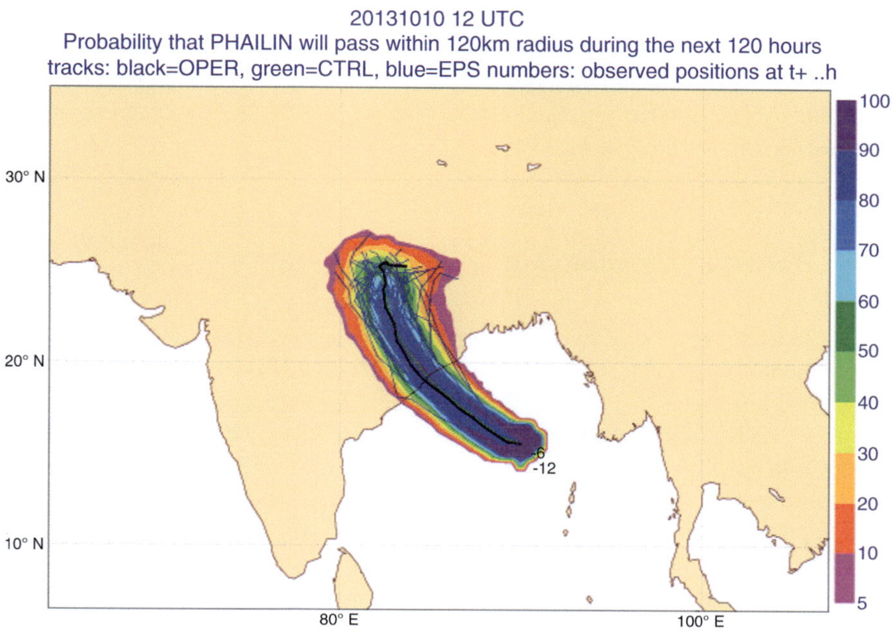

Fig. 9 A sample probabilistic TIGGE -Ensemble Prediction guidance for landfall of TC Phailin (08–14 October 2013) based on 10th October 2013/1200 UTC initial conditions

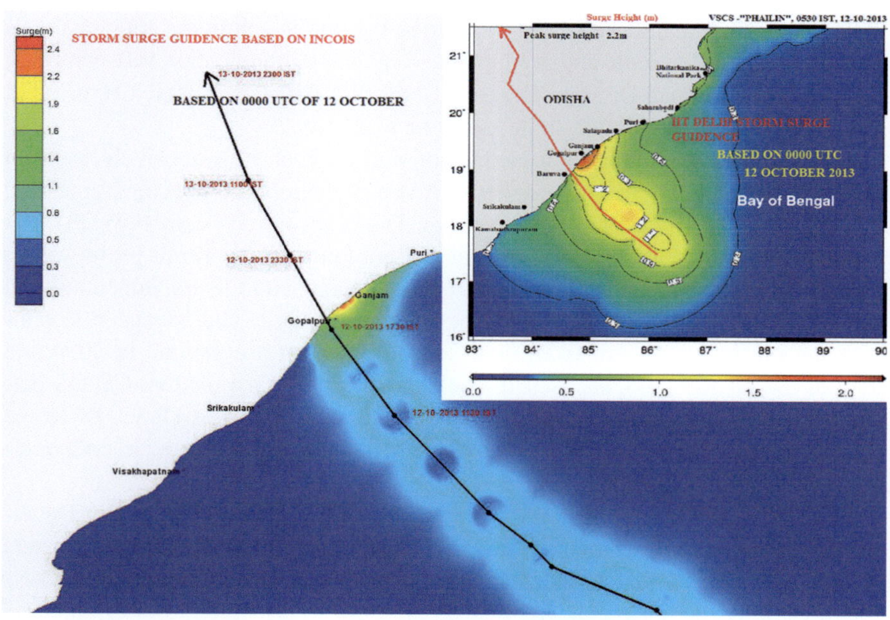

Fig. 10 Sample storm surge guidance by IIT-Delhi and INCOIS in association with landfall of TC Phailin (08–14 October 2013)

Collaboration between IMD and INCOIS-Hyderabad has resulted in implementation of coastal inundation forecast since October 2013. Thus, it is evident that our international and national collaborators for NWP guidance, which is presently the backbone of TC forecasting, have played a vital role in IMD's progress in TC forecasting and warning processes.

3.4 Warning Dissemination

For early dissemination of cyclone warnings, wireless telegraph (W/T) facility of Department of Posts and Telegraph was used in the past. However, the all important early dissemination of warnings to the public came by way of collaboration with the All India Radio as radio was considered to be the most dependable medium of speedy dissemination during the 1970s (Koteswaram 1971b) and subsequently with the Doordarshan. Cyclone warnings are communicated to ships navigating in the NIO region through the port authorities also. On receipt of IMD's cyclone warnings, the port authorities hoist the requisite danger signal so that the ships would avoid the danger areas and take alternate routes. Police wireless communication network, microwave telecommunication links of Railways and Aeronautical Fixed Telecom Network (AFTN) have also been used widely when the situation warranted. Currently, IMD's bulletins and warnings pertaining to TC forecasting are disseminated to the State and Central government civil administrators and the national disaster management authorities (NDMA) by fax, email, SMS and phone without any time delay. This is in addition to the public communication channels such as internet, electronic and print media. Regular communication between IMD, NDMA and National Disaster Response Force (NDRF) is a vital exercise for disaster mitigation within the short lead time available.

For warning the public, radio, television and internet are extensively used. In order to disseminate warnings under extreme conditions of telecommunication failure, a satellite based Cyclone Warning Dissemination System (CWDS) was implemented in late 1980s along coastal Andhra Pradesh and Tamil Nadu was an indigenous effort by ISRO. The success of the scheme led to its implementation in other cyclone prone areas also. With the advent of internet and advanced technological facilities, IMD embarked upon a modernisation programme in 2009 and state-of-the-art technologies used in cyclone warning dissemination. CWDS network is being replaced by Digital CWDS. Also satellite based Direct-To-Home (DTH) dissemination of cyclone warnings is also being implemented in collaboration with Doordarshan.

Further, bulk SMS are generated using a tool developed by Centre for Development of Advanced Computing (CDAC) and disseminated to registered users through INCOIS warning system and using the system of dissemination of farmer's bulletin through Kisan portal. Regular press briefings are given to electronic and print media and press release indicating the salient features of the TC is also made as and when the situation demands. In the case of TC *Hudhud* during 08–14 October

2014, 4 press conferences were held and 2 press releases were made. Hourly updates on the cyclone situation were issued to the press on 12th October, the day of landfall. Such interactions are tremendously improving the role of media in disseminating TC warnings.

For the benefit of WMO/ESCAP member countries in the NIO region, RSMC, New Delhi, has launched a dedicated website (www.rsmcnewdelhi.imd.gov.in) for TCs over NIO to communicate the TC warnings without any time delay. Further, RSMC bulletins are disseminated to the meteorological departments of the member countries through GTS also.

IMD's quantitative precipitation forecasts (QPF) for river catchments/sub-catchments serve as vital inputs to the Central Water Commission (CWC) to monitor and mitigate flooding along river beds.

3.5 Research and Development

Recognising the importance of research and development for improving TC forecasting, IMD has been undertaking R&D projects on TCs over NIO. IMD scientists carry out research pertaining to TCs of NIO and publish their papers in various national and international journals. IMD also started a quarterly journal in 1950, which is presently named *Mausam*, wherein several research results on various meteorological aspects are published. IMD is involved in joint research projects with academic institutions. IMD scientists are nominated for M.Tech programmes by CAS, IIT New Delhi. IMD and CAS, IIT New Delhi are jointly working on storm surge guidance since 1979. Further, based on Prof. Koteswaram's Cyclone Distress Mitigation Committee recommendations, IMD established a dedicated TC research unit, Cyclone Warning Research Centre (CWRC) in 1972 which is functioning from RMC Chennai. Software which presents tracks and statistics of cyclones and depressions over NIO from the year 1891 onwards, namely, *Cyclone eAtlas*, developed at this centre, is presently available freely in the web. It is frequently used by TC forecasters, civil administrators, disaster managers and research communities at large.

Recognising the importance of scientific and technical cooperation in Earth Observations and Earth Sciences a Memorandum of Understanding (MoU) was signed on the Earth sciences and Observations by the Ministry of Earth Sciences, and NOAA in April 2008. The MoU is an enabling mechanism for undertaking joint activities between India and the US to use the combined scientific and technical skills in enhancing the observations of the Earth and use the information most effectively for the benefit of the society. Tropical Cyclone research is one of the implementation agreements under this MoU. The MoU has been further renewed in 2013 for next 5 years.

3.5.1 Forecast Demonstration Project

Under INDO-US joint collaboration on TC research, a multi-institutional campaign on TC forecasting, viz., Forecast Demonstration Project for TCs over BOB, aimed to demonstrate improvement in forecasting of cyclogenesis, intensification and movement of TCs over NIO with enhanced observations over data sparse oceanic region by way of satellite based high-resolution microwave products, Sagar Kanya cruise observations, buoy network data, coastal AWS data etc. is conducted every year since 2008 during the post monsoon cyclone season. As per the science plan of this project, continuous monitoring of environmental conditions is undertaken for analysing and forecasting cyclogenesis and intense observation periods are declared when a depression forms over the BOB for taking additional observations. The data are assimilated in NWP models for improved NWP guidance and TC forecasting. The data collected during the campaign is archived and used for TC research.

3.6 Capacity Building

Several MoUs are in vogue for sharing of knowledge and expertise in the field of earth system sciences (www.moes.gov.in). Following are the noteworthy international collaborations:

MoUs signed with NCEP, USA and UKMO for implementation of NWP models in IMD and NCMRWF have contributed immensely towards capacity building in the field of NWP in India. An agreement has been signed by the MoES, India on the cooperation for Regional Integrated Multi-Hazard Early Warning System for Afro-Asian region (RIMES) (functioning from Thailand) in February 2011. An MoU for cooperation of weather climate and geophysics sciences and early warning of coastal hazards has been signed between MoES, India and Agency for Meteorology, Climatology and Geophysics, Indonesia in January 2011. An MoU for cooperation in the field of earth sciences and services has been signed between MoES, India and Korean Meteorological Administration in September 2010.

In an effort towards sharing of knowledge and expertise, RSMC New Delhi also conducts training programmes for meteorologists of WMO/ESCAP member countries every year.

4 Conclusions and Future Scope

Progress achieved so far is no reason for any complacency on our part towards our strife for improved TC forecasts and disaster risk reduction. Despite overall improvements in TC track and intensity prediction, there have been lapses too and there is further scope for improvement. Prediction of unusual TCs is still challenging especially, the recurving TCs, rapidly or slowly moving TCs, stationary TCs, rapidly intensifying or weakening TCs over the sea. The track prediction of TC, Viyaru over the BOB during 10–16 May 2013, which crossed Bangladesh coast had large landfall time error, as it crossed coast well before the predicted time (RSMC, New Delhi 2014). It was mainly due to rapid movement (about 40–50 kmph) of the TC 12 h prior to landfall. Also, intensity forecast of TCs Lehar (23–28 November 2013) and Nilofar (25–31 October 2014) had large errors, since, their rapid weakening over the sea could not be predicted well in advance. Aside from improving our understanding and knowledge of TCs of NIO, these issues also point towards need for a collaborative, co-ordinated mechanism for accurate and early warning of TCs over the NIO. Our thrust areas now include location specific heavy rainfall prediction, coastal inundation modelling and forecasting.

An important collaborative effort being undertaken presently is to include the ocean component in the IMD-HWRF run. The Ocean Model and Ocean Coupler require customization of Ocean Model for Indian Seas. IMD is expecting to implement the Ocean coupling in collaboration with INCOIS, Hyderabad.

Scarcity of observational data near the TC centre, leads to poor initial condition in NWP models. It further leads to increase in forecast error of NWP models, which are the backbone for official track forecast of TCs. Hence, crucial data collection from the TC field, need to be undertaken by aircraft probing. Scientific and technical know-how for this challenging task need to be tapped from international expert communities through collaborative efforts.

Acknowledgements The authors thank the officials of Cyclone Warning Division, IMD New Delhi for their help in data collection and preparation of the manuscript.

References

Das, P.K. and Bose, B.L. (1958). Numerical prediction of the movement of Bay depressions. *IJMG*, **9**, 225–232.

Das, P.K. (1972). A prediction model of storm surges in the Bay of Bengal. *Nature*, **239**, 211–213.

Das, P.K. (1994). Prediction of storm surges in the Bay of Bengal. *Proceedings of the Indian National Science Academy*, **60**, 513–533.

Das, P.K, Sinha, M.C. and Balasubramanyan, V. (1974). Storm surges in the Bay of Bengal. *Quarterly Journal of the Royal Meteorological Society*, **100**, 437–449.

Das, S. (2008). Impressions about the development of NWP in India. *In*: Tyagi, A and Pattanaik, D.R. (eds.) Five Decades of Numerical Weather Prediction in India. IMD New Delhi, India.

Dube, S.K, Jain, I, Rao, A.D. and Murthy, T.S. (2009). Storm surge modelling for the Bay of Bengal and Arabian Sea. *Natural Hazards*, **51**, 3–27.

Dube, S.K, Rao, A.D, Sinha, P.C. and Chittibabu, P. (1994). A real time storm surge prediction system: An application to east coast of India. *Proceedings of the Indian National Science Academy*, **60**, 157–170.

Datta, R.K, Sinha, M.C, Chhabra, B.M. and Prasad, K. (2008). Five decades of numerical weather prediction in India Meteorological Department: Early days and evolution. *In*: Tyagi, A. and

Pattanaik, D.R. (eds.), Five Decades of Numerical Weather Prediction in India. IMD New Delhi, India.
Ghosh, S.K. (1977). Prediction of storm surges on the coast of India. *Indian Journal of Meteorology and Geophysics*, **28**, 157–168.
India Met. Dept. (IMD) (2002). Damage potential of Tropical Cyclones. IMD Pune, India.
India Met. Dept. (IMD) (2013). Cyclone Warning in India – Standard Operation Procedure, IMD New Delhi, India.
Kotal, S.D. and Bhattacharya, S.K. (2013). Tropical cyclone Genesis Potential Parameter (GPP) and it's application over the north Indian Sea, *Mausam*, **64**, 149–170.
Kotal, S.D, Battacharya, S.K, Bhowmik, S.K.R. and Kundu, P.K. (2014). Development of NWP-based cyclone prediction system for improving cyclone forecast service in the country. *In:* Kamaljit Ray, Mohapatra, M, Bandyopadhyay, B.K. and Rathore, L.S. (eds.) High Impact Weather Events over the SAARC Region, Springer, Netherlands and Capital Publishing Co., New Delhi, India, pp. 113–129.
Koteswaram, P. (1971a). A decade of satellite meteorology in India, *Indian Journal of Meteorology and Geophysics*, **22**, 273–278.
Koteswaram, P. (1971b). Reports of the Cyclone Distress Mitigation Committee – Andhra Pradesh and Orissa, Govt. of India.
Krishnamurti, T.N, Kumar, A, Yap, K.S, Dastoor, A.P, Davidson, N. and Sheng, J. (1990). Performance of a high resolution mesoscale tropical prediction model, *Advances in Geophysics*, **32**, Academic Press Inc., 133–286.
Mathur, M.B. (1991). The national Meteorological center's Quasi-Lagrangian model for hurricane prediction. *Monthly Weather Review*, **109**, 1419–1447.
Mitra, A.K, Bohra, A.K, Rajeevan, M.N. and Krishnamurti, T.N. (2009). Daily Indian Precipitation Analysis from a Merge of Rain-gauge data wutg TRMM TMPA Satellite-derived rainfall estimates. *Journal of the Meteorological Society of Japan*, **87A**, 265–279.
Mohanty, U.C, Paliwal, R.K, Tyagi, A. and John, A. (1989). Evaluation of a multi-level primitive equations limited area model for short range prediction over Indian region. *Mausam*, **40**, 34–42.
Mohapatra, M, Mandal, G.S, Bandyopadhyay, B.K, Tyagi, A. and Mohanty, U.C. (2012a). Classification of cyclone hazard prone districts of India. *Natural Hazards*, **63** (**3**), 1601–1620.
Mohapatra, M, Bandyopadhyay, B.K. and Tyagi, A. (2012b). Best track parameters of tropical cyclones over the North Indian Ocean: A review. *Nature Hazards*, **63** (**3**), 1285–1317.
Mohapatra, M, Nayak, D.P. and Bandyopadhyay, B.K. (2012c). Evaluation of cone of uncertainty in tropical cyclone track forecast over north Indian ocean issued by India Meteorological Department. *Tropical Cyclone Research and Review*, **2**, 331–339.
Mohapatra, M, Sikka, D.R, Bandyopadhyay, B.K. and Tyagi, A. (2013a). Outcomes and challenges of Forecast Demonstration Project (FDP) on landfalling cyclones over the Bay of Bengal, *Mausam*, **64**, 1–12.
Mohapatra, M, Nayak, D.P, Sharma, R.P. and Bandyopadhyay, B.K. (2013b). Evaluation of official tropical cyclone track forecast over north Indian Ocean issued by India Meteorological Department. *Journal of Earth System Science*, **122**, 589–601.
Mohapatra, M, Nayak, D.P, Sharma, R.P. and Bandyopadhyay, B.K. (2013c). Evaluation of Official Tropical Cyclone intensity forecast over north Indian Ocean Issued by India Meteorological Department. *Natural Hazards*, **68**, 433–451.
Mohapatra, M, Bandyopadhyay, B.K. and Tyagi, A. (2014). Construction and quality of best track parameters for study of climate change impact on tropical cyclones over the north Indian Ocean during satellite era. *In*: Mohanty, U.C, Mohapatra, M, Singh, O.P, Bandyopadhyay, B.K. and Rathore, L.S. (eds.) Monitoring and Prediction of Tropical Cyclones in the Indian Ocean and Climate Change, Springer, Netherlands and Capital Publishing Co., New Delhi, India, pp. 3–17.
Mohapatra, M. and Sharma, M. (2015). Some characteristics of surface wind structure of tropical cyclones over the north Indian Ocean. *Journal of Earth System Science*, **124**, 1573–1598.

Rao, Y.P. (1976). Southwest Monsoon, IMD, Met. Monograph No. Synoptic Meteorology – 1/76.
Rao, A.D, Murty, P.L.N, Jain, I, Kankara, R.S, Dube, S.K. and Murty, T.S. (2012). Simulation of water levels and extent of coastal inundation due to a cyclonic storm along the east coast of India. *Natural Hazards*, doi: 10.1007/s11069-012-0193-6.
RSMC, New Delhi (2014). Report on cyclonic disturbances over the north Indian Ocean during 2013, published by RSMC, IMD, New Delhi
Sikka, D.R. (1975). Forecasting the movement of tropical cyclones in the Indian seas by non-divergent barotropic model. *Indian Journal of Meteorology and Geophysics*, **26**, 323–325.
Sikka, D.R. and Suryanarayana, R. (1972). Verification of forecasts of storms and depressions using computer oriented technique based on climatology and persistence. *Indian Journal of Meteorology and Geophysics*, **23**, 247–257.
Sikka, D.R. (2006). Major advances in understanding and prediction of tropical cyclones over north Indian Ocean – a perspective. *Mausam*, **57**, 165–196.
Tyagi, A, Mohanty, U.C. and Ramesh, K.J. (1994). Planetary boundary layer structure in the monsoon trough region. *Mausam*, **45**, 213–222.
Tyagi, A, Hatwar, H.R, Mukhopadhyay, B, Pattanaik, D.R, Bhowmik, S.R.K, Rama Rao, Y.V. and Pai, D.S. (2008). Numerical Weather Prediction at IMD: Present status, Ongoing activities and Future plans. *In*: Tyagi, A. and Pattanaik, D.R. (eds.), Five Decades of Numerical Weather Prediction in India. IMD New Delhi, India.

Hydro-Meteorological Aspects of Tropical Cyclone Phailin in Bay of Bengal in 2013 and the Assessment of Rice Inundation due to Flooding

Kamaljit Rays, M. Mohapatra, K. Chakravarthy, S.S. Ray, S.K. Singh, A.K. Das, B.A.M. Kannan, and B.K. Bandyopadhyay

1 Introduction

Tropical Cyclones (TCs), one of the most destructive of all the natural disasters, are capable of causing loss of life and extensive damage to property. The Bay of Bengal is a potentially energetic region for the development of cyclonic storms and approximately 7 % of the global annual tropical storms form over this region with two cyclone seasons in a year (Gray 1968). Much of the TC related damage is attributed to storm surges, high winds, damage associated with strong thunderstorm complexes and TC-induced heavy rainfall. Predicting rainfall associated with TCs is a major operational challenge. Over the last few decades flooding from TCs at landfall has become a threat to human lives in India. Although track-forecasts continue to improve, quantitative precipitation forecasts (QPF) for TCs have shown little skill. One of the uncertainties in QPF is a lack of precipitation data over the open oceans to evaluate and validate numerical weather prediction (NWP) model results. TC rainfall forecasting techniques are lagging behind those of the track forecast. However, significant progress has been made in recent years due to the advance in remote sensing observations and the improvement of mesoscale models and data assimilation techniques. Until relatively recently, TC rainfall prediction was carried out mainly using empirical methods and subjective experience on the part of the forecaster. However, advanced techniques for Quantitative Precipitation Estimate (QPE) are currently employed in operational applications in some major forecasting centres, which already have greatly improved the

K. Rays (✉) • M. Mohapatra • K. Chakravarthy • A.K. Das • B.A.M. Kannan
B.K. Bandyopadhyay
India Meteorological Department, Mausam Bhawan, Lodi Road, 110003 New Delhi, India
e-mail: kamaljit_ray@rediffmail.com

S.S. Ray • S.K. Singh
Mahalanobis National Crop Forecast Centre, Pusa Campus, 110012 New Delhi, India

© Capital Publishing Company 2017

M. Mohapatra et al. (eds.), *Tropical Cyclone Activity over the North Indian Ocean*,
DOI 10.1007/978-3-319-40576-6_2

forecasting for TC-related rainfall. Minakshi Devi et al. (2014) have shown predicted tracks of a few cyclonic events such as SIDR (Nov, 2007), Aila (May, 2009) and Laila (May, 2010) along with their contribution to precipitation in the NE India. Recent studies have indicated that some high resolution dynamical model simulations are capable of capturing the rainfall pattern of TCs. Lee and Choi (2010) investigated the torrential rainfall associated with Typhoon Rusa in South Korea in 2002 through numerical simulation using Weather Research Forecast (WRF) model. Haggag and Yamashita (2009) studied the hydro-meteorological features of TC Gonu using coupled atmosphere, ocean and land surface modelling with an atmospheric component based on the MM5 model. There are other studies on performance of models including, Raju et al. (2012), Osuri et al. (2012), Abhilash et al. (2012), Routray et al. (2013) and Srivastava et al. (2011). All these studies indicate that the high resolution along with the improved data assimilation, especially DWR and satellite based data can improve the rainfall forecast by the models. Though heavy rainfall prediction is still a challenge, Hurricane WRF (HWRF) is a promising model for this purpose. A number of studies have examined the track, intensity, structure and genesis of TCs, however very few studies have considered the rainfall dynamics associated with TCs. The Hydro-meteorological aspects of TC have been dealt mainly in terms of coastal storm surge and inundation studies only.

Recent TC Phailin was the most intense cyclone that crossed India coast on 12 October 2013 after Odisha Super Cyclone of 29 October 1999. It affected around 12 million people. The cyclone prompted India's biggest evacuation in 23 years with more than 5,50,000 people moved up from the coastline in Odisha and Andhra Pradesh to safer place. At the time of landfall on 12 October, maximum sustained surface wind speed in association with the cyclone was about 115 knots (215 kmph) and estimated central pressure was 940 hPa. It caused very heavy to extremely heavy rainfall over Odisha leading to floods. Maximum rainfall occurred over northeast sector of the system centre at the time of landfall and 36 h thereafter. Maximum 24 h cumulative point rainfall was 38 cm in Odisha. The NWP and dynamical statistical models provided good guidance with respect to its genesis, track and intensity. However, there was spatiotemporal displacement and large intensity error in the forecast rainfall with respect to actual rainfall.

The main objective of this paper is to investigate the heavy rainfall caused by TC Phailin. In this paper, we have analysed the rainfall associated with TC Phailin before and after landfall. We also use the observed manual data after the landfall and also buoy and satellite merged data for sea area to arrive at the behavior of rainfall during the Cyclone period. In this study, we use HWRF model to study the rainfall predicted daily for 24 and 48 h along the track of cyclone and its comparison with the observed rainfall.

2 TC Phailin Track

The very severe cyclonic storm (VSCS) Phailin originated from a remnant cyclonic circulation from the South China Sea. The cyclonic circulation lay as a low pressure are over Tenasserim coast on 6 October 2013. It lay over North Andaman Sea as a

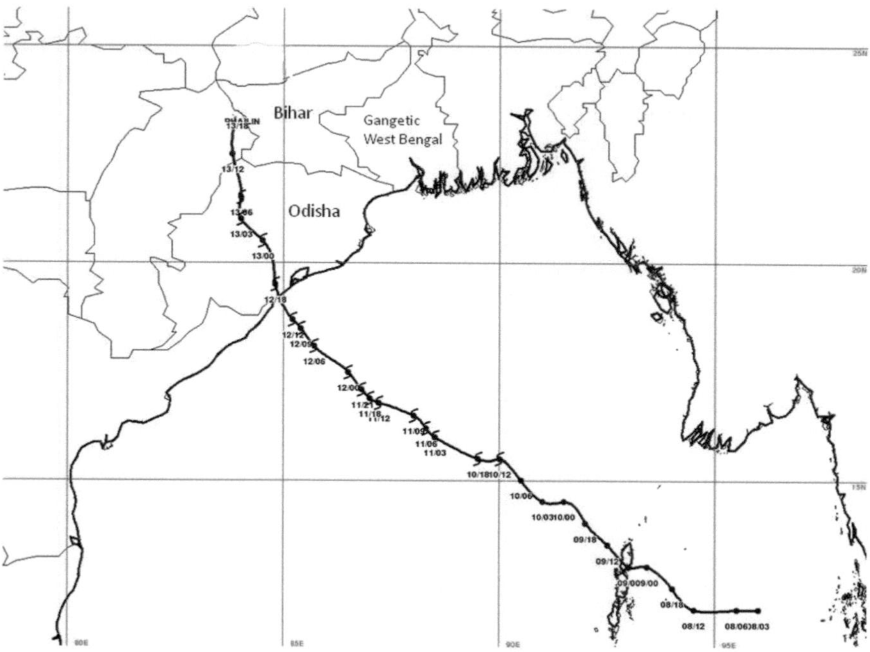

Fig. 1 Track of VSCS Phailin in Bay of Bengal (North Indian Ocean) (8–13 October 2013)

well marked low pressure area on 7 October. It concentrated into a depression over the same region on 8 October near latitude 12.0°N and longitude 96.0°E. Moving west-northwestwards, it intensified into a deep depression on 9 morning and further into cyclonic storm (CS), 'Phailin' in the same day evening. Moving northwestwards, it further intensified into a severe cyclonic storm (SCS) in the morning and into a VSCS in the forenoon of 10 October over east central Bay of Bengal. The VSCS, Phailin crossed Odisha and adjoining North Andhra Pradesh coast near Gopalpur (Odisha) around 2230 h IST of 12 October 2013 with a sustained maximum surface wind speed of 200–210 kmph gusting to 220 kmph. The track of the TC is shown in Fig. 1.

3 Rainfall due to TC, Phailin

3.1 Rainfall Monitoring Process of IMD

At the genesis stage, the rainfall was monitored mainly with satellite observations, supported by meteorological buoys, coastal and island observations. As the system entered into the east central Bay of Bengal moving away from Andaman and Nicobar Islands, it was mainly monitored by satellite observations, supported by buoys. On 12 October, when the system lay within the radar range, the DWR at Visakhapatnam was utilised and continuous monitoring by this radar started from 0100 UTC of 12 October when the system was at approximately 310 km east-southeast of Visakhapatnam coast and continued till 1800 UTC of that date. In addition to the observations from satellite and Radar, data from conventional observatories and Automatic Weather Stations (AWS) were used.

India Meteorological Department (IMD) has a synoptic observatory network of 552 stations, 2868 raingauges under Daily Rainfall Monitoring Scheme and 675 Automatic Weather Stations which record rainfall. In addition State Governments are maintaining over 3540 raingauge stations whose data are made available to IMD from Indian Railways, Forest and Agriculture Departments and other organisations which maintain about 5039 non-reporting raingauge station to meet their specific needs and also to provide data to IMD.

3.2 Rainfall Estimates Based on Point Raingauge Stations

The VSCS Phailin caused very heavy (13–24 cm/day) to extremely heavy rainfall ($\geq$25 cm/day) over Andaman and Nicobar Islands and Odisha and isolated ($<$25 % of area), heavy (7–12 cm) to very heavy rainfall over adjoining eastern India. The daily precipitation on 13 and 14 October averaged for the state as a whole was 92 % of the monthly rainfall of October and 71 % of the post-monsoon seasonal (October–November–December) rainfall for Odisha state.

Maximum 24 h cumulative rainfall of 380 mm was reported by an automatic raingauge in Banki in Cuttack district of Odisha. The highest 24 h cumulative average rainfall of 187 mm was recorded in Nayagarh district on 13 October (ending at 0830 h IST on 13 October) and highest 48 h cumulative rainfall of 232 mm was recorded in Mayurbhanj district on 14 October (ending at 0830 h IST of 14 October 2013). The districts in the north of Odisha (Balasore, Bhadrak and Mayurbhanj) recorded as much as 300–400 mm and other parts of Odisha received between 200 and 300 mm as Phailin made landfall on the night of 12 October 2013. Figure 2 depicts the 24 h cumulative rainfall as on 13 October (ending 0300 UTC) averaged districtwise over the Odisha state. Hourly rainfall data of the self recording rain gauge (SRRG) at Gopalpur recorded observations until 1600 UTC after which data is not available due to blowing away of the instrument due to the cyclone. Till 1600 UTC, it recorded maximum hourly rainfall of 17 mm at 1430 h IST. The maximum hourly rainfall (80 mm), directly attributable to the cyclone occurred at Rairangangpur station of Mayurbhanj district in Odisha at 0300 h IST of 13 October (5 h after the landfall). The isohyetal analysis of 24 h accumulated rainfall as on 13 October, 0300 UTC based on automatic weather and rain gauge stations is shown in Fig. 3. Maximum rainfall occurred over northeast sector of the system centre at the time of landfall. Maximum 24 h cumulative rainfall of 38 cm was reported over Banki in Cuttack district of Odisha.

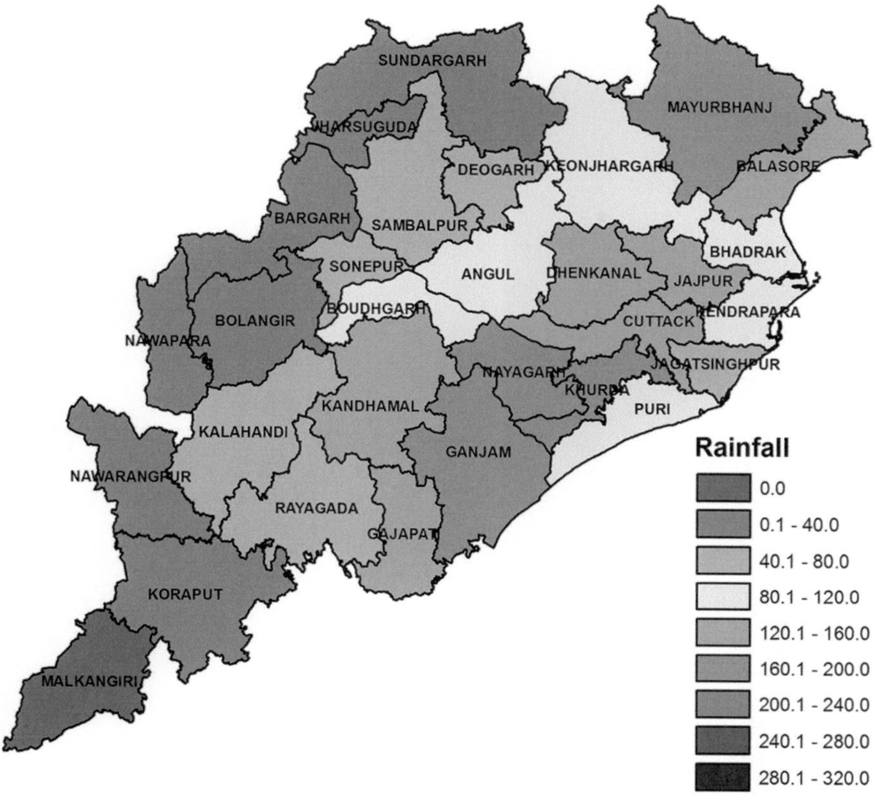

Fig. 2 24-h accumulated, districtwise rainfall (in mm) over Odisha based on rain gauge observations recorded as on 13 October 2013, 0300 UTC

3.3 QPE Based on Merged Data Set of Raingauge and Satellite

IMD has also developed a new high spatial resolution ($0.25 \times 0.25°$) daily IMD-NCMRWF merged TRMM TMPA 3B42 Satellite derived rainfall dataset which includes Land and North Indian Oceanic regions (Mitra et al. 2009). An example of this product for TC Phailin is shown in Fig. 4.

The daily Indian precipitation analysis formed from a merge of IMD rain-gauge data with the TRMM TMPA satellite-derived rainfall estimates showed Cyclone Phailin's heaviest rainfall ($\geq$800 mm) occurred over open waters of east central Bay of Bengal. The merged data was used to create a map of rainfall generated by Cyclone Phailin as it marched through Bay of Bengal from 8 October to 12 October. Phailin gave maximum cumulative rainfall of around 740 mm in the Southwest sector of the track over east-central Bay of Bengal near Andaman Islands, when it intensified from a deep depression into a cyclonic storm. It was seen that the rainfall

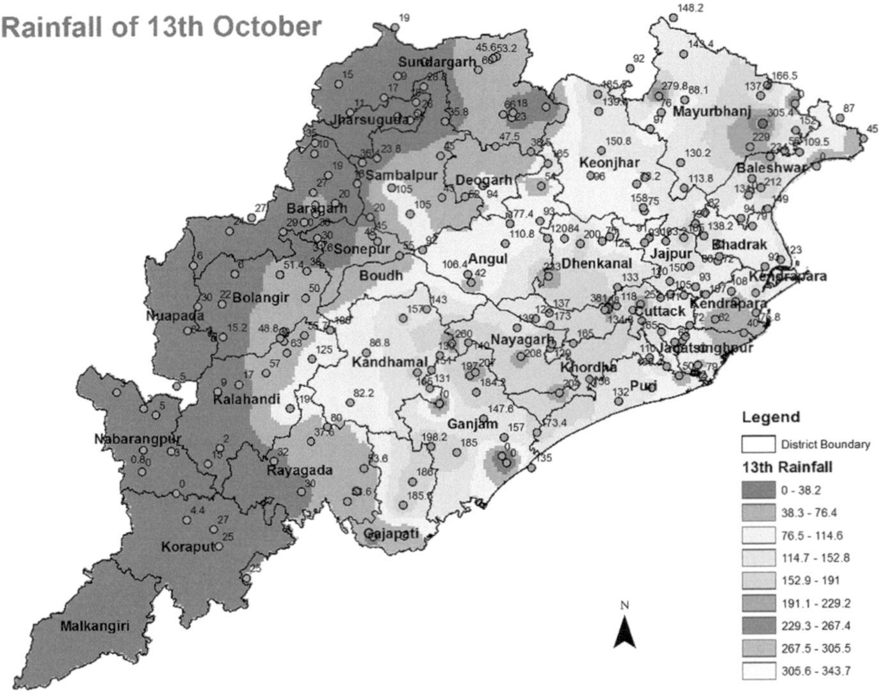

Fig. 3 Isohyetal analysis of 24 h accumulated rainfall (in mm) over Odisha from 12 October/0300 UTC to 13 October/0300 UTC based on various surface rainfall observations

was generally confined up to 400 kms south–southwest and 200 km north–northeast of the storm centre till the TC was in the sea. As it approaches coast, the rainfall sector shifts to within 400 km N and NE of the storm centre. The area of heaviest rainfall was within 200–300 kms in the SW sector from the storm centre and within 100 km in the N–NE sector from the storm centre.

3.4 Radar Based QPE

Radar Images from Vishakhapatnam are shown in Fig. 5a. The eye of the cyclone was beyond 250 km range of the DWR and thus the maximum rainfall zone was not captured. How-so-ever at 13:31 UTC of 12 October a narrow rainband, very well developed in the outer wall cloud region of the Cyclone near Srikakulam station of Vishakhapatnam extended east west across the Andhra Pradesh Coast. The rainband was stationary and separate from the spiral rainfall system of TC and gave 10 cm of rainfall to that district of Andhra Pradesh along with thunder squalls, giving an impression of Cyclone crossing, few hours before the actual landfall (Figs. 5a and 5b). At 1551 UTC, the rainfall intensity was maximum to the

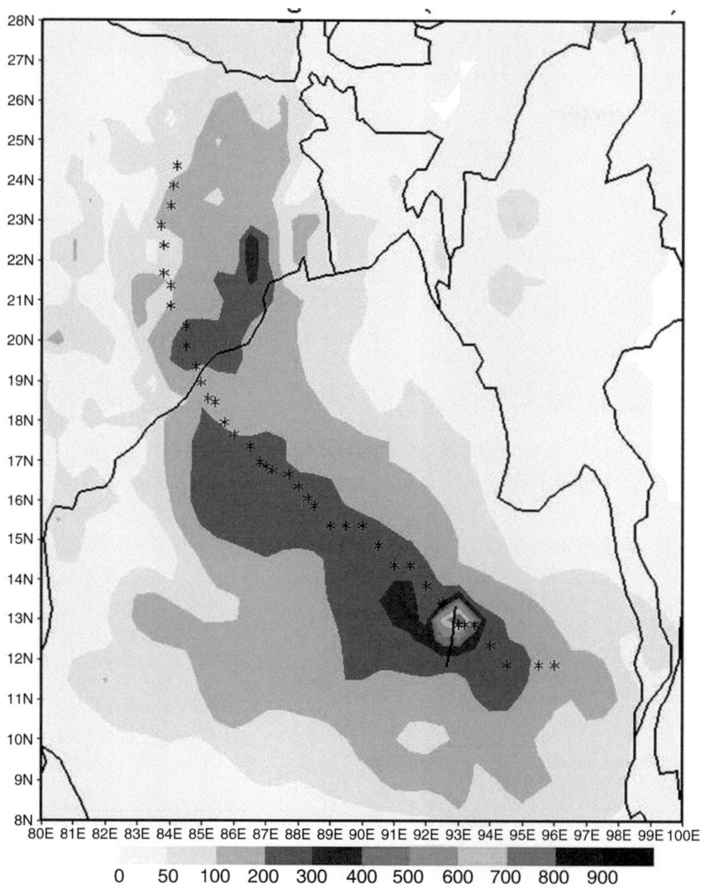

Fig. 4 Total accumulated rainfall during the life period of TC Phailin (08 October to 14 October 2013) based on IMD-NCMRWF's satellite (TRMM TMPA 3B42) – gauge merged rainfall (in mm) data. Dots indicate IMD's best track positions of TC Phailin

Southeast to Southwest sector of the eye. The rainfall intensity had decreased considerably in the southern periphery of the TC by 2000 UTC as the maximum rainfall intensity had shifted north–eastwards.

4 Synoptic Environment Related to Rainfall due to Phailin

We analyse the NCEP reanalysis data to investigate the synoptic environment associated with TC Phailin. Figure 6 shows the wind vectors at 850 hPa at 0000 UTC of 12 October and 13 October. As the TC was approaching the coast, the moist Southerly to SE'ly winds with wind speed more than 18 m/s from Bay of Bengal were directing massive moisture flux to coastal Odisha, Bihar and adjoining

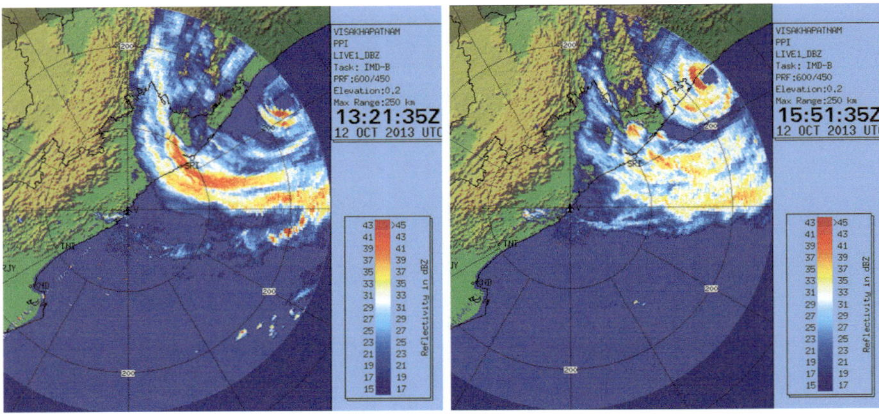

Fig. 5a Reflectivity of Cyclone Phailin before and during landfall as seen from nearest Doppler weather Radar at Visakhapatnam

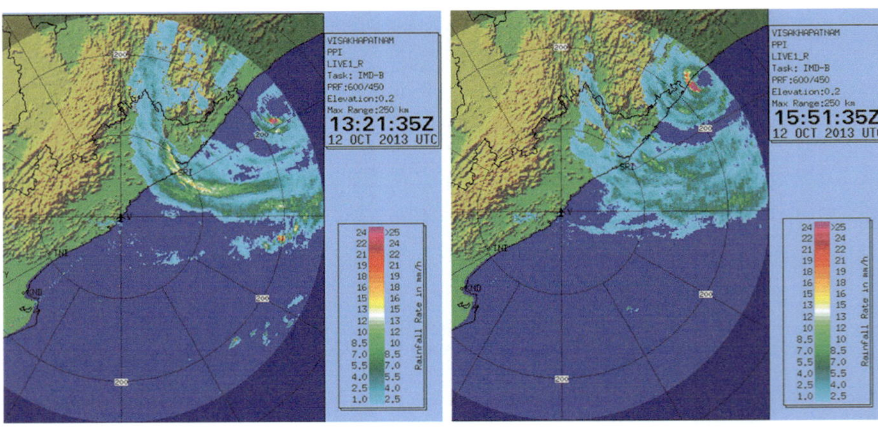

Fig. 5b Surface rainfall intensity of Cyclone Phailin before and during landfall as seen from nearest Doppler weather Radar at Vishkhapatnam

Gangetic West Bengal. The low level southerly flow brought in massive moisture convergence in the NE sector of the TC, which led to very heavy rains over Bihar Jharkhand and adjoining West Bengal (Eastern India). Figure 7 depicts the vertical velocity (Omega) on 12 October and 13 October. The upward motion of the air is supported by the windward lifting of the southerly flow in association with TC.

It is observed that orography influenced the rainfall distribution and intensity in addition to the factors as discussed above. The east coast of India and the neighbouring countries surrounding the Bay of Bengal have diverse features which include mountainous ranges, plains and deltas. A study by Debnath and Mandal (2012) on the orographic effect on rainfall shows that orographic effect is very much prominent in producing heavy rainfall in association with passage of a

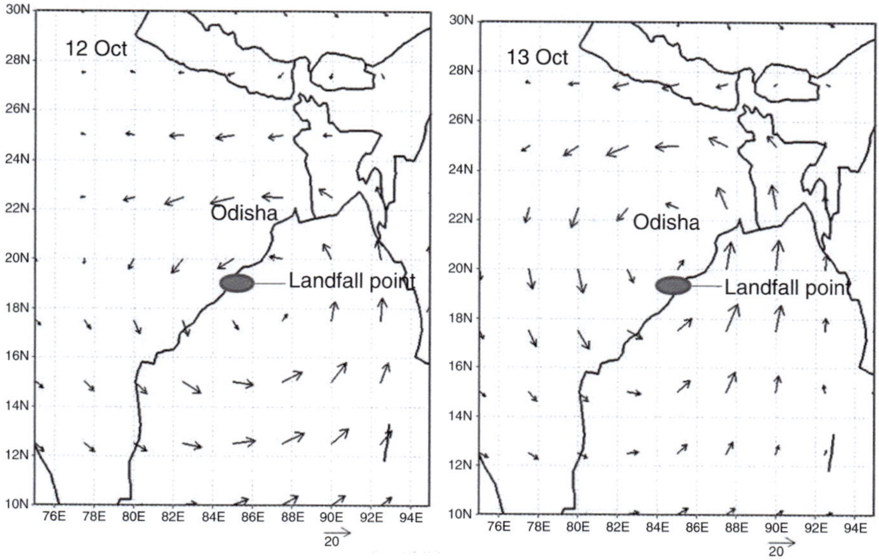

Fig. 6 850 mb mean vector wind (m/s) on 12 October and 13 October 2013

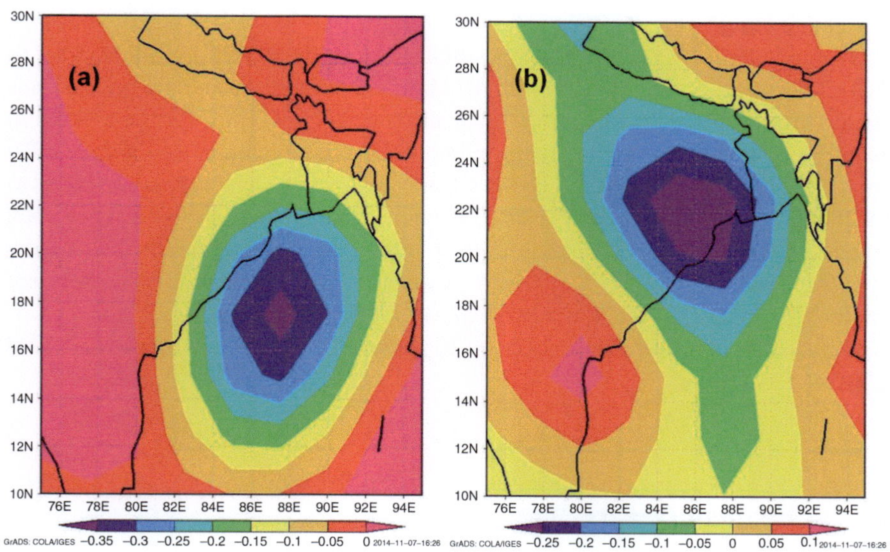

Fig. 7 500 mb mean vertical velocity (Pa/s) on (**a**) 12 October and (**b**) 13 October 2013

TC. In case of Phailin, the rainfall at the time of landfall was also influenced by orography as the interior part of north Odisha is a plateau region with a few hill peaks in Balasore and Mayurbhanj district, which helped in orographic lifting of moist air from north coastal Odisha and hence higher rainfall over north interior Odisha.

5 HWRF Model Performance

Recently, under Indo-US joint collaborative program, IMD adopted Hurricane Weather Research and Forecast system (HWRF) model for TC track and intensity forecast for North Indian Ocean (NIO) region for its operational requirements. The basic version of the model HWRF (V3.2+) which was operational at Environmental Monitoring Centre (EMC), National Centre for Environmental Prediction (NCEP), USA was ported in IMD's IBM P-6/575 machine with nested domain of 27 and 9 km horizontal resolution and 42 vertical levels with outer domain covering the area of $80 \times 80°$ and inner domain $6 \times 6°$ with centre of the system adjusted to the centre of the observed cyclonic storm. The outer domain covers most of the North Indian Ocean including the Arabian Sea and Bay of Bengal and the inner domain mainly covering the cyclonic vortex moving along the movement of the system. The model has special features such as vortex initialisation coupled with ocean model to take into account the changes in SST during the model integration and diagnostic software to provide the graphic and text information on track, intensity and rainfall. The operational version of the model is run incorporating vortex re-location and moving nesting procedure on real time twice a day based on 00 and 12 UTC initial conditions to provide 6-h track, intensity and rainfall forecasts valid up to 120 h. The model uses IMD GFS-T574L64 analysis/forecast as initial and boundary conditions.

Figure 8 shows the error in 24 h Rainfall forecast by HWRF from 9 October to 12 October 2013, in comparison with the observed rainfall. The model forecast was on the higher side with an average error of more than 300 mm for the maximum rainfall. Table 1 gives the maximum rainfall observed in comparison with HWRF forecast. The maximum rainfall observed on 9 October was 168 mm as compared to the model forecast of 647 mm (error of 447 mm). The model indicated a lag in track forecast by 12–24 h (average land fall point error was 144 km) and thus a location error of around 100 km for the daily maximum rainfall. The model also could not capture the rainfall in the northeast sector of the TC after crossing. The error was more than 100 mm over Odisha state on 12 October for 24 h forecast based on 0000 UTC of 12 October.

6 Assessment of Rice Inundation Area

The heavy rainfall and the storm surge of 2 m along the coast associated with the Phailin, led to flooding in coastal Odisha districts. The water could not recede back due to swollen sea and thus led to damage of standing rice crop. Rice being the major crop during this period in Odisha, an attempt was made to assess the Rice crop inundation due to flood using microwave remote sensing data from Indian SAR (Synthetic Aperture Radar) satellite RISAT-1. This satellite provides SAR data in C band at 25 m resolution (in MRS mode) and 24-day repetivity (Misra et al. 2013).

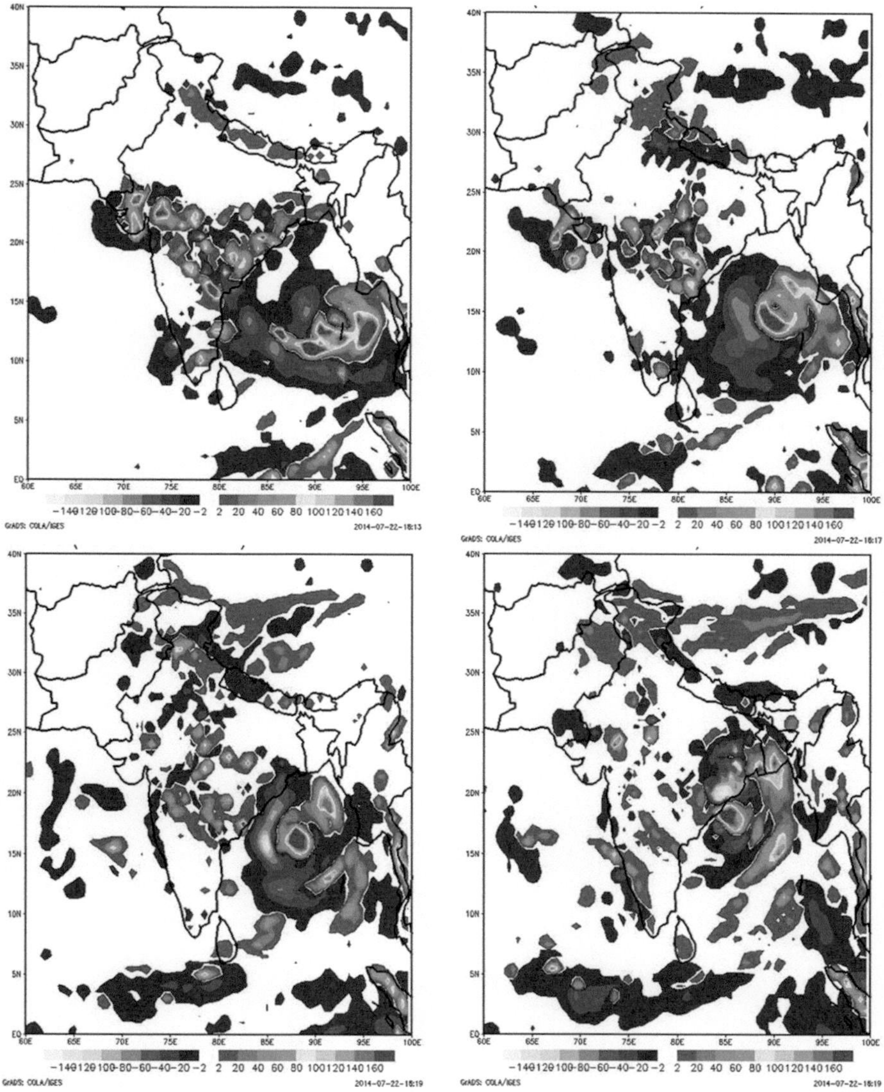

Fig. 8 Error in 24 h rainfall forecast by HWRF during 09–12 October 2013, in comparison with the observed rainfall

Table 1 Observed and 24-h HWRF forecast of maximum rainfall during 09–12 October 2013

Date	Observed rainfall in mm (recorded on next day at 0830 h)	HWRF model rainfall (24 h forecast)	Error in mm
9 October	168	647	479
10 October	128	508	380
11 October	175	711	536
12 October	216	512	296

The SAR, an active sensor, transmits pulses of microwave and detects echo, which carries information about the surface. Due to relatively long wavelengths in microwave, radar signals are capable of penetrating clouds in the atmosphere and are independent of sunlight. These characteristics of SAR are particularly useful in monitoring floods over large areas, while accurate flood mapping using optical imagery are limited, because of cloud cover associated with cyclone. Satellite-derived flood inundation maps in near-real time are invaluable to state or national agencies for disaster monitoring and relief efforts. Precise mapping of the maximum flood extent is also required for detecting deficiency in existing flood control measures and for arbitrating damage claims late. Due to unique signature of water correspondence to other features, flood affected area has been extracted through the detailed analysis of the SAR image. SAR appears to be an ideal sensor for detecting flooding in extensive areas, since the backscattering signature is so distinctive for water, compared to the vegetation.

Rice area of Odisha state was mapped using multi-date (3 date) C-band SAR data of RISAT-1 following a logical classification approach (Chakraborty et al. 1997). This was overlaid by the flood inundation map received from National Remote Sensing Centre, which had been prepared using SAR data. From the Fig. 9,

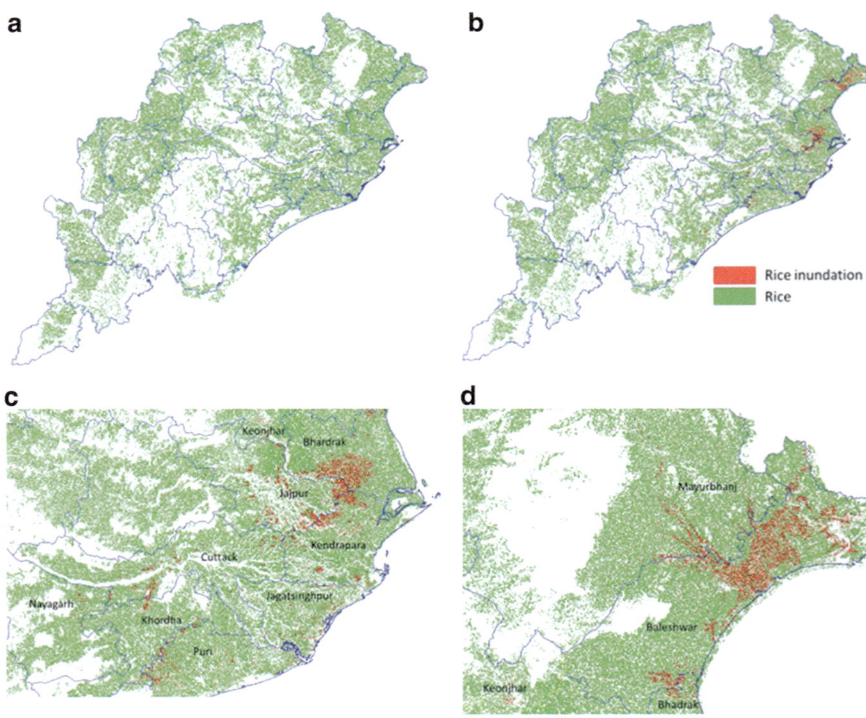

Fig. 9 Map showing the (**a**) rice cropped area, (**b**) flood inundated rice area of Odisha, and zoomed images showing rice inundation in (**c**) Bhadrak district, and (**d**) Balasore and Mayurbhanj districts

Table 2 Districtwise rice area inundated over Odisha in association with passage of TC Phailin

District	Rice cropped area (ha)	Rice inundation area (%)
Baleshwar	160569.1	22.3
Bhadrak	115012.5	17.8
Jajpur	98380.2	16.4
Kendrapara	100548.3	15.7
Puri	100596.8	11.2
Jagatsinghpur	58852.0	12.2
Cuttack	121855.8	5.9
Mayurbhanj	302197.8	2.3
Ganjam	199323.6	2.3
Khordha	96846.0	4.2
Keonjhar	188705.5	0.5
Nayagarh	86855.0	1.4
Dhenkanal	117850.1	0.6
Boudh	65364.4	0.02

it is clearly seen that coastal areas have maximum rice coverage mainly in Puri, Bhadrak, Kendrapara, Balasore and Jagatsinghpur districts. The maximum rice area inundated due to floods is in Balasore, Bhadrak, Jajpur, Jagatsinghpur and Kendrapara districts (Fig. 9 and Table 2). Among these the highest rice inundated area was in Balasore followed by Bhadrak district. The total rice inundated area was 132.4 thousand hectares. A field visit was carried out to check accuracy of the inundated area mapping. Out of 18 sites visited, 16 were found to be correctly mapped using remote sensing data. This showed the efficacy of microwave SAR data in rice inundated area mapping.

7 Conclusions

Another low pressure area that formed over the Bay of Bengal led to extremely heavy rainfall over Odisha during 21–28 October. Due to previous standing water over the region due to TC, Phailin, the rainfall due to low pressure area caused unprecedented flood over Odisha. As a result the floods in association with low pressure area caused more human death and economic loss than the flood due to TC, Phailin. It also happened in case of cyclone, Nilam (28 October–1 November 2012) over Bay of Bengal. In the case of CS Nilam (28 October–1 November 2012), at the time of landfall, the cloud mass was significantly sheared to the northeast of the system during its dissipation stage leading to heavy rainfall activity over entire Andhra Pradesh and adjoining Odisha. The prediction of heavy rainfall during slow movement or practical stationarity of TC near the coast is a challenge for the forecasters. Further, the forecasters are proved to be over-warning in case of TCs rapidly weakening near the coast and also the TCs moving rapidly just before landfall. Though heavy rainfall prediction is still a challenge, HWRF is a promising model for this purpose (although it showed over estimation of intensity in case of Phailin). Assimilation of DWR data can improve the potential of this model to accurately predict the heavy rainfall during landfall. Microwave remote sensing could be successfully used to map the rice inundated area post-Phailin cyclone.

Acknowledgement The authors from MNCFC wish to thank Director, NRSC for providing inundation area map, which was used for rice inundation mapping.

References

Abhilash, S, Sahai, A.K, Mohanakumar, K, George, J.P. and Das, S. (2012). Assimilation of Doppler Weather Radar Radial Velocity and Reflectivity Observations in WRF-3DVAR system for Short-Range Forecasting of Convective Storms. *Pure and Applied Geophysics*, **169**, 2047–2070.

Chakraborty, M, Panigrahy, S. and Sharma, S.A. (1997). Discrimination of rice crop grown under different cultural practices using temporal ERS-1 SAR data. *ISPRS Journal of Photogrammetry and Remote Sensing*, **52**, 183–191.

Debnath, G.C. and Mandal, P. (2012). The role of orography in producing extremely on the role of orography in producing extremely heavy rainfall induced by cyclonic storm Aila – A WRF model analysis for disaster management. *In*: WMO Technical Document on Proceedings of Second International Conference on Indian Ocean Tropical Cyclones and Climate Change (IOTCCC-II) held at New Delhi, India during 14–17 Feb, 2012, WWRP 2013–4.

Gray, M.W. (1968). Global view of the origin of tropical disturbances and storms. *Monthly Weather Review*, **96**, 669–700.

Haggag, M. and Yamashita, T. (2009). Environmental simulator application to the analysis of the tropical cyclone Gonu in 2007. *Jl. of Int. Development and Cooperation*, **15**, 1–2, 47–63.

Lee, D.K. and Choi, S.J. (2010). Observation and numerical prediction of torrential rainfall over Korea caused by Typhoon Rusa (2002), *J. Geophys. Res.*, **115**, D12105. doi: 10.1029/2009JD012581.

Minakshi Devi, Kalita, S, Das, S, Goswami, H. and Barbara, A. (2014). Model prediction of cyclonic tracks over Bay of Bengal and resultant precipitation in the north-east region. Geomatics, *Natural Hazards and Risk*, **5:2**, 93–114. doi: 10.1080/19475705.2013.775186.

Mitra, A.K, Bohra, A.K, Rajeevan, M. and Krishnalmurti, T.N. (2009). Daily Indian precipitation analysis formed from a merge of rain-guage data with TRMM TMPA satellite derived rainfall estimates. *Journal of the Meteorological Society of Japan*, **87A**, 265–279.

Misra, T, Rana, S.S, Desai, N.M, Dave, D.B, Jyoti, R, Arora, R.K, Rao, C.V.N, Bakori, B.V, Neelakanthan, R. and Vachchani, J.G. (2013). Aperture Radar payload on-board RISAT-1: Configuration, technology and Performance. *Current Science*, **104**, 446–461.

Osuri, K.K, Mohanty, U.C, Routray, A. and Mohapatra, M. (2012). The impact of satellite-derived wind data assimilation on track, intensity and structure of tropical cyclones over the North Indian Ocean. *International Journal of Remote Sensing*, **33**, 5.

Raju, P.V.S, Potty, J. and Mohanty, U.C. (2012). Prediction of severe tropical cyclones over the Bay of Bengal during 2007–2010 using high-resolution mesoscale model. *Natural Hazards*, **63**, 1361–1374, doi: 10.1007/s11069-011-9918-1.

Routray, A, Mohanty, U.C, Osuri, K.K. and Kiran Prasad, S. (2013). Improvement of monsoon depressions forecast with assimilation of Indian DWR data using WRF-3DVAR analysis system. *Pure and Applied Geophysics*, **170**, 2329–2350. doi: 10.1007/s00024-013-0648-z.

Srivastava, K, Rashmi, B. and Roy Bhowmik, S.K. (2011). Assimilation of Indian Doppler Weather Radar observations for simulation of mesoscale features of a land-falling cyclone. *Natural Hazards*, doi: 10.1007/s11069-011-9835-3.

Spatial Verification of Rainfall Forecasts During Tropical Cyclone 'Phailin'

K. Sharma, R. Ashrit, G.R. Iyengar, A. Mitra, B. Ebert, and E.N. Rajagopal

1 Introduction

During October 2013 Bay of Bengal (BOB) tropical cyclone (TC) 'Phailin' hit east coast of India. This was the most intense cyclone that made landfall over India after the Odisha Super Cyclone (29 October 1999). This TC originated from a remnant cyclonic circulation from the South China Sea. It intensified into a cyclonic storm on the 9 October 2013 and moved northwestwards. It further intensified into a very severe cyclonic storm on 10 October 2013 over east central BOB. It crossed Odisha coast near Gopalpur around 2230 h IST of 12 October 2013 with a sustained maximum surface wind speed of 200–210 kmph gusting to 220 kmph. Some of its unique features included the rapid intensification of the system from 10 October to 11 October 2013 resulting in an increase of wind speed from 83 to 215 kmph. Also, at the time of landfall on 12 October, maximum sustained surface wind speed in association with the cyclone was about 215 kmph and estimated central pressure was 940 hPa with pressure drop of 66 hPa at the center compared to surroundings (RSMC, New Delhi, 2014).

This TC caused heavy rainfall over Odisha leading to floods and storm surge leading to coastal inundation in the state of Odisha. Based on post-cyclone survey report, maximum rainfall occurred over northeast sector of the system centre at the time of landfall. Banki in Cuttack district of Orissa received a maximum 24 h cumulative rainfall of 38 cm.

The verification of TC track predicted by NCMRWF's Unified model (NCUM) and NCMRWF's Global Forecast System (NGFS) in terms of track error with respect to observed track based on Joint Typhoon Warning Center (JTWC) shows

K. Sharma (✉) • R. Ashrit • G.R. Iyengar • A. Mitra • B. Ebert • E.N. Rajagopal
National Centre for Medium Range Weather Forecasting, A-50 Sector-62, 201309 Noida, Uttar Pradesh, India
e-mail: kuldeep@ncmrwf.gov.in

© Capital Publishing Company 2017
M. Mohapatra et al. (eds.), *Tropical Cyclone Activity over the North Indian Ocean*,
DOI 10.1007/978-3-319-40576-6_3

that the forecast track error is lowest in NCUM forecasts (Ashrit et al. 2013). Here in the present study, an attempt has been made to evaluate the skill of NCUM and NGFS in terms of rainfall after the landfall of the very severe cyclonic storm (VSCS) 'Phailin' which hit the Orissa coast near Gopalpur on the evening of 12 October 2013 with a sustained maximum surface wind speed of 215 kmph. Available at: http://www.imd.gov.in/section/nhac/dynamic/phailin.pdf.

This manuscript is divided in to following sections: Section 2 deals with the details and description of Numerical Weather Prediction (NWP) models used in the present study. Section 3 covers the data and methodology used for rainfall verification. The tracks produced by NCUM and NGFS along with the observed track have been presented in Sect. 4. Section 5 describes the results obtained from the present study. Finally, in Sect. 6 conclusions based on the current study are presented.

2 NWP Models at NCMRWF

In this section, we briefly discuss about some noticeable differences between the formulations of the two models. Details about the deterministic models operational at NCMRWF can be found at (Rajagopal et al. 2007; Prasad et al. 2011) for NGFS and (Rajagopal et al. 2012) for the Unified Model (NCUM). Table 1 gives a brief

Table 1 Details and description about NCUM and NGFS model

	NGFS	NCUM
Horizontal resolution	Spectral truncation of 574 waves in the zonal direction (T574) with a Gaussian grid of 1760 × 880 points (~23 km resolution near equator)	N512 (~25 km at mid-latitudes) with a EW-NS grid of 1024 × 769 points
Vertical levels	Hybrid sigma-pressure (64 levels). The hybrid coordinate system is terrain following in the lower levels and transforming to pure pressure levels in the upper levels	70 vertical levels
Model time step	2 min	10 min
Forecast lead time	10 days	10 days
Data assimilation	Grid point Statistical Interpolation – GSI Wu et al. (2002)	4D var Data assimilation System Rawlins et al. (2007)
Dynamics	Spectral, Hybrid sigma-p, reduced Gaussian grids	Non-hydrostatic dynamics with deep atmosphere. Height vertical coordinates with levels transitioning from terrain following to height. Global latitude–longitude Grids
Time integration	Leapfrog/semi-implicit	Semi-implicit integration with 3D semi-Lagrangian advection

overview of the main features of the two models. The differences in the formulations of two models arise due to several factors including: horizontal and vertical resolutions, physical parameterisations, different time integration methods, as well as data assimilation schemes etc. (Table 1). The most important among these are the different data assimilation schemes. NGFS utilises the Grid point Statistical Interpolation (GSI; Wu et al. 2002) which is based on the three dimensional variational data assimilation (3D-VAR) system whereas NCUM uses four dimensional variational data assimilation (4D VAR) system (Rawlins et al. 2007) for data assimilation. 4D-VAR is a simple generalisation of 3D-VAR and it takes into account the temporal evolution processes which lead to improved representation of synoptic systems in the initial conditions. Extensive studies have been conducted, at various meteorological organisations (UKMet Office, Meteorological Service of Canada, National Centre for Environmental Prediction (NCEP), USA etc.), for comparing the respective skills of 4D-VAR and 3D-VAR in assimilating data and forming the initial conditions for different models. All these studies have helped in forming a consensus that the 4D-VAR performs better than the 3D-VAR scheme (Lorenc and Rawlins 2005; Laroche et al. 2005). This is reflected in the better estimation of observed synoptic systems in the initial conditions (analysis) of the models using 4D-VAR for data assimilation. Additional details about the models' configuration and forecast products are summarised in Table 1.

3 Data and Methodology

Rainfall forecasts produced by NGFS and NCUM model has been verified against IMD-NCMRWF merged satellite gauge (NMSG) data (Mitra et al. 2009, 2013). This rainfall data is a merged product of satellite estimates (Tropical Rainfall measuring mission (TRMM)) and IMD rain gauge observations. The rainfall data from NMSG and from the two models are 24 h accumulated valid at 03 UTC. The model forecast rainfall, originally at a higher grid resolution, is interpolated at 0.5° grid resolution to match with the grid and resolution at which NMSG data is available. The landfall occurred on 12 October evening 1700 UTC followed by the movement in north-northeast direction on 13 and 14 October which led to heavy rain and flooding. So here in the present study, analysis has been performed for 14 October (rainfall received between 03 UTC of 13 October and 03 UTC of 14 October) and 15 October 2013 (rainfall received between 03 UTC of 14 October and 03 UTC of 15 October).

Spatial verification of the rainfall forecasts after the landfall of the cyclone in the present study is carried out using the contiguous rain area (CRA) method. This method was developed for estimating the systematic errors in the rainfall forecasts (Ebert and McBride 2000; Stefano and Marco 2008; Ebert and Gallus 2009). It was one of the first methods to measure errors in predicted location and to separate the total error into components due to errors in location, volume and pattern. The steps involved in the CRA technique are described in Ebert and McBride (2000). The CRA method is an object-oriented verification procedure suitable for gridded quantitative precipitation forecasts (QPFs). In the CRA framework a weather system is defined as a region bounded by a user specified isohyet (entity) of precipitation in the union of the forecast and observed rain field. This technique is then simply based on a pattern matching of two contiguous areas (entities), defined as the observed and forecast precipitation areas delimited by the chosen isohyet. The forecast and observed entities need not overlap, but they must be associated with each other, which means that they should be close to each other. The best match between the two entities can be determined either: (a) by maximising the correlation coefficient, (b) by minimising the total mean squared error, (c) by maximising the overlap of the two entities, or (d) by overlaying the centers of gravity of the two entities. For a good forecast, all the methods should give very similar location errors. In the present study, the best match is determined by maximising the correlation.

For each entity that can be identified in forecast and observations, the CRA method determines the location, volume and pattern errors, which are then combined in the form of a total mean squared error (MSE). To estimate the location error, the forecast field is horizontally translated over the observed field until the best match is obtained. The location error is then simply the vector displacement of the forecast. MSE and its decomposition (location error, volume error and pattern error) are shown below:

$$\mathrm{MSE}_{\mathrm{Total}} = \mathrm{MSE}_{\mathrm{Displacement}} = +\mathrm{MSE}_{\mathrm{Volume}} + \mathrm{MSE}_{\mathrm{Pattern}} \tag{1}$$

where the component errors are estimated as

$$\begin{aligned} \mathrm{MSE}_{\mathrm{Displacement}} &= 2S_{\mathrm{F}}S_{\mathrm{O}}(r_{\mathrm{OPT}} - r) \\ \mathrm{MSE}_{\mathrm{Volume}} &= (F' - O')^2, \\ \mathrm{MSE}_{\mathrm{Pattern}} &= 2S_{\mathrm{F}}\mathrm{S}_{\mathrm{O}}(1 - r_{\mathrm{OPT}}) + (S_{\mathrm{F}} - S_{\mathrm{O}})^2 \end{aligned} \tag{2}$$

In the above expressions F' and O' are the mean forecast and observed precipitation values after shifting the forecast to obtain the best match, s_{F} and s_{O} are the standard deviations of the forecast and observed precipitation, respectively, before shifting. The spatial correlation between the original forecast and observed features (r) increases to an optimum value (r_{OPT}) in the process of correcting the location via pattern matching.

Displacement and pattern errors are associated with errors in dynamics (predicted flow) while volume error is associated with errors in physics (moisture) treatment. These components provide guidance for model developers when the statistics of error components are studied for large sample of cases.

4 Observed and Forecast Track

TC 'Phailin' originated from a remnant cyclonic circulation from the South China Sea. It intensified into a cyclonic storm on 9 October 2013 and moved northwestwards. It further intensified into a very severe cyclonic storm on 10 October 2013 over east central Bay of Bengal (BOB).

The tracks predicted, with different initial conditions, by deterministic models, NCUM and NGFS along with observed track based on Joint Typhoon Warning Center (JTWC) are shown in Fig. 1a–d. Forecast positions based on NGFS are shown at 6-h interval while the forecast positions based on NCUM is shown at 24-h interval. Figure 1a–d shows the observed and forecasts tracks based on 00 UTC of 09, 10, 11 and 12 October 2013, respectively. The forecasts indicate landfall over Andhra Pradesh and Odisha border. On one hand, Fig. 1a, b shows that the tracks produced, based on initial conditions of 9 and 10 October, by two models, are close to observed track at the time of landfall. On the other hand, tracks produced, based on initial conditions of 00 UTC of 11 and 12 October (Fig. 1c, d), by NCUM show

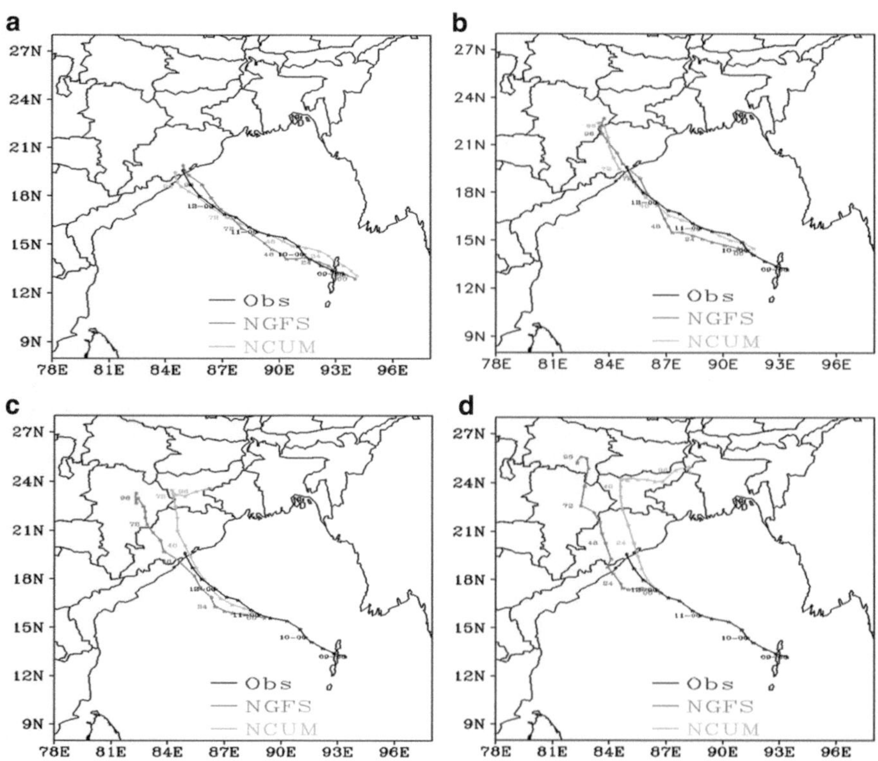

Fig. 1 (**a–d**). Tracks of Tropical Cyclone 'Phailin' predicted by NCUM and NGFS based on Initial Conditions of 00 UTC of 9 (**a**), 10 (**b**), 11 (**c**) and 12 (**d**), October 2013 along with observed track-based Joint Typhoon Warning Center (JTWC)

the movement of the cyclone in north-northeast direction while NGFS predicted tracks move in north-northwest direction (Fig. 1c, d).

5 Results and Discussion

First, qualitative inter-comparison of spatial pattern of wind and rainfall produced by NGFS and NCUM will be discussed. Further, to quantify the forecast biases in both the models, verification of quantitative precipitation forecast (QPF) using the CRA has been used.

5.1 Wind and Rainfall After the Landfall of TC

TC 'Phailin' hit Orissa coast near Gopalpur around 2230 h IST of 12 October 2013. So, the rainfall associated with this cyclone over the land experienced on 14 October 2013 (rainfall received during 03 UTC of 13 October and 03 UTC of 14 October 2013) as well as 15 October 2013 (rainfall received during 03 UTC of 14 October and 03 UTC of 15 October 2013) are verified. Figures 2 and 3 show the observed

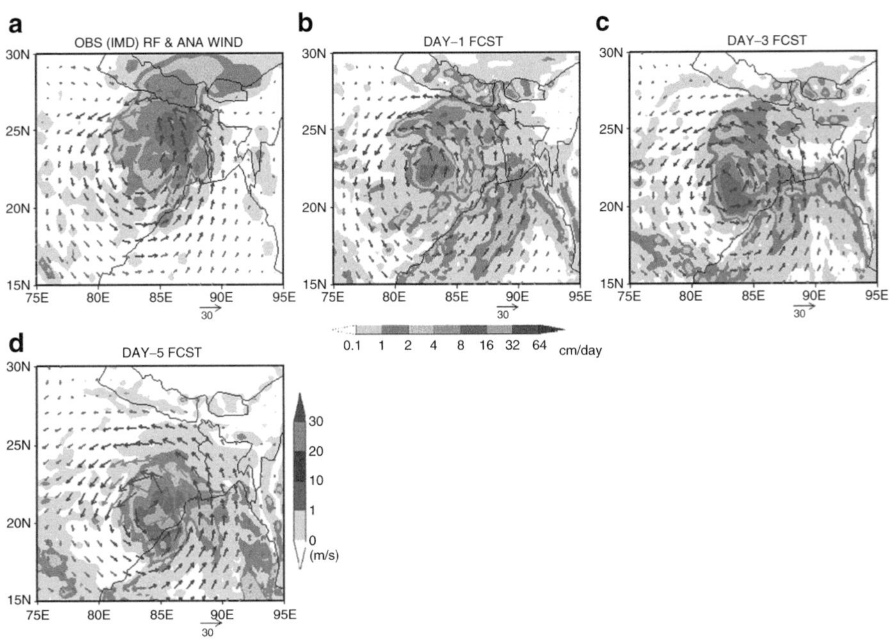

Fig. 2 Observed and NGFS Model Predicted rainfall in cm/day (horizontal colour bar) and 850 hPa circulation in m/s (vertical colour bar) over Indian region. valid for 03Z/14 October 2013

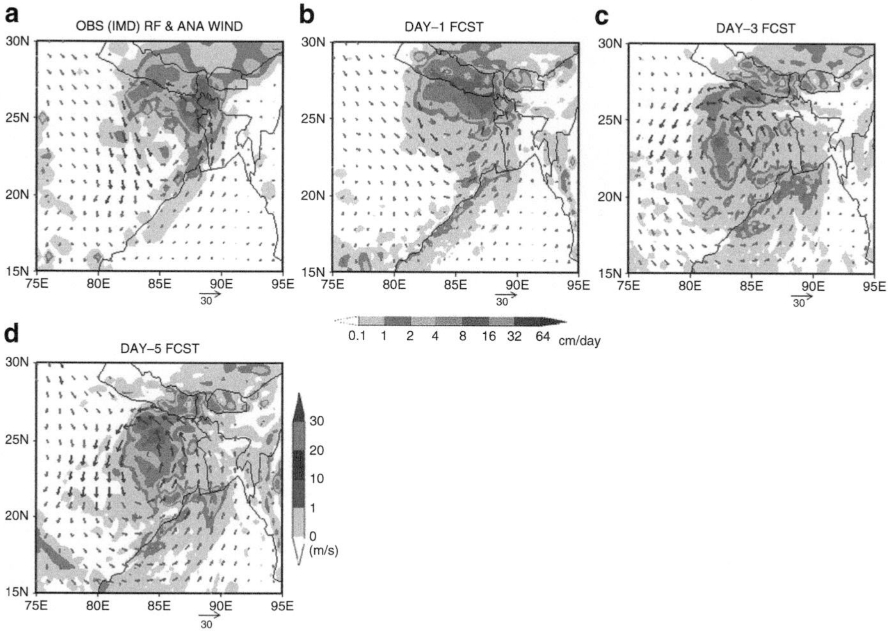

Fig. 3 Observed and NGFS Model Predicted rainfall in cm/day (horizontal colour bar) and 850 hPa circulation in m/s (vertical colour bar) over Indian region valid for 03Z/15 October 2013

and predicted rainfall along with 850 hPa wind (wind vectors are valid at 00 UTC) over eastern part of India valid on 14 and 15 October produced by NGFS. Day-1 forecast valid for 14 October 2013 produced by NGFS captures the rainfall band with peak rainfall much to the southwest with respect to observations while Day-3 and Day-5 rainfall forecast with peak of the rainfall band can be seen much to the south as compared with observation. Wind circulation at 850 hPa shows the movement of the cyclone is slower as compared to observation in all Day-1, Day-3 and Day-5 forecasts valid on 14 October 2013. Rainfall in Day-1 forecast shown by NGFS valid for 15 October 2013 matches well with the peak rainfall and is slightly to the north compared to observations. Day-3 and Day-5 rainfall forecasts also capture the rainfall band very well with slight underestimation in the peak rainfall amounts valid for 15 October 2013. Similar plots valid for 14 and 15 October 2013 produced by NCUM are presented in Figs. 4 and 5. NCUM predicts rainfall band and circulation in Day-1 and Day-3 forecast close to observations. However, peak rainfall amount and associated circulation is located much to the south with respect to observations in Day-5 forecast valid for 14 October 2013. Day-1 and Day-3 forecasts valid on 15 October 2013 match well as compared with observation with slight overestimation in peak amount of rainfall in Day-1 forecast. However, Day-5 forecast predicts the rainfall band slightly to the south of the observed locations. Thus, the rainfall and circulation predicted by NCUM are close to observation while NGFS prediction is slightly in the south with respect to

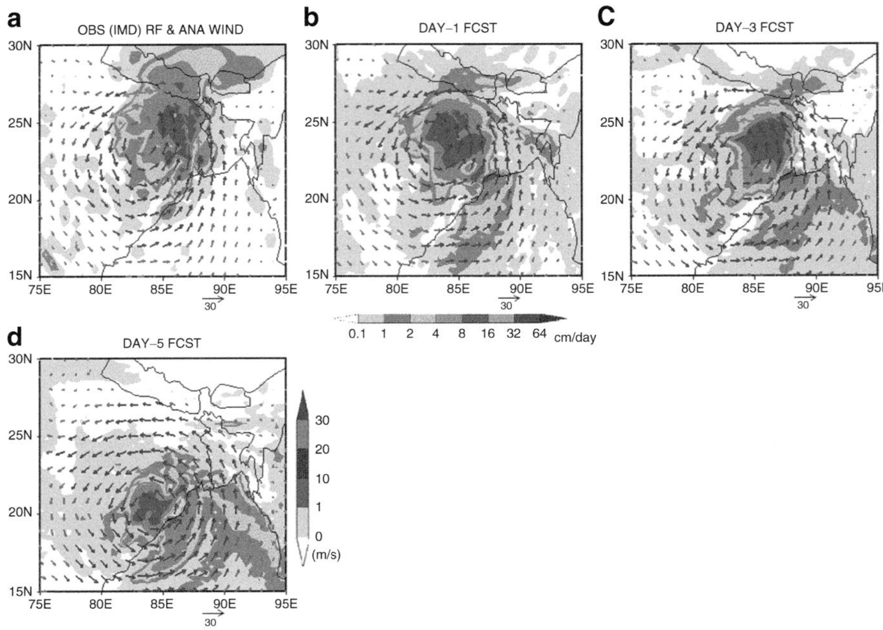

Fig. 4 Observed and NCUM Model Predicted rainfall in cm/day (horizontal colour bar) and 850 hPa circulation in m/s (vertical colour bar) over Indian region valid for 03Z/14 October 2013

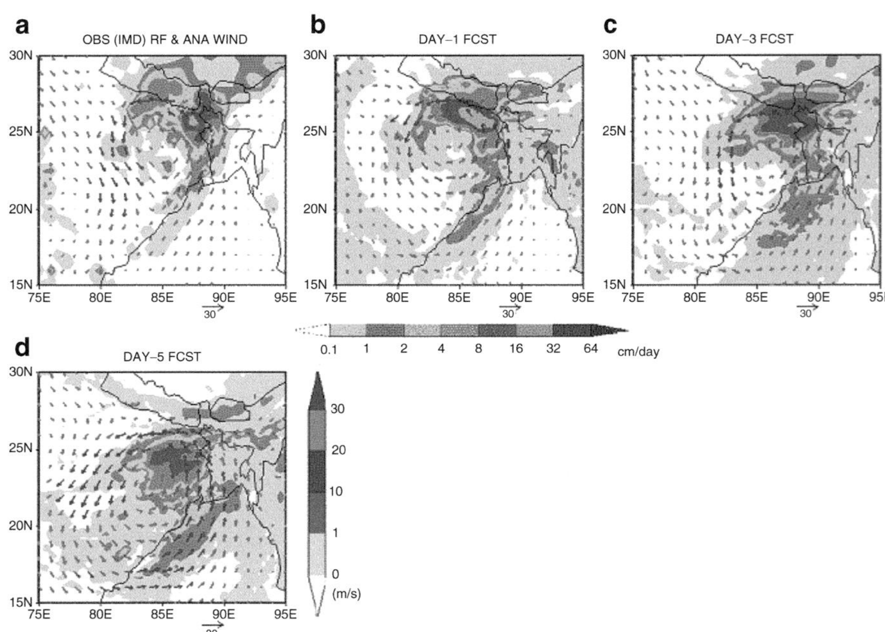

Fig. 5 Observed and NCUM Model Predicted rainfall in cm/day (horizontal colour bar) and 850 hPa circulation in m/s (vertical colour bar) over Indian region valid for 03Z/15 October 2013

observations on 14 October. However, rainfall and wind predicted by both models in Day-1 and Day-5 forecast valid on 15 October are close to observations while NCUM are much near to observations in Day-3 relatively to NGFS.

5.2 Verification of Rainfall Forecasts During 14 and 15 October Using CRA Method

Since TC 'Phailin' crossed Orissa coast as a very severe cyclonic storm, it produced extremely heavy rainfall after landfall. So, CRAs defined by higher thresholds of 10, 20, 40 and 80 mm/day are used to isolate the events corresponding to the high rainfall region due to 'Phailin'. As a next' step, a pattern matching technique is used for estimating the location error. In this case, the best match between the forecast and observed entities is done by maximising the correlation coefficient (Pearson correlation) between the forecast and observed fields. Figures 6 and 7 show the CRA verification isohyets for Day-1, Day-3 and Day-5 forecasts from NCUM and NGFS valid for 14 and 15 October.

Table 2 shows MSE and root mean square error (RMSE) (before and after the CRA procedure) obtained from CRA verification for Day-1–Day-5 forecasts valid on 14 and 15 October, respectively. The comparison of MSE between NCUM and NGFS reflect that NGFS has larger error than NCUM which implies the performance of NCUM is better than NGFS in terms of matching the displacement, volume and pattern of the forecast precipitation entities valid on 14 October. However, NCUM shows slightly larger MSE only in Day-1 and Day-5 forecast as compared to NGFS valid on 15 October. RMSE is also shown in the table for the original and shifted rainfall (i.e., before and after the CRA method). It is known as a direct indicator of the forecast accuracy. When we compare RMSE before and after applying CRA procedure, we find NCUM show lower value than NGFS which leads to better forecast than NGFS valid on 14 October 2013. However, Day-1 forecast valid on 15 October 2013 produced by NGFS has 23 % lower RMSE as compared with NCUM while in Day-5 forecast errors are close to each other before and after the shift.

5.3 Verification of QPF Statistics

Quantitative verification of rainfall associated with TC Phailin after the landfall is also made using CRA method. The rainfall forecast after the landfall has been categorised as 'hits', 'misses' depending upon whether the position and intensity were well predicted.

We have evaluated the CRA statistics for a number of rainfall thresholds ranging from 10 to 80 mm/day. Based on these thresholds, the components of contingency

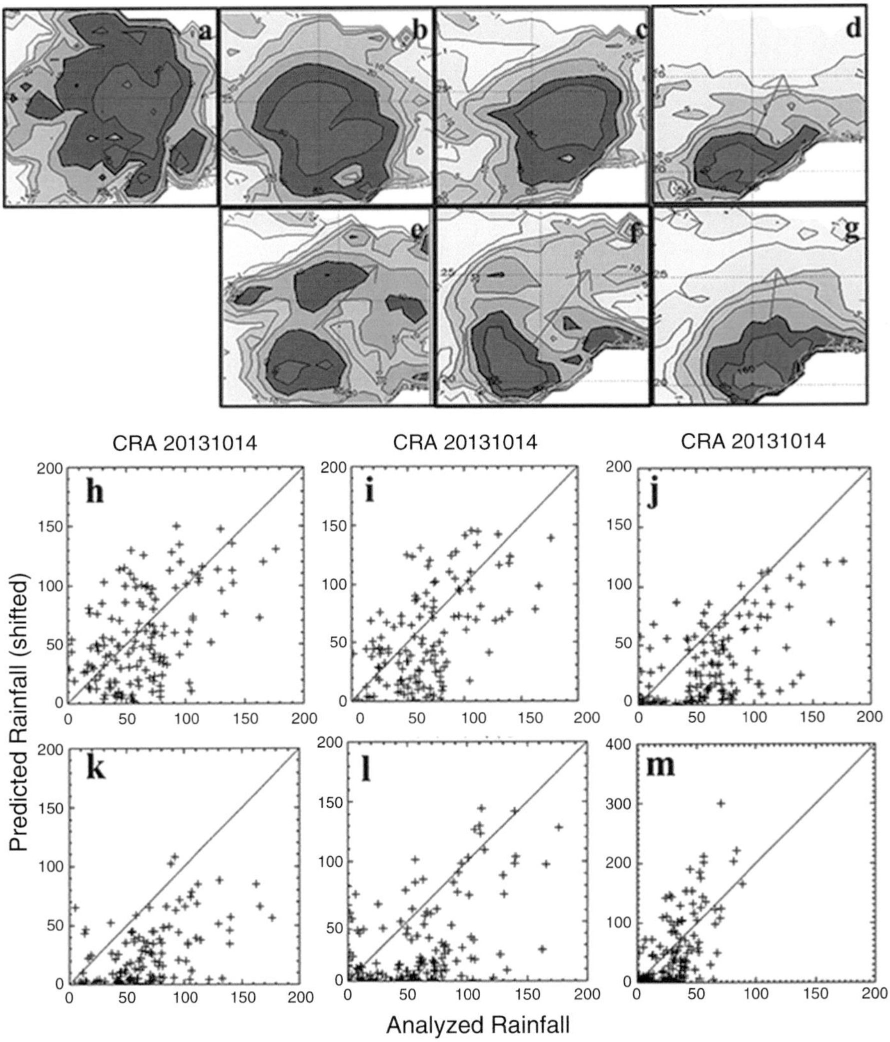

Fig. 6 Isohytes in mm over eastern coast of India (**a**) Analysis, (**b**) Day-1,(**c**) Day-3, (**d**) Day-5, from NCUM and (**e**) Day-1,(**f**) Day-3, (**g**) Day-5, from NGFS. Scatter plots show the number of matching grid points between model forecast and observation (analysis) valid on 14 October 2013. Plots (**h**–**m**) represents Day-1, Day-3 and Day-5 of NCUM and NGFS, respectively

table (number of hits, number of misses, false alarms and number of correct negatives) have been calculated during CRA analysis. Frequency of number of hits and misses valid for 14 and 15 October is shown in Fig. 8a–d for different rainfall thresholds (<1, 1–10, 10–20, 20–40 and 40–80 mm/day). The rainfall forecast from Day-1 to Day-5 predicted by NCUM shows higher number of hits and lower number of misses as compared with NGFS for all rainfall thresholds valid

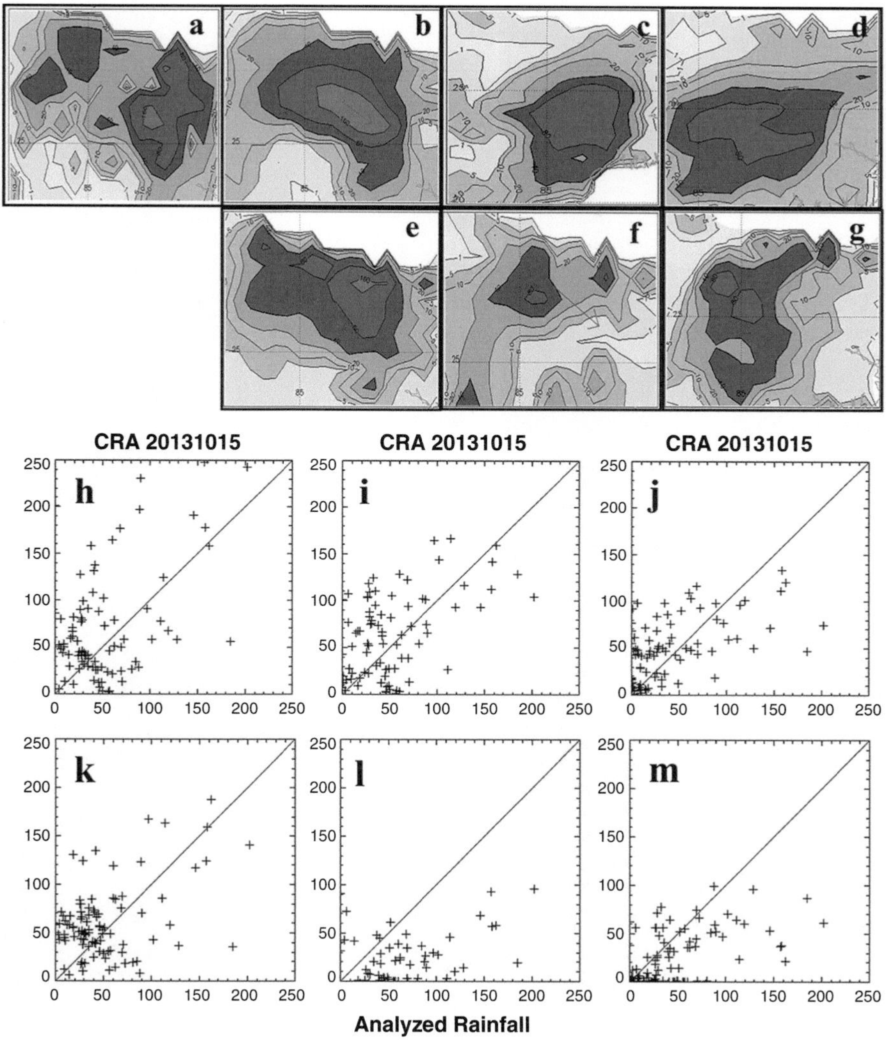

Fig. 7 Isohytes in mm over eastern coast of India (**a**) Analysis, (**b**) Day-1, (**c**) Day-3, (**d**) Day-5 from NCUM and (**e**) Day-1, (**f**) Day-3, (**g**) Day-5 from NGFS. Scatter plots show the number of matching grid points between model forecast and observation (analysis) valid on 15 October 2013. Plots (**h–m**) represents Day-1, Day-3 and Day-5 of NCUM and NGFS, respectively

for 14 October. But the number of hits is nearer in Day-1 forecast at all thresholds for the rainfall valid for 15 October. Further, we have computed equitable threat score (ETS), probability of detection (POD) and Hanssen & Kuipers (HK) score by using the components of the contingency table. The scores are shown in Figs. 9 and 10 in the form of bar graph from Day-1 to Day-5 for two rainfall thresholds (10–20 mm/day: left panel and 20–40 mm/day: right panel) for both the models valid for 14 and 15 October.

Table 2 MSE, initial and shifted RMSE for Day-1–Day-5 forecasts obtained from the CRA verification for rainfall forecast from NCUM and NGFS valid on 14 and 15 October 2013

	Mean Squared Error (MSE)		Initial RMSE		Shifted RMSE	
Date	NCUM	NGFS	NCUM	NGFS	NCUM	NGFS
20131014_24	2251.0	3431.6	47.4	58.5	37.8	51.8
20131014_48	2596.5	3866.9	50.9	62.1	40.4	52.6
20131014_72	2635.8	4733.4	51.3	68.8	38.3	47.8
20131014_96	4294.1	5559.2	65.5	74.5	43.7	41.7
20131014_120	5912.7	7319.7	76.8	85.5	43.0	45.8
20131015_24	3677.3	2177.7	60.6	46.6	56.4	43.8
20131015_48	3958.1	6388.5	62.9	79.9	47.4	73.9
20131015_72	2857.8	4791.9	53.4	69.2	46	59.9
20131015_96	3225.3	9203.5	56.7	95.9	46.1	95.7
20131015_120	4437.7	4200.6	66.6	64.8	39.8	42.8

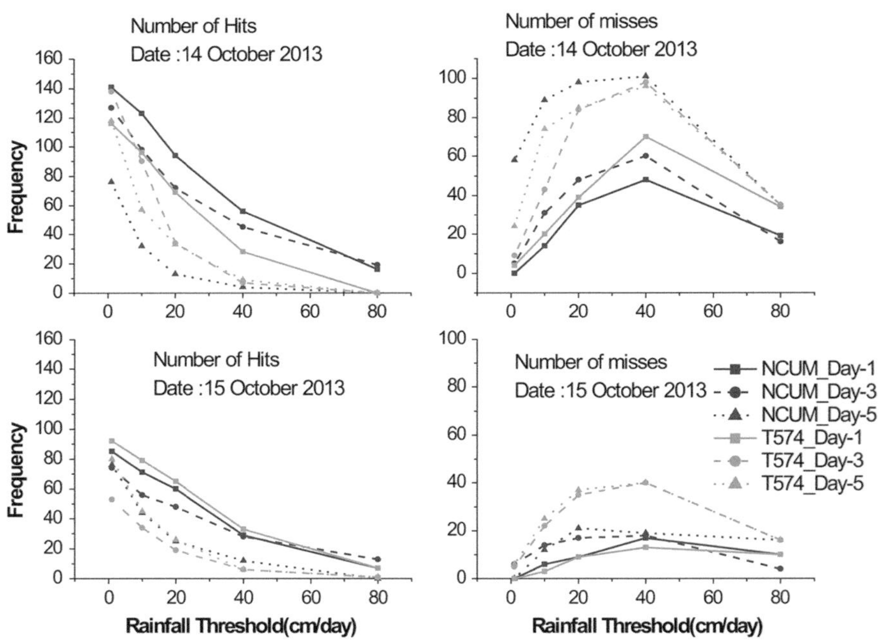

Fig. 8 Number of hits and misses based on the event verification during CRA procedure. Upper Panel correspond to number of hits and misses for Day-1, Day-3 and Day-5 forecasts valid on 14 October and Lower panel corresponds to the hits and misses for Day-1, Day-3 and Day-5 forecasts valid on 15 October 2013. These are based on <1, 1–10, 10–20, 20–40 and 40–80 mm/day rainfall thresholds

ETS measures the fraction of events that are correctly predicted accounting for hits by random chance. High ETS would imply that there is a large number of correctly predicted forecast entities near to the location of the matching observed

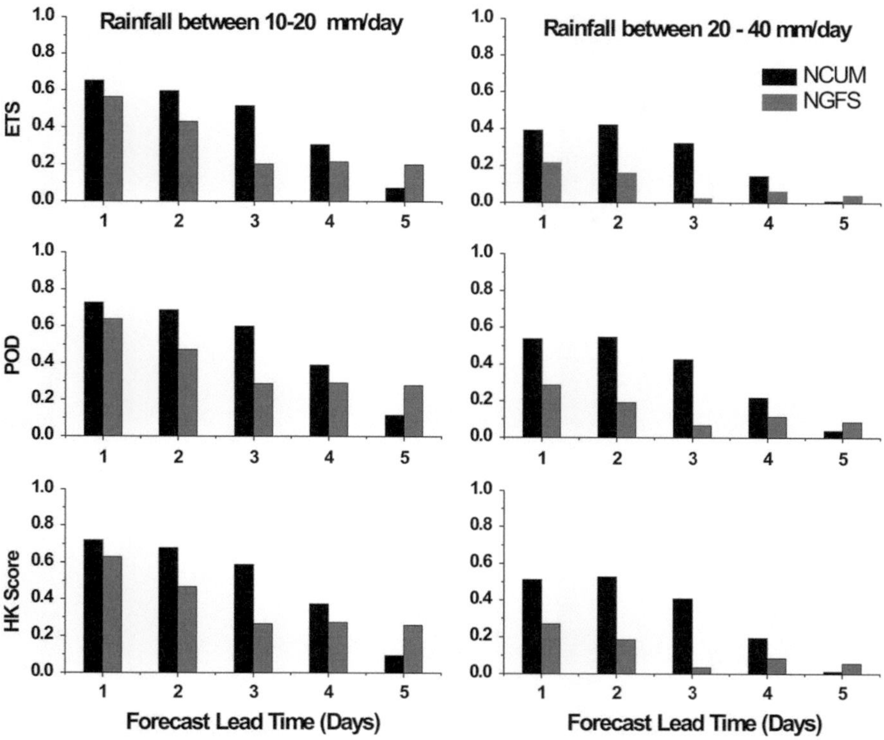

Fig. 9 ETS, POD, and HK score for NCUM and NGFS from Day-1 to Day-5 forecasts valid for 14 October 2013 based on 10–20 mm/day (*left panel*) and 20–40 mm/day (*right panel*)

entities (hits) and lesser number of forecast entities far away from the observations (misses and false alarms). The top panel of Figs. 9 and 10 show ETS from Day-1 to Day-5 forecast for two rainfall thresholds (10–20 mm/day: left panel and 20–40 mm/day: right panel) valid for 14 October and 15 October, respectively. NCUM shows higher ETS from Day-1 to Day-4 for both the rainfall threshold (10–20 and 20–40 mm/day) for 14 October. For 15 October, NGFS shows slightly higher ETS only in Day-1 for both the thresholds. However, for Day-2–Day-5, NCUM shows higher ETS than NGFS for both the thresholds.

HK score, also known as the True Skill Score (TSS), is defined as the difference between the hit rate and the false alarm rate (Hanssen and Kuipers 1965). A high HK score indicates more hits relative to false alarms. The bottom panel of Figs. 9 and 10 show HK score from Day-1 to Day-5 forecast for two rainfall thresholds (10–20 mm/day: left panel and 20–40 mm/day: right panel) valid for 14 and 15 October, respectively. Here in this present case of TC 'Phailin', NCUM shows higher HK score than NGFS till Day-4 while in Day-5, score is slightly higher NGFS for 14 October for both rainfall thresholds. For 15 October, NGFS and NCUM are close to each other in Day-1 forecast for both the threshold. However,

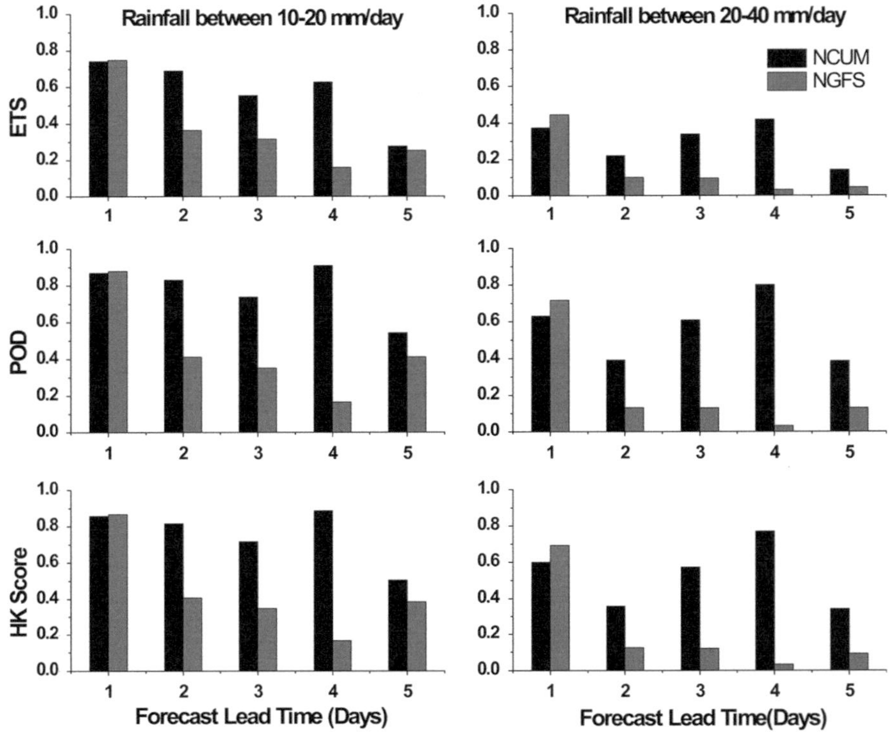

Fig. 10 ETS, POD, and HK score for NCUM and NGFS from Day-1 to Day-5 forecasts valid for 15 October 2013 based on 10–20 mm/day (*left panel*) and 20–40 mm/day (*right panel*)

NCUM shows higher HK score from Day-2 to Day-5 as compared with NGFS for both the thresholds.

POD is defined as the fraction of observed events that are correctly predicted. Therefore a high POD indicates good forecast skill of a model. In the current case, a high POD would imply that many forecast entities with intensities approximately matching the observations are lying close enough to the observed entities hence having a higher number of hits (Fig. 9). From the middle panels of Figs. 9 and 10, it is seen that POD is consistently higher for NCUM (Day-1–Day-4) for the forecast but in Day-5 NGFS shows higher POD than NCUM valid for 14 October for both thresholds. On one hand, rainfall forecast valid on 15 October, NGFS and NCUM shows comparable skill in Day-1 forecast similar to HK score and ETS. On the other hand, NCUM shows higher POD for Day-2–Day-5 forecast as compared to NGFS for both rainfall thresholds from 10–20 to 20–40 mm/day. Thus, based on the above statistics, we can say that NCUM has higher skill in predicting the rainfall due to TC 'Phailin' as compared to NGFS.

6 Summary

In this paper, the comparison of skills of NCUM and NGFS in predicting the rainfall associated with the TC 'Phailin' after the landfall has been carried out. After hitting the coast, it caused heavy to extremely heavy rainfall over Odisha, Coastal Andhra Pradesh, Jharkhand, Bihar, Sub-Himalyan West Bengal and some parts of Sikkim on 14 and 15 October 2013. The findings of the paper are as follows:

- The forecast tracks produced by both models indicate landfall over Andhra Pradesh and Odisha border. However, after the landfall, NCUM-predicted track shows the movement of the cyclone in north-northeast direction while NGFS track drifted the system in north-northwest direction.
- On one hand, the forecast produced by NCUM was successfully able to capture the intensity of circulation and rainfall. On the other hand, NGFS shows the circulation and associated movement of rainfall band slower and representing rainfall much to the south with respect to observations.
- The verification of spatial pattern of rainfall indicates the underes-timation of average rain rate and the rain volume produced by NGFS as compared to NCUM which shows lower percentage errors and total MSE and RMSE.
- Event verification indicates NGFS has lower number of hits and higher number of misses than NCUM. Also, a comparison of statistical scores (POD, HK score and ETS) between the two models again confirms that NCUM performs better than NGFS in predicting the rainfall associated with this cyclone specifically after the landfall.

Acknowledgements The authors are thankful to Director, NCMRWF for constant support and encouragement. Inputs and suggestions provided by Scientists, NCMRWF are gratefully acknowledged.

References

Ashrit, R, Amit Ashish, Kuldeep Sharma, Anumeha Dube, Indira Rani. and Dasgupta M. (2013). Verification of Predicted Tropical Cyclone Tracks by NCMRWF Global Forecast Models. NMRF/RR/04/2014.

Ebert, E.E. and McBride, J.L. (2000). Verification of precipitation in weather systems: Determination of systematic errors. *Journal of Hydrology*, **239**, 179–202.

Ebert, E.E. and Gallus, Jr, W.A. (2009). Towards better understanding of Contiguous Rain Areas (CRA) method of spatial verification. *Weather and Forecasting*, **24**, 1401–1415.

Hanssen, A.J. and Kuipers, W.J. (1965). On the relationship between the frequency of rain and various meteorological parameters. Koninklijk Neferlands Meteorologist Institua Meded. Verhand, 81, pp. 2–25.

Lorenc, A.C. and Rawlins, F. (2005). Why does 4D-Var beat 3D-Var? *Quarterly Journal of the Royal Meteorological Society*, **131**, 3247–3257.

Laroche, S, Gauthier, P, Tanguay, M, Pellerin, S, Morneau, J, Koclas, P. and Ek, N. (2005). Evaluation of the operational 4D-Var at the Meteorological Service of Canada. *In*: Proceedings

of the Fourth WMO International Symposium on Assimilation of Observations in Meteorology and Oceanography, Prague, Czech Republic, World Meteorological Organization, pp. 139.
Mitra, A.K, Bohra, A.K, Rajeevan, M.N. and Krishnamurti, T.N. (2009). Daily Indian precipitation analysis formed from a merge of rain-gauge data with the TRMM TMPA satellite-derived rainfall estimates. *Journal of the Meteorological Society of Japan*, **87A**, 265–279.
Mitra, A.K, Momin, I.M, Rajagopal, E.N, Basu, S, Rajeevan, M.N. and Krishnamurti, T.N. (2013). Gridded daily Indian monsoon rainfall for 14 seasons: Merged TRMM and IMD gauge analyzed values. *Journal of Earth System Science*, **122**, 1173–1182.
Prasad, V.S, Mohandas, S, Das Gupta, M, Rajagopal, E.N. and Kanti Dutta, S. (2011). Implementation of Upgraded Global Forecasting Systems (T382L64 and T574L64) at NCMRWF: NCMR/TR/5/2011.
Rajagopal, E.N, Das Gupta, M, Mohandas, S, Prasad, V.S, George, John, P, Iyengar, G.R. and Preveen Kumar, D. (2007). Implementation of T254L64 Global Forecast System at NCMRWF, NCMRWF Technical Report, pp. 1–42.
Rajagopal, E.N, Iyengar, G.R, George, J.P, Das Gupta, M, Mohandas, S, Siddharth, R, Gupta, A. and Chourasia, M. (2012). Implementation of Unified Model based Analysis-Forecast System at NCMRWF. NMRF/TR/2/2012, 45p.
Rawlins, F, Ballard, S, Bovis, K, Clayton, A, Li, D, Inverarity, G, Lorenc, A.C. and Payne, T. (2007). The met office global four-dimensional variational data assimilation scheme. *Quarterly Journal of the Royal Meteorological Society*, **133**, 347–362.
Stefano, M. and Marco, C. (2008). Forecast Verification: A summary of common approaches and examples of application, FORALPS Technical Report 5, Uni-versità degli Studi di Trento, Dipartimento di Ingegneria Civilee Ambientale, Trento, Italy, 60 pp.
Wu, W.S, Purser, R.J. and Parrish, D.F. (2002). Three-dimensional variational analysis with spatially inhomogeneous covariances. *Monthly Weather Review*, **130**, 2905–2916.

Diagnostics of Upper Level Dynamics and Rainfall Asymmetry of Very Severe Cyclonic Storm *MADI* (2013)

S. Balachandran and B. Geetha

1 Introduction

Intensity and structural changes of a Tropical Cyclone (TC) are known to be associated with environmental factors such as vertical wind shear (VWS), sea surface temperature (SST) etc. as well as TC specific factors such as its location, speed and direction of motion.

The influence of environment on TC intensity change is effected through complex interaction of multi-scale physical and dynamical processes. Rhome and Raman (2006) made a review on some of the well known processes identified to be associated with TC intensity change such as VWS, SST, upper level divergence, land interactions, eddy angular momentum fluxes etc. Several studies have been undertaken to understand the large / synoptic scale eddy forcings on TC intensification (Molinari and Vollaro 1989; DeMaria et al. 1993; Yu and Kwon 2005). McBride and Zehr (1981), based on a study of composite TC parameters, found that developing storms contained large inward eddy fluxes of cyclonic angular momentum at outer radii, which were confined almost entirely to the outflow layer of the composite storms. However, non-developing tropical disturbances contained much weaker and more diffused cyclonic eddy momentum fluxes. DeMaria et al. (1993) showed that upper level outer Eddy Momentum Flux (EMF)/Eddy Flux Convergence (EFC) at the inner radii can serve as catalyst to organise diabatic sources through secondary radial circulations which excite internal instabilities in tropical cyclones.

Different types of environmental interactions have been brought out by various case studies. Eddy forcing in the form of westerly trough interaction has been well

S. Balachandran (✉)
India Meteorological Department, Chennai, India
e-mail: balaimd@gmail.com

B. Geetha
India Meteorological Department, Mausam Bhawan, Lodi Road, 110003 New Delhi, India

© Capital Publishing Company 2017

M. Mohapatra et al. (eds.), *Tropical Cyclone Activity over the North Indian Ocean*,
DOI 10.1007/978-3-319-40576-6_4

documented for Atlantic hurricanes (Molinari and Vollaro 1989; Hanley et al. 2001) and Pacific typhoons (Yu and Kwon 2005). Ramage (1974) and Rucker (1992) noted that Pacific typhoon *Joan* (October 1970) and *Flo* (September 1990) intensified dramatically under the influence of weakening of the upper anti-cyclone.

Regarding the environmental effects on structural changes, several studies on asymmetric rainfall structures with reference to VWS have been carried out in the past. Rodgers and Adler (1981), Marks (1985) and Burpee and Black (1989) have worked on radial profiles of rainfall associated with some Pacific typhoons and Atlantic Hurricanes using satellite passive microwave imager and airborne radar observations and have brought out the complex and asymmetric rainfall structures associated with TCs. Lonfat et al. (2004) determined the asymmetry pattern in rainfall distribution for TCs of global oceanic basins and attributed vertical wind shear and variations in the translational speeds of TCs of various basins as important factors amongst others for the evolution of different asymmetry patterns over various oceanic basins. Chen et al. (2006) brought out the roles of VWS and TC motion in rainfall asymmetric structures over various oceanic basins.

In the present study, the environmental influences on the intensity and structural changes of a very severe cyclonic storm (VSCS; maximum sustained surface wind speed (MWS): 64–119 knots) *MADI* that formed and affected the North Indian Ocean (NIO) basin during 6–13 December 2013 are analysed. The VSCS *MADI* formed as a depression (D; MWS: 17–27 knots) over the Bay of Bengal (BOB) on 6 December and initially moved northwards until 9 December during which time it intensified into a VSCS. On 9 and 10 it underwent several structural and intensity changes, started weakening and recurved southwestwards. The study addresses the intensity changes of *MADI* from the context of synoptic/large scale eddy forcings. Variations in the precipitation structures during the intensification phase as well as under strong/weak VWS scenario are also brought out. The results obtained are comparable to those obtained by the earlier workers for other oceanic basins.

2 Data and Methodology

The methodology of analysis comprises of two parts – (1) Intensity change analysis and (2) Study of evolution of asymmetric structural patterns. The intensity changes associated with the TC *MADI* are analysed with respect to VWS, SST and EFC. Structural patterns evolving during the intensification phase of *MADI* as well as under different VWS conditions are brought out by determining and analysing Fourier first order wave number-1 asymmetry (Lonfat et al. 2004) in the rainfall structures.

The VWS data at 6-h intervals (area averaged over 600 km from the TC centre at the time of observation) is obtained from the US National Oceanic and Atmospheric Administration (NOAA), Cooperative Institute for Research in the Atmosphere (CIRA) website. Daily SST data is taken from NCEP reanalysis dataset (NOAA high resolution blended analysis). The basic data for computation of EFC is NCEP's $1 \times 1^{\circ}$ FNL dataset. *MADI*'s position and intensity are taken from IMD's best track data and its speed of motion is determined using the same data (RSMC, New Delhi 2014).

The methodology of computation of EFC is similar to the one followed by the earlier workers (Molinari and Vollaro 1989) and is detailed below. From the conventional zonal and meridional wind data, radial (u) and tangential (v) velocities are computed for the entire life period of *MADI* at 6-h intervals in a storm relative Lagrangian frame of reference. For this purpose, the storm movement, determined from the IMD's best track data at 6-h intervals, is subtracted from the actual radial and tangential velocities at each grid point.

The EFC at the 200 hPa level is computed at 6-h intervals for the entire life period of *MADI* in storm relative cylindrical co-ordinate system up to 12° from the centre with radial interval $\Delta r = 1°$ and azimuthal interval $\Delta\phi = 15°$ using bilinear interpolation according to following equation (Molinari and Vollaro 1989).

$$EFC = \frac{1}{r^2}\frac{\partial}{\partial r} r^2\overline{u'v'}$$

where u and v refer to storm relative radial and tangential velocities and prime refers to the deviations from the azimuthal mean, r is the radial distance from the TC centre.

For determining the asymmetry in the rainfall structure associated with *MADI* Tropical Rainfall Measuring Mission (TRMM) based $0.25 \times 0.25°$ rain rate data at 3-h intervals (3B42V7) is downloaded from the NASA website (Available at: http://disc.sci.gsfc.nasa.gov/). The Fourier first order wave number-1 asymmetry is calculated using the method followed by Lonfat et al. (2004). The computations are carried out in a TC centric moving co-ordinate system with the direction of motion of the TC as the reference direction. First, the mean rain rate in each 1° wide annulus ($\approx$10 km) around the TC centre is computed upto 5° radial distance (50 annuli). For each annulus, the first order Fourier coefficients are computed using all individual rain rates as

$$a_{\mathrm{i}} = \sum_{\mathrm{i}} [R_{\mathrm{i}} \cos\theta_{\mathrm{i}}] \qquad \text{and} \qquad b_{\mathrm{i}} = \sum_{\mathrm{i}} [R_{\mathrm{i}} \sin\theta_{\mathrm{i}}]$$

where R_{i} represents individual rain rates and θ_{i} the phase angle of the corresponding grid point relative to the direction of motion of the TC. The spatial structure of the first order asymmetry can then be represented by

$$M_1 = \frac{\lfloor a_1 \cos\theta + b_1 \sin_\theta \rfloor}{R}$$

where R is the mean rain rate calculated over the entire annulus. The asymmetry amplitudes are then normalised to the ambient mean rain rate of each annulus so that amplitude near unity implies that the wave number signal at that point is as strong as the axi-symmetric average.

3 Results and Discussions

3.1 Synoptic History of VSCS MADI

A low pressure area that emerged into south Andaman Sea on 01 December 2013 morning from south China Sea concentrated into a depression near latitude 10.0 °N/ longitude 84.0 °E on 6 December. It intensified into a cyclonic storm (CS; MWS: 34–47 knots) *MADI* with centre near latitude 10.5 °N/longitude 84.0 °E at 0000 UTC of 07 and into a severe cyclonic storm (SCS; MWS:48–63 knots) over the same region at 0900 UTC of 07. It then further intensified into a VSCS at 0600 UTC of 08 near latitude 12.3 °N/longitude 84.7 °E. It then moved to the north and weakened into an SCS at 1200 UTC of 09 December and lay centered near latitude 14.6 °N/longitude 84.7 °E. It continued to move slowly north-northeastwards till 0900 UTC of 10 December as an SCS up to latitude 15.7 °N/longitude 85.3 °E under influence of upper tropospheric steering ridge which moved northward along with northward movement of *MADI*. However, due to gradual weakening of *MADI*, the steering level changed from upper troposphere to lower and middle troposphere. The influence of the upper tropospheric anti-cyclonic circulation to the east of system centre decreased and that of lower and middle level anti-cyclonic circulation lying to the west of the system centre (over central India) increased. As a result, the SCS recurved westwards initially and then southwestwards commencing from 0900 UTC of 10. Under combined impact of colder SST, low Ocean thermal energy, high VWS and incursion of cold and dry air from central India into the core of the cyclone, it gradually weakened into a CS on 11 and crossed Tamil Nadu coast close to Vedaranyam as D around 1330 UTC of 12 December (RSMC, New Delhi 2014). Figure 1 presents the track of VSCS *MADI*.

The intensity changes associated with the VSCS, *MADI* and the role of EFC in the intensity changes are discussed in the following sections.

3.2 Environmental Influences on the Intensity of VSCS MADI – *VWS and SST*

The environmental VWS between 850 and 200 hPa layer and the intensity of the TC as depicted by the Estimated Central Pressure (ECP) during the life period of *MADI* are shown in Fig. 2a, b respectively. It is noted that while VWS direction remained more or less southerly to southwesterly, the magnitude varied significantly from about 4 m/s to 16 m/s. During the initial stages (from 06/06–07/18 UTC) the VWS was in the range 6–9 m/s and was favourable for intensification of *MADI* from the depression stage to SCS stage. However, during the period 08/00 TC to 10/06 UTC, the VWS was in the range 10–15 m/s and yet the system intensified into VSCS and sustained for about 2 days.

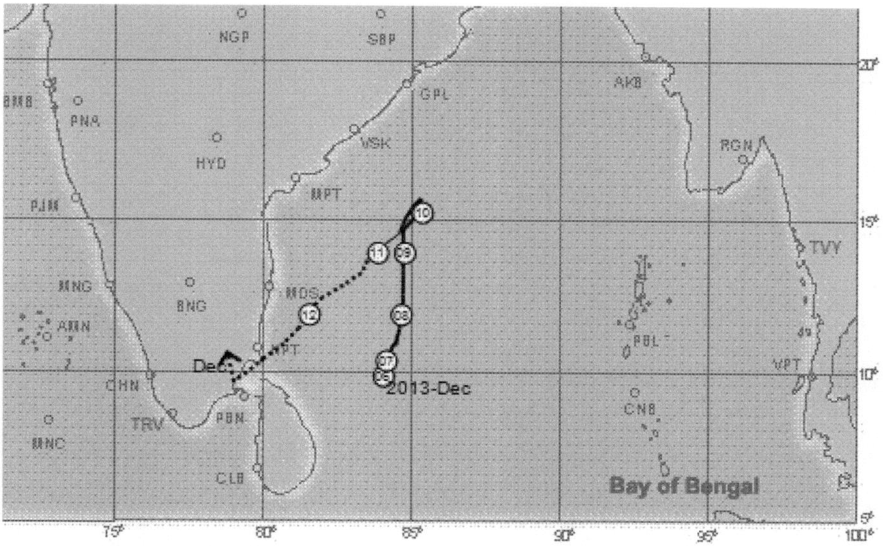

Fig. 1 Track of VSCS *MADI*: 6–13 December 2013. Source: (www.rmcchennaieatlas.tn.nic.in)

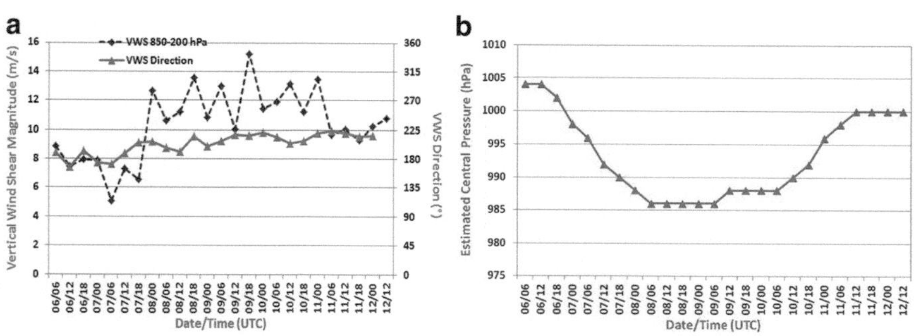

Fig. 2 (**a** and **b**). Time series of vertical wind shear (VWS) (**a**) and the estimated central pressure (**b**) during the life period of VSCS *MADI*

The SST anomaly averaged over the BOB region 80 °E–90 °E & 10–20 °N and SST over the TC centre (Fig. 3a, b) show that the system was in mostly under unfavorable SST conditions. Thus, both VWS and SST could not contribute for the sustenance of the TC as a SCS on 8 and 9. In this context, the role of another parameter known to influence the intensity change of a TC (Molinari and Vollaro 1989), viz., synoptic/large scale eddy forcing is examined for further analysis.

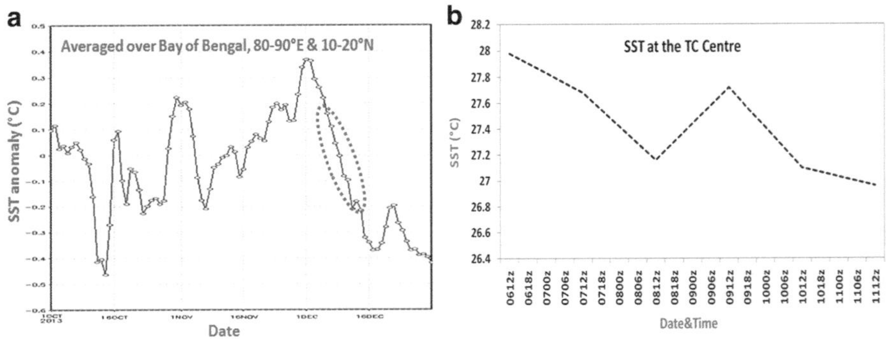

Fig. 3 (**a** and **b**). SST anomaly over the Bay of Bengal between 80–90 °E and 10–20 °N during the period 1 October–31 December 2013 (**a**) and SST at the TC centre during the life period of *MADI* (**b**). Dotted circle in (**a**) indicates the period of life cycle of *MADI*

3.3 EFC Changes vis-a-vis *Intensity Change of VSCS* MADI

The large/synoptic scale environmental interaction feature responsible for the eddy forcing is illustrated by the streamline flow and isotach analysis at 200 hPa and corresponding satellite pictures during 07/06 and 09/1200 UTC in Fig. 4a, b. Progression of a shortwave trough from west to east with increase of wind speed over northern side of *MADI*'s centre and northeastward asymmetric outflow may be noted. This synoptic situation provided enhanced poleward asymmetric outflow channel above the TC centre in the upper troposphere influencing the intensification of the system.

Figure 5a, b present the azimuthally averaged EFC (in units of m/s/day) and radial velocity (in m/s) during the life period of the TC. The x-axis of the plots depict the radial distance from the TC centre in degrees ($1° \approx 110$ km) and the y-axis corresponds to the date and time (UTC). During the initial stages of intensification (06/18–08/06 UTC) there was inward propagation of EFC (10 m/s/day) from about 700 to 400 km towards TC centre which favoured *MADI*'s intensification. Concomitantly, the outflow maximum (Fig. 5b), which was initially at about 600 km from the *MADI* centre, shifted inwards and was located near the TC centre during the mature stage also. During the recurving phase (09/12–10/00 UTC), EFC maximum (10–15 m/s/day) shifted outward from about 200 to 400 km due to deformation of equatorward anticyclonic outflow (Fig. 4b). However, at the same time strong EFC (15–25 m/s/day) was noticed at outer radii 800–1100 km which could be attributed to the influence of upper anti-cyclone in maintaining the poleward outflow. During the period 10/06–11/00 UTC strong negative EFC was noticed over the inner radii (upto 400 km) which contributed to the weakening of *MADI* as can be noted from the inflow in the radial velocity during 10/12–11/12 UTC indicating filling of the system. Even though there was strong positive EFC (about 10–15 m/s/day) at the outer radii (300–800 km) it could not help in maintaining the system further.

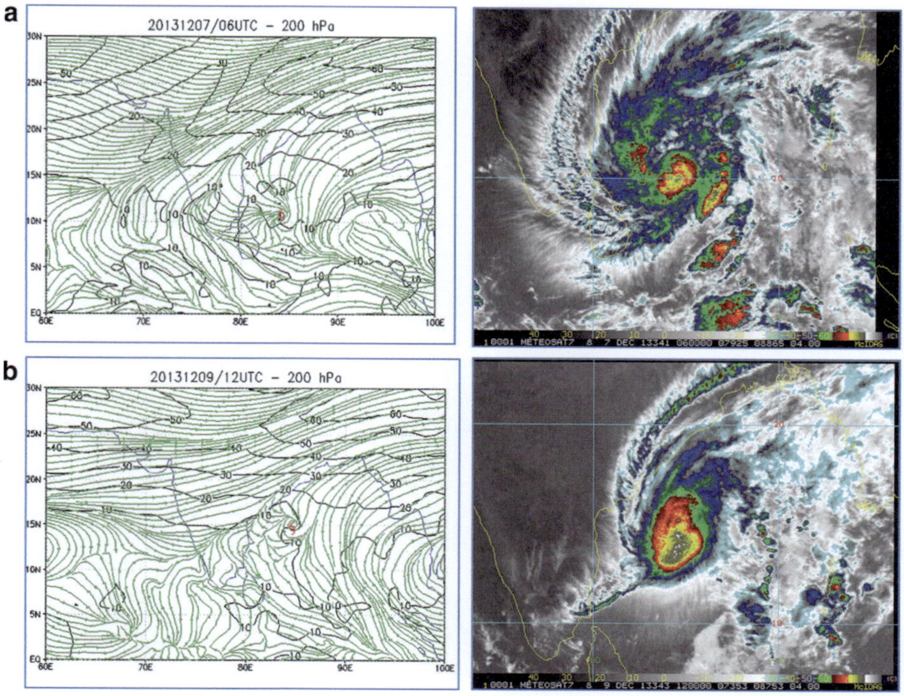

Fig. 4 (**a** and **b**). Streamline flow pattern and isotach analysis at 200 hPa during 07/06 UTC (**a**) and 09/1200 UTC (**b**) and corresponding satellite imageries (METEOSAT-7 with overlaid Cloud Top Temperatures; *Source*: rammb.cira.colostate.edu/). TC centre is indicated in red symbol in the wind analysis charts

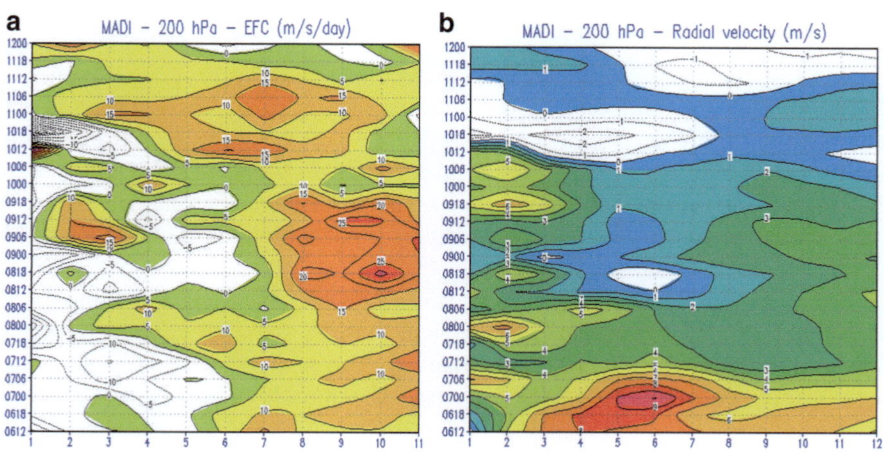

Fig. 5 (**a** and **b**). Azimuthally averaged EFC (m/s/day) (**a**) and radial velocity (m/s) (**b**) during the life period of *MADI*. *X*-axis corresponds to radial distance from the TC centre in degrees and *y*-axis corresponds to the date and time (UTC)

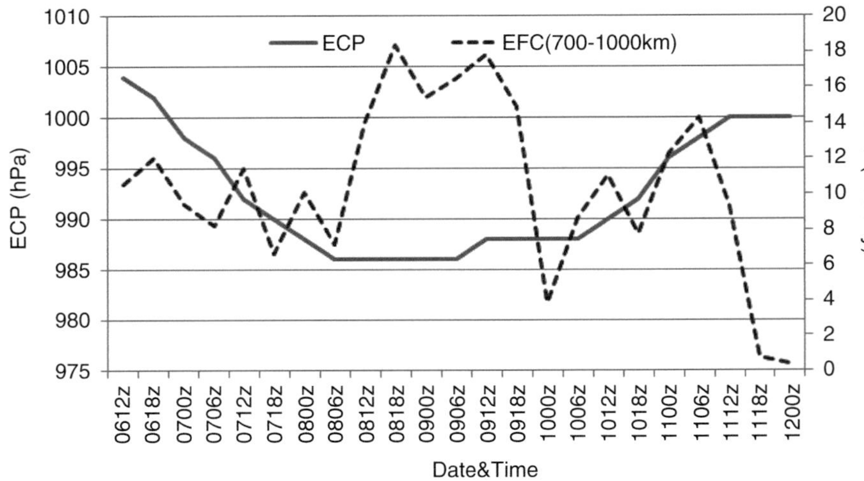

Fig. 6 Variation of EFC averaged over the radial band 700–1000 km with the TC intensity (represented by the ECP) at 6-h intervals during the life period of *MADI*

The role of the outer eddy forcing is further reiterated by analysing the variation of ECP with EFC (averaged over the 700–1000 km band) and the same is presented in Fig. 6. It may be noted that the value of EFC is greater than 10 m/s/day during intensification and the mature stage (06/12–09/12 UTC) but it decreased drastically during 09/12–09/18 UTC) which contributed to the weakening of *MADI*.

3.4 Structural Variations of VSCS MADI *– Rainfall Asymmetry Analysis*

In this section, structural variations associated with environmental influences are analysed in the precipitation field of *MADI* upto 300 km from its centre. Asymmetric rainfall structures associated with *MADI* are examined in the context of VWS and recurving motion using Wave number-1 asymmetry analysis. Figure 7 presents the asymmetry in rainfall structures during the intensification phase of *MADI*.

It may be noted that there was a cyclonic shift of wave number-1 asymmetry during the intensification phase. During the depression stage, the asymmetry maximum is noticed in the front left quadrant which moved to back left quadrant as system intensified into a CS which further moved to front right quadrant as it intensified further into a SCS. Also it is noted that relative magnitude of asymmetry decrease (increase) within (beyond) 200 km as TC intensity increases.

As mentioned earlier, during the recurving phase of *MADI*, VWS was high at about 12–14 m/s and there was a change in the relative orientation between VWS and the TC motion vectors. Figure 8 presents the asymmetry in rainfall structures

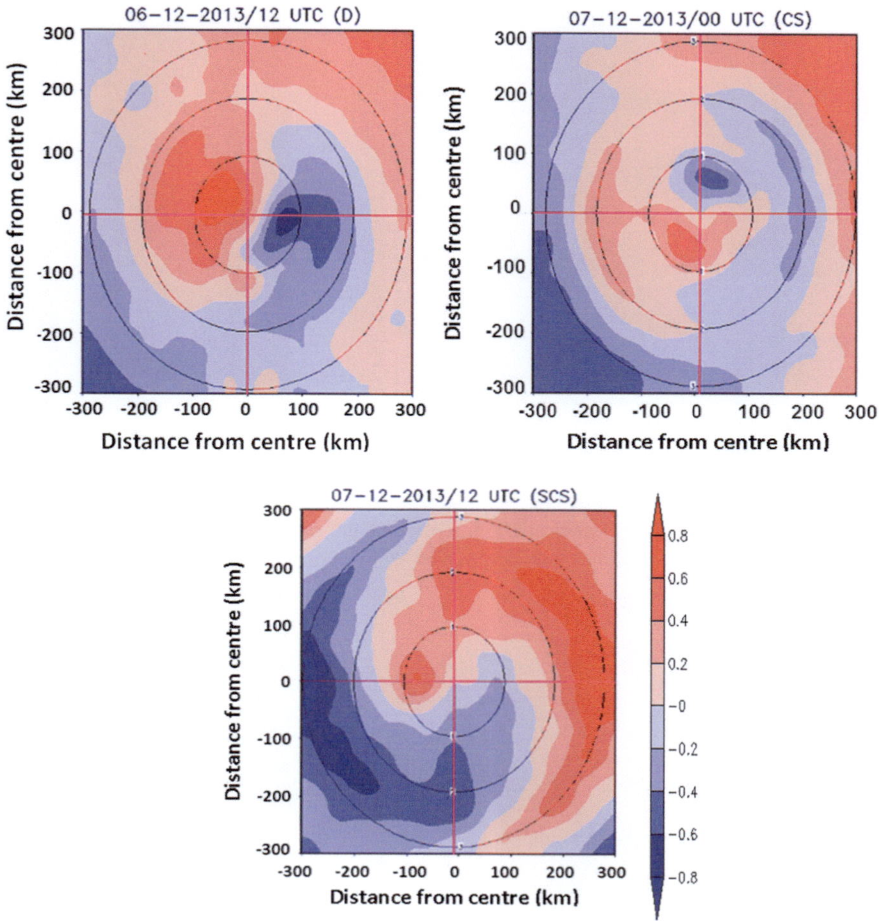

Fig. 7 Asymmetry in radial rainfall distribution during the intensification phase of VSCS *MADI* (06th/12 UTC, 7th/00 UTC and 7th/12 UTC), December 2013. Intensity at the respective times are indicated in *brackets* (See text for notations)

during the period of recurvature, viz., 10/06 and 10/12 UTC. It may be noted that the VWS and TC motion are in phase (180° out of phase) during its northward (southeastward) movement at 10/06 UTC (10/12 UTC). In both cases, the wave-1 asymmetry maxima are in the down shear – left quadrant indicating the influence of VWS in determining rainfall asymmetric structure.

However, during the intensification stage, between 07/00 and 07/18 UTC, VWS was less (about 5–7 m/s), but the speed of movement of the TC varied considerably. At 00 UTC of 7, it was 3 m/s and 12 UTC of 7, it was 7 m/s. Under weak VWS conditions, the effects of TC motion dominates and accordingly, a front asymmetry is noted (Fig. 9). These results are in line with those obtained by earlier workers (Chen et al. 2006).

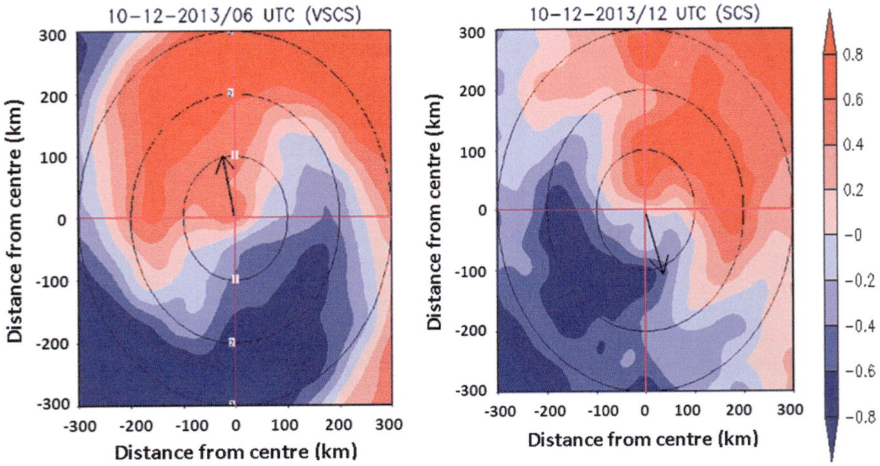

Fig. 8 Asymmetry in radial rainfall distribution at the time of recurvature of VSCS *MADI* – 10th/06 UTC and 10th/12 UTC, December 2013. *Arrows* at the centre indicate the direction of the vertical Wind Shear relative the direction of motion of the TC (shown to coincide with the positive-Y direction)

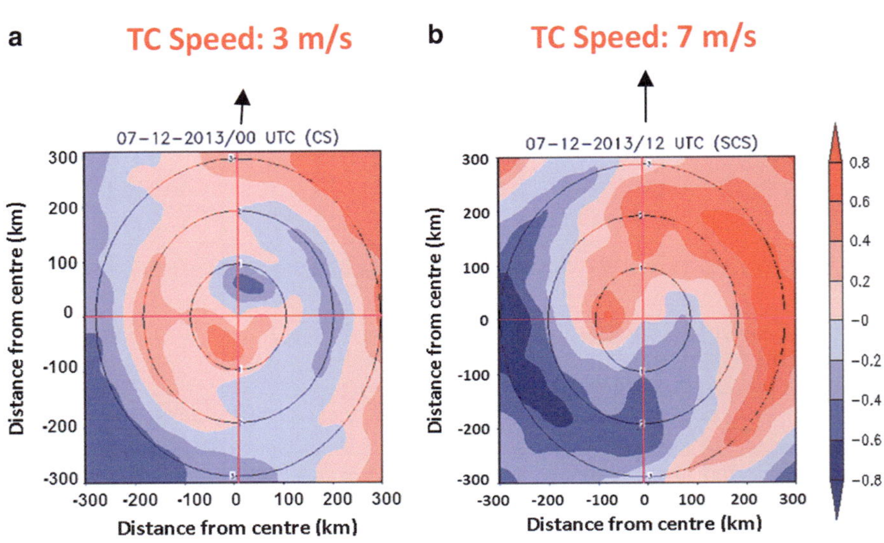

Fig. 9 Asymmetry in radial rainfall distribution in respect of VSCS *Madi* during 07th/00 UTC (**a**) and 07th /12 UTC, December 2013 (**b**) when the VWS was low (4–6 m/s) and oriented along the direction of motion of the TC. *Arrows* at the centre indicate the direction of the VWS relative the direction of motion of the TC (positive Y-direction). *Arrows* outside the plots indicate the TC motion vectors

4 Conclusion

Intensity and structural variations during the life cycle of VSCS, *MADI* that formed and affected the North Indian Ocean basin during 06–13 December 2013 are analysed in the context of environmental influences and following results are obtained:

1. During the life period of VSCS, *MADI*, the VWS and the SST over the Bay of Bengal were detrimental to its growth save for a lower VWS during the initial growth of the TC.
2. Environmental eddy forcings, through transport of angular momentum flux (quantified by eddy angular momentum convergence at the inner radii) have contributed to the sustenance of *MADI* during its mature stage. The eddy interaction is manifested in the intensity change with shift of positive EFC (10 m/s/day) towards inner radius and concomitant inward shift of radial outflow maxima during the intensification phase.
3. During the Mature stage, high values of EFC (15–25 m/s/day) were noticed at the outer radii which helped to maintain the poleward outflow.
4. A drastic decrease in the EFC is noticed at the start of the weakening phase.
5. Structural analysis carried out on rainfall asymmetry in the context of VWS and TC motion shows that wave number-1 asymmetry shifts cyclonically from front left to front right quadrant with increasing intensity and the relative magnitude of asymmetry decrease (increase) within (beyond) 200 km as TC intensity increases.
6. During recurving stage, VWS was high and there was a change in the relative orientation between VWS and TC motion vectors. Under strong shear conditions asymmetry maximum was noted in down shear – left quadrant.
7. During the initial intensification stage, when the VWS was less, the role of TC speed manifested on the rainfall structure with front asymmetry.

While the present analysis is a case study, similar computations for more TC cases would help in understanding the characteristics and quantifying the environmental influence on the intensity and structural variations of TCs of NIO.

Acknowledgements The authors thank the Deputy Director General of Meteorology, Regional Meteorological Centre, Chennai for providing facilities to undertake this study.

References

Burpee, R.W. and Black, M.L. (1989). Temporal and spatial variations of rainfall near the centres of two tropical cyclones. *Monthly Weather Review*, **117**, 2204–2218.

Chen, S.S, Knaff, J.A. and Marks, F.D. (2006). Effects of Vertical Wind Shear and Storm Motion in Tropical Cyclone Rainfall Asymmetries deduced from TRMM. *Monthly Weather Review*, **134**, 3190–3208.

DeMaria, M, Baik, J.J. and Kaplan, J. (1993). Upper-Level Eddy Angular Momentum Fluxes and Tropical Cyclone Intensity Change. *Journal of Atmospheric Sciences*, **50**, 1133–1147.

Hanley, D, Molinari, J. and Keyser, D. (2001). A composite study of the interactions between tropical cyclones and upper-tropospheric troughs. *Monthly Weather Review*, **129**, 2570–2584.

Lonfat, M, Marks, F.D. Jr. and Chen, S.S. (2004). Precipitation distribution in tropical cyclones using the Tropical Rainfall Measuring Mission (TRMM) microwave imager: A global perspective. *Monthly Weather Review*, **132**, 1645–1660.

Marks, F.D. Jr. (1985). Evolution of the structure of precipitation in Hurricane Allen (1980). *Monthly Weather Review*, **113**, 909–930.

McBride, J.L. and Zehr, R. (1981). Observational analysis of tropical cyclone formation. Part II: Comparison of non-developing versus developing systems. *Journal of Atmospheric Sciences*, **38**, 1132–1151.

Molinari, J. and Vollaro, D. (1989). External influences on hurricane intensity. Part I: Outflow layer Eddy angular momentum fluxes. *Journal of Atmospheric Sciences*, **46**, 1093–1105.

Ramage, C.S. (1974). The Typhoons of October 1970 in the South China Sea: Intensification, Decay and Ocean Interaction. *Journal of Applied Meteorology*, **13**, 739–751.

Rhome, J.R. and Raman, S. (2006). Environmental influences on tropical cyclone structure and intensity: A review of past and present literature. *Indian Journal of Geo-Marine Sciences*, **35**, 61–74.

Rodgers, E.B. and Adler, R.F. (1981). Tropical cyclone rainfall characteristics as determined from a satellite passive microwave radiometer. *Monthly Weather Review*, **109**, 506–521.

RSMC, New Delhi (2014). Report on Cyclonic Disturbances over north Indian Ocean during, 2013, published by RSMC, New Delhi, pp. 1–236.

Rucker, J.H. (1992). Upper-Tropospheric Forcing on the Intensification Rates of Tropical Cyclones Flo and Ed based on TCM-90 observations", Masters' Thesis – Naval Postgraduate School, Monterey, California.

Yu, H. and Kwon, H.J. (2005). Effect of TC-Trough Interaction on the Intensity Change of Two Typhoons. *Weather Forecasting*, **20**, 199–211.

The Role of Information System in Data/Product/Warning Dissemination and Future Improvements

S.L. Singh and Kuldeep Srivastava

1 Introduction

Communication plays a significant role in (1) the data collection from different observatories all over Indian region and other countries, (2) dissemination of data to different agencies for processing and generating information useful for forecaster, (3) exchanging the processed data and product among forecasters and other users and (4) reception of forecast and warnings from forecasters and dissemination to various agencies involved in disaster management.

Meteorological telecommunication of India Meteorological Department (IMD) consists of an integrated network of point-to-point circuits and multipoint circuits which interconnect meteorological centers within the country and the world for receiving meteorological data/products and relaying it selectively. It is mainly organized in two levels viz, the meteorological telecommunication network within the Global Telecommunication System (GTS) of World Weather Watch (WWW) program of World Meteorological Organization (WMO), and the National Meteorological Telecommunication Network. A new latest state-of-the-art Automatic Message Switching System (AMSS) at Regional Telecommunication Hub (RTH), New Delhi is functional from mid 2009. As per WMO mandate, mirror RTH as a disaster recovery system (DRS) and Global Information System Centre (GISC) has been installed at Pune under WMO Information System (WIS) programme of WMO. With the WIS implementation, IMD status has come up from Regional level to Global data dissemination centre.

The role of communication is very important in forecasting weather events. With the advent of new technologies day by day, the communication has changed rapidly from very old Telegram, very low speed (few baud) Teleprinter and telex to high

S.L. Singh (✉) • K. Srivastava
India Meteorological Department, Lodi Road, 110003 New Delhi, India
e-mail: sl.singh@imd.gov.in

© Capital Publishing Company 2017
M. Mohapatra et al. (eds.), *Tropical Cyclone Activity over the North Indian Ocean*,
DOI 10.1007/978-3-319-40576-6_5

Table 1 Milestones of IMD's telecommunication facilities

Year	System	High speed circuits	Maximum speed handling	Speed of other circuits	Data exchange per day	Type of data
1974	Philips DS-714	2	2400 bps	50–300 bps	Few KB	Alphanumeric
1988	VAX-11/750	8	9600 bps	50–300 bps	3 MB	Scanned file + alphanumeric
2000	NETSYS SUN E-250 (weatherman)	128	128 kbps	50–128 kbps	4 GB	Charts + files + alphanumeric
2009	TRANSMET (operational RTH till date)	No limit; any practical no. of circuits	Present – 8 Mbps, supports any practical speed	64 Kbps to 8 Mbps, supports any practical speed	400 GB	Different modes/formats like audio, video, binary ASCII, pdf etc.
2013	Messir-Comm and Messir-WIS (Mirror RTH and GISC under framework of WIS till date)	No limit; any practical number of circuits	Present – 8 Mbps, supports any practical speed	64 Kbps to 8 Mbps, supports any practical speed	400 GB	Different modes/formats like audio, video, binary ASCII, pdf etc.

speed (Gbps) internet, Multiprotocol Label Switching Virtual Private Network (MPLS VPN), mobile etc. with continuous growth. Table 1 gives various milestones of IMD's automated telecommunication facilities briefly since its inception.

2 Information System of IMD

Information exchange system of IMD consists of two AMSS (one for national data and the other for International data) at Headquarter (HQ), New Delhi, six AMSS at six Regional Meteorological Centres (RMCs), Mirror RTH system and GISC at Pune for meteorological data collection and dissemination system. The data flow within IMD is given in Fig. 1. All the systems are running in redundant hot standby mode to avoid any single point failure.

Under modernization of IMD, a new project named VARSAMANA was taken up to augment the observational, communication and forecasting system. The various systems under the project and their data flow are shown in Fig. 2. Central Information and Processing System (CIPS) acts as task centre for creation/development/production of various value added meteorological products according to needs of the forecasters. Synergy systems (forecasters' workstation) have been installed at 14 locations (HQ and important field stations) for overlay of various meteorological parameters one over the other, charting, plotting and drawing the

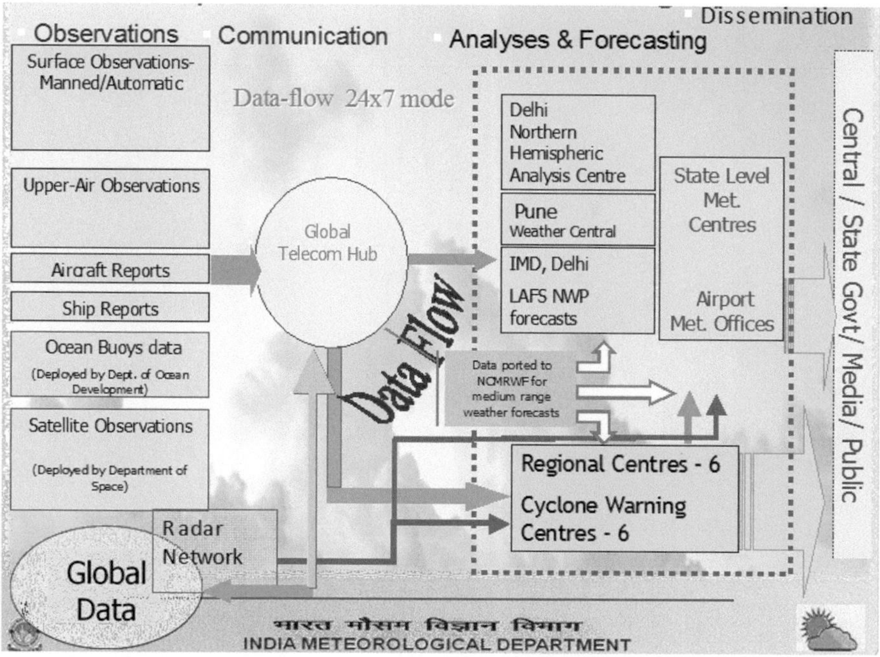

Fig. 1 IMD's data flow for weather monitoring, forecasting and warning services

isolines for better visualisation/analysis of weather phenomena and issue of weather forecasts/warnings/alerts etc.

Public Weather Service (PWS) system, installed at HQ, RMC New Delhi and Kolkata, consists of MeteoFactory, Visumet and TV systems. Meteo-Factory system can automatically produce the weather forecast bulletins into desired format, schedule and disseminate it through various means like ftp, email, SMS, voice etc at right time to the end users. MeteoFactory system ingests meteorological raw data, processes these data and produces automatically tailored products and services. MeteoFactory generates all India daily weather forecast, city forecast and rainfall distribution, sub-division wise city forecast, national daily weather report, warning product and Video capsule for Doordarshan (National Television (TV)). Visumet system is used to display various forecast products and warnings on display screens located at various places. The TV system is used to produce the capsules for Doordarshan (National TV) news bulletins.

In addition to above, a Video Wall system is being used by forecasters to display various products/forecasts/weather events etc. in dynamic mode to the visitors and media persons.

IMD has Video Conferencing (VC) system at nine locations including five Regional Meteorological Centers (Mumbai, Chennai, Kolkata, Nagpur and Guwahati), Weather forecasting Center (Pune), National Weather Forecasting Centre (NWFC, New Delhi), Meteorological Center (Bhubaneswar) and

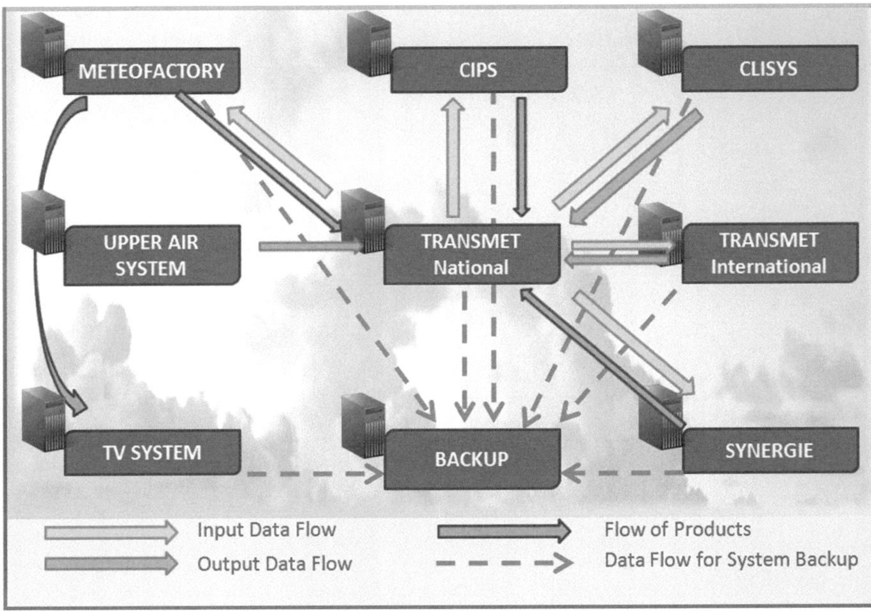

Fig. 2 VARSAMANA system in IMD and their data flow

Meteorological Training Institute (Pune). Forecasters of HQ and RMCs find video conferencing system very much useful to discuss among themselves the daily weather forecast as well as during disastrous weather events such as cyclone, heavy rainfall, cloud burst, flash flood etc. Further, IMD is using Multi conferencing unit (RMX 2000) of M/s Polycom installed at IMD Head quarter (NWFC) for organising multi-conferencing.

Various types of communication media and facilities used in India Meteorological Department for collection and dissemination of weather data, forecast products and cyclone warning/other weather alerts etc. are as follows.

1. Telephone/Fax
2. Digital Meteorological Data Dissemination (DMDD)
3. Very Small Aperture Terminal (VSAT)
4. Internet (two ISPs + National knowledge network (NKN)
5. Multi Protocol Label Switching-Internet Protocol Virtual Private Network (MPLS IP-VPN)
6. IMD websites, RSMC New Delhi website
7. Short Message Service (SMS)
8. Email
9. Navigation Telex (NAVTEX)

3 Bulletin Issued During Cyclone Period

Regional Specialised Meteorological Centre (RSMC), New Delhi prepares and disseminates the following bulletins (IMD 2013).

3.1 Tropical Weather Outlook

Tropical Weather Outlook is issued daily at 0600 UTC based on 0300 UTC observations in normal weather for use of the member countries of WMO – Economic and Social Commission for Asia and the Pacific (ESCAP) Panel on tropical cyclones. This contains description of synoptic systems over NIO along with information on significant cloud systems as seen in satellite imageries. It also provides probabilistic genesis forecast (formation of depression) over Bay of Bengal and Arabian sea separately for day 1 (up to 24 h), day 2 (24–48 h) and day 3 (48–72 h). The forecast is issued in probabilistic terms like nil, low, fair, moderate and high probability corresponding to expected probability of occurrence of 00, 01–25, 26–50, 51–75 and 75–100 %. This forecast has been introduced on 1 June 2014.

3.2 Special Tropical Weather Outlook

The Special Tropical Weather Outlooks are issued at 0600 and 1500 UTC based on 0300 and 1200 UTC observations, respectively when a tropical depression forms over NIO. The special tropical outlook indicates discussion on various diagnostic and prognostic parameters. The 120-h track and intensity forecasts are issued from the stage of deep depression. The track and intensity forecast are issued for +06, +12, +18, +24, +36, +48, +60, ... 120 h or till the system is likely to weaken into a low-pressure area. These bulletins contain the current position and intensity, past movement, central pressure of the cyclone, description of satellite imageries, cloud imageries, expected direction and speed of movement, expected track and intensity of the system up to 120 h. It also includes the description of sea condition. The time of issue of this bulletin is HH + 03 h. The cone of uncertainty in the track forecast is also included in the graphical presentation of the bulletin (Fig. 3). Tropical weather outlooks are transmitted to WMO/ESCAP panel member countries through GTS and are also made available on real time basis through internet at IMD's website: www.rsmcnewdelhi.imd.gov.in.

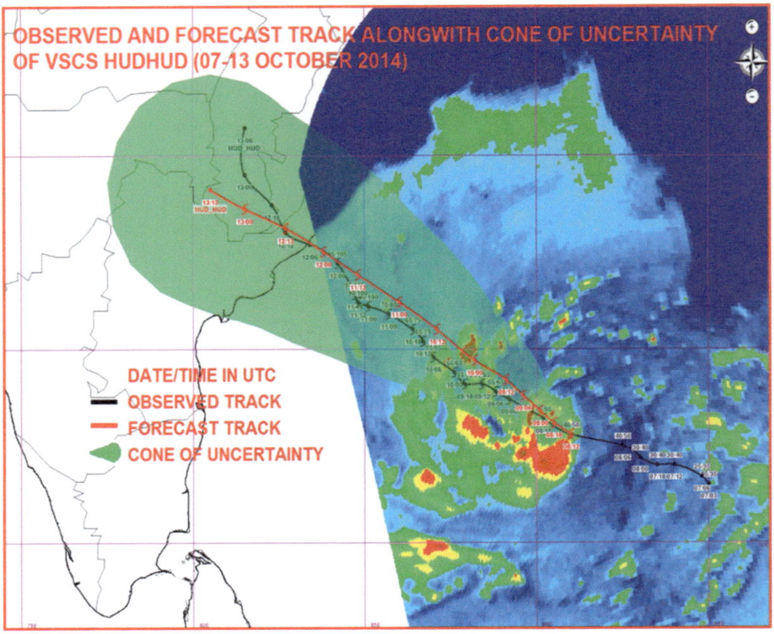

Fig. 3 A typical example of observed and forecast track of TC HUDHUD during October 2014

3.3 Tropical Cyclone Advisories

Tropical cyclone advisory bulletin is issued when a deep depression intensifies into a tropical cyclone (TC, wind speed = 34 knots or more). It replaces the 'Special Tropical Weather Outlook' bulletin. Tropical cyclone advisories are issued at 3-h intervals based on 00, 03, 06, 09, 12, 15, 18 and 21 UTC observations. The time of issue is HH + 03 h. These bulletins contain the current position and intensity, past movement, central pressure of the cyclone, description of satellite imageries, cloud imageries, expected direction and speed of movement, expected track and intensity of the system up to 120 h. The expected point and time of landfall, forecast winds, squally weather and state of the sea in and around the system are also mentioned. Storm surge guidance is provided in the bulletin as and when required. Tropical cyclone advisories are transmitted to panel member countries through GTS and are also made available on real time basis through internet at IMD's website: www.rsmcnewdelhi.imd.gov.in.

3.4 Global Maritime Distress and Safety System (GMDSS)

Under Global Maritime Distress and Safety System (GMDSS) Scheme, India has been designated as one of the 16 services in the world for issuing sea-area bulletins for broadcast through GMDSS for MET AREA VIII (N), which covers a large portion of NIO. As a routine, two GMDSS bulletins are issued at 0900 and 1800 UTC. During cyclonic situations, additional bulletins (up to 4) are issued for GMDSS broadcast. In addition, coastal weather and warning bulletins are also issued for broadcast through NAVTEX transmitting stations.

3.5 Tropical Cyclone Advisories for Aviation

Tropical Cyclone Advisories for aviation are issued for international aviation as soon as any disturbance over the NIO attains or is likely to attain the intensity of cyclonic storm (maximum sustained surface wind speed $\geq$34 knots) within next 6 h. These bulletins are issued at 6-h intervals based on 00, 06, 12 and 18 UTC synoptic charts and the time of issue is HH + 03 h. These bulletins contain present location of cyclone in latitude/longitude, maximum sustained surface wind (in knots), direction of past movement and estimated central pressure, forecast position in latitude/longitude and forecast winds in knots valid at HH + 6, HH + 12, HH + 18 and HH + 24 h in coded form. The tropical cyclone advisories are transmitted on real time basis through GTS and AFTN channels to designated International Airports of the region prescribed by International Civil Aviatin Organisation (ICAO) and by ftp to WMO's Aviation Disaster Risk Reduction (ADRR) at Hong Kong in coded form. It is also being sent in graphics (png format) through GTS for above users.

3.6 Bulletin for India Coast

These bulletins are issued from the stage of depression onwards. During the stage of depression/deep depression; it is issued based on 00, 03, 06, 12 and 18 UTC observations.

When the system intensifies into a cyclonic storm over NIO, these bulletins are issued at 00, 03, 06, 09, 12, 15, 18 and 21 UTC (every 3-h interval) based on previous observations. This bulletin contains present status of the system i.e. location, intensity; past movement and forecast intensity and movement for next 120 h or till the systems weaken into a low pressure area, likely landfall point and time and likely adverse weather including heavy rain, gale wind and storm surge. Expected damage and action suggested are also included in the bulletins. This bulletin is completely meant for national users and these are disseminated through various modes of communication including All India Radio, Doordarshan (National TV), Telephone/Fax, SMS Print and electronic media. It is also posted on RSMC New Delhi website.

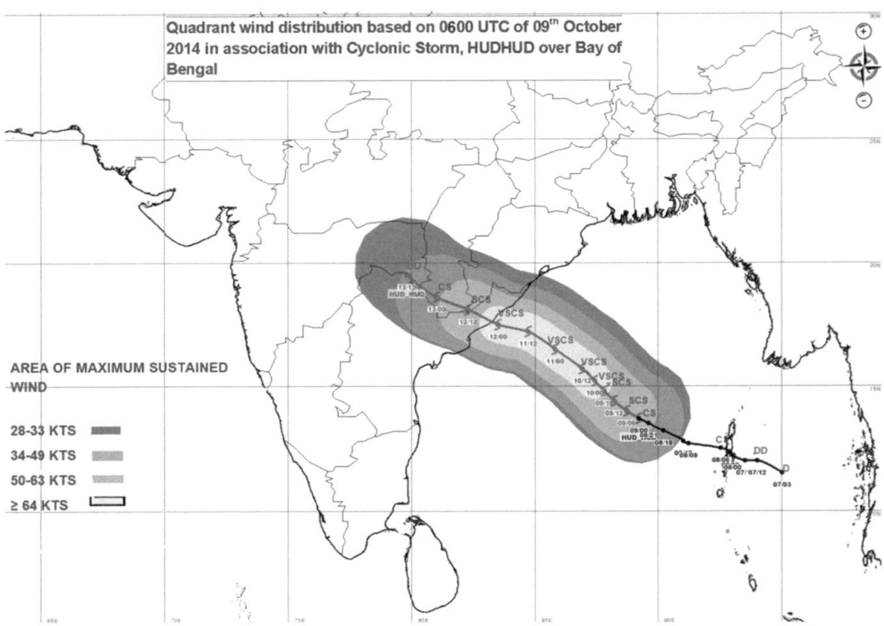

Fig. 4 A typical graphical presentation of cyclone wind forecast during TC HUDHUD in October 2014

3.7 Wind Forecast for Different Quadrants

The forecast of radius of maximum sustained wind in four quadrants northeast (NE), northwest (NW), southeast (SE) and southwest (SW) of a cyclone commenced with effect from cyclone, GIRI during October 2010. In this forecast, the radius of 28, 34, 50 and 64 knot winds are given for various forecast periods like +06, +12, +18, +24, +36, +48, +60, … 120 h. A typical graphical presentation of this forecast is shown in Fig. 4. This quadrant wind forecast is issued as bulletin from the deep depression stage onwards to various users through global telecommunication system.

3.8 TC Vital

The TC Vital is issued by RSMC, New Delhi to various NWP Centers in coded form for their use in creating the synthetic vortex in NWP models and running storm surge and coastal inundation model. It is issued 4 times a day based on 00, 06, 12 and 18 UTC. This bulletin contains the information on location (latitude/ longitude), intensity (MSW and estimated central pressure), movement (speed/ direction), size, radius of maximum wind and wind radii of 34 knot wind in four geographical quadrants namely northeast (NE), northwest (NW), southeast (SE) and southwest (SW) quadrants etc. This system has been introduced in 2012.

3.9 Cone of Uncertainty Forecast

The Cone of uncertainty (COU) represents the probable position of a CD/TC's Circulation Center, and is made by drawing a set of circles centered at each forecast point – 06, 12, 18, 24, 36, 48, 60, 72, 84, 96, 108 and 120 h for a 5-day forecast. The radius of each circle is equal to the average official track forecast errors of 10, 20, 30, 45 60, 80, 100, 120, 135, 150, 160, 170 and 180 nautical miles for 06, 12, 18, 24, 36, 48, 60, 72, 84, 96, 108 and 120 h forecasts, respectively. The radii of circle to construct cone of uncertainty have been changed based on the average error of 2009–2013. The new radii have been introduced with effect from cyclone, Hudhud in October 2014.

4 SMS

RTH system transmits the cyclone warnings/alerts to administrators and authorities engaged in disaster mitigation selectively depending upon the likely affected coastal areas. Extracts of the Bulletin for Indian coast are to be sent to disaster management officials by SMS using 'Mobile Seva' under Digital India Programme of Government of India. The messages are to be sent to IMD and disaster management officials at New Delhi and the states likely to be affected as well as to registered users of the respective states. The content of the message will be of 200 characters including space between words. SMS messages are sent to farmers through Kisan portal, to fishermen through Indian National Centre for Ocean Information Services (INCOIS) SMS system and to registered public by RSMC New Delhi.

5 Access/Utilization of Information System During Cyclone Period

All the information systems at IMD have been utilised optimally during cyclone periods, which resulted efficient data/products collection/exchange in almost near real time utilising the automated communication system and high speed communication links, their analysis through High Power Computing System (HPCS), task development tool of CIPS, value addition by various information systems, visualisation through synergie systems etc. It resulted in issue of accurate cyclone track

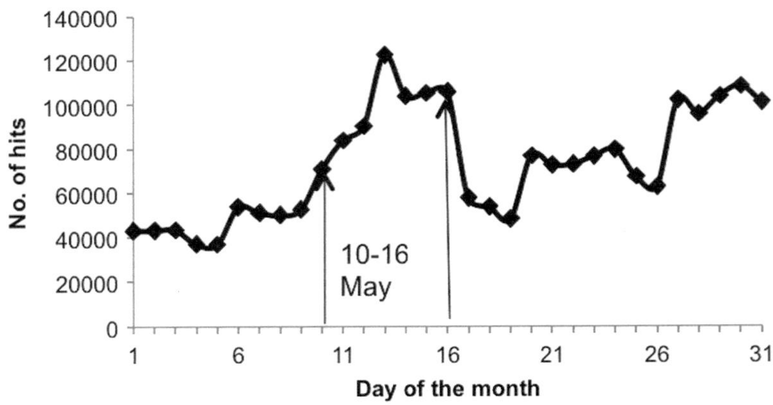

Fig. 5 Hits on IMD website during CS Viyaru (10–16 May 2013, Landfall occurred 1330 h IST of 16 May)

prediction/forecast/alerts for the public and authorities engaged in disaster mitigation. Such systems thus led to minimise the loss of valuable life and national property.

During cyclone period, starting from its genesis, maturity and even beyond landfall stage, the public, authorities and scientists had constant watch on the weather and its effect which is well evident from access/hits on the IMD's public web portal www.imd.gov.in during the period. Such analysis for each cyclone during the year 2013 is given in Figs. 5, 6, 7 and 8.

Figures 5, 6, 7 and 8 show the number of hits on IMD website www.imd.gov.in during (a) cyclonic storm (CS) Viyaru, (b) very severe cyclonic storm (VSCS), Phailin, (c) severe cyclonic storm (SCS) Helen and VSCS Lehar and (d) VSCS Madi respectively. Details about these cyclones are available in RSMC, New Delhi (2014). As shown in Fig. 5, the period of cyclone was 10–16 May 2013 with landfall around 1330 h IST of 16 May 2013. The hits on IMD website started increasing from 9 May onwards reaching a peak value of 122,671 hits on 13 May and then started decreasing after 16 onwards. The hits again increased after 26 May and remained high (between 96,029 and 108,456) in comparison to monthly average (71,229 for May 2013) till 31 May due to depression in Bay of Bengal from 29 to 31 May 2013. During the VSCS 'Phailin' (08–14 October 2013), the access to the website started increasing from 7 October onwards reaching to a peak value (~2.2 lacs) on 12 October (landfall at 2230 h IST) with decreasing trend afterwards with daily number of hits being more than 1 00 000 during the entire cyclone period (Fig. 6). During SCS Helen (19–23 November 2013, Landfall 1400 h IST of 22 November) and VSCS Lehar (23–28 November 2013, Landfall 0630 h IST of 25 November and again 1400 h IST of 28 November), it clearly indicates that the access to the website has gone up during the cyclone periods with a comparatively higher value during the VSCS period. The increased number of hits during 13–-17 November may be attributed to a depression during the period in Bay of Bengal

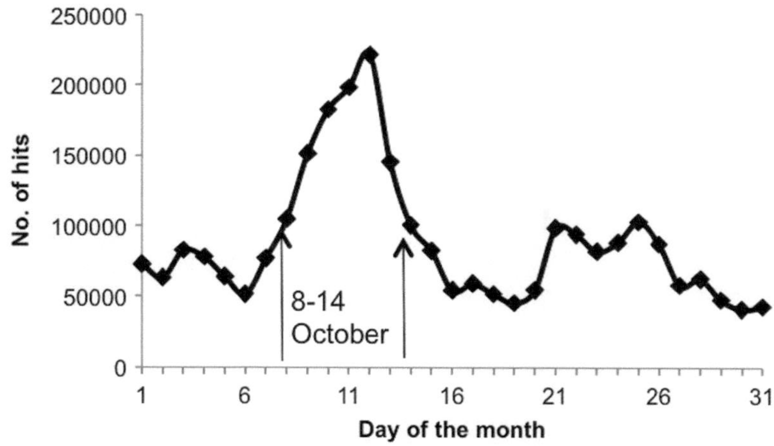

Fig. 6 Hits on IMD website during VSCS Phailin (8–14 October 2013, Landfall occurred 2230 h IST of 12 October)

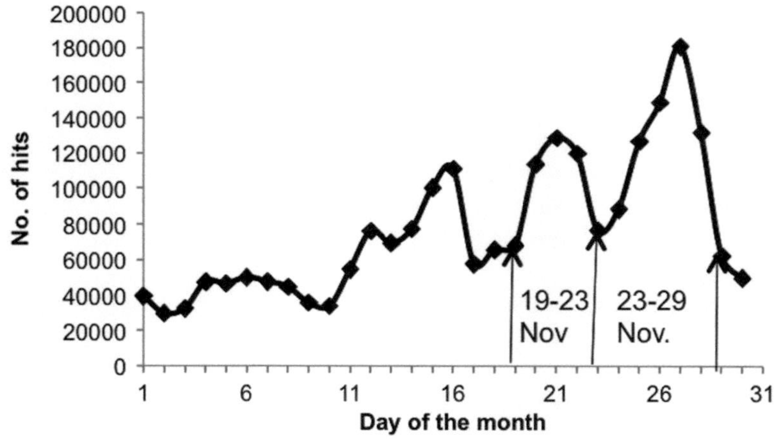

Fig. 7 Hits on IMD website during SCS Helen (19–23 November 2013, Landfall occurred at 1400 h IST of 22 November) and VSCS Lehar (23–28 November 2013, landfall occurred at 0630 h IST of 25 November and again at 1400 h IST of 28 November)

(Fig. 7). Hits on IMD website during VSCS Madi (6–13 December 2013, Landfall at 1900 and 2230 h IST of 12 December) indicate the peaks on the initial maturity date (7 December) and landfall date (Fig. 8).

The access to IMD website during the cyclone period for all the cyclones in the year 2013 have been compared with monthly average access during the respective cyclone months and yearly average access as depicted in Fig. 9. It clearly indicates that the peak access to the IMD website is higher for VSCS compared to CS and SCS, except for VSCS 'Madi' for the cyclones during the year 2013. The lower peak access in case of VSCS Madi may be due to its unique track with near

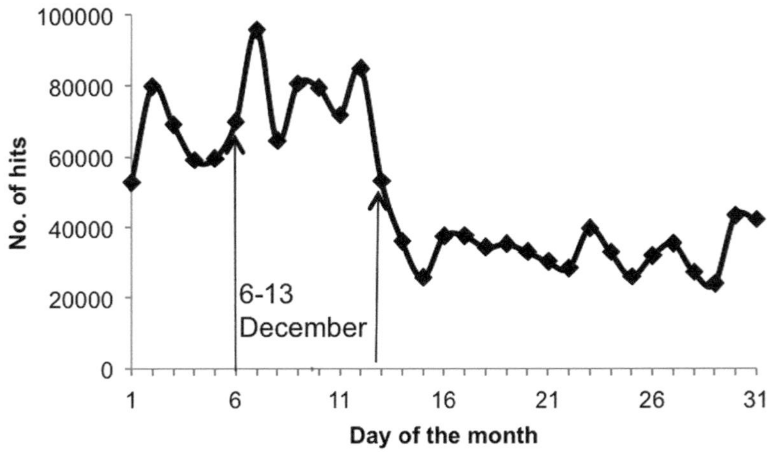

Fig. 8 Hits on IMD website during VSCS Madi (6–13 December 2013, landfall occurred at 1900 and 2230 h IST of 12 December)

northward movement till 15.7° N. Its speed peaked up gradually after recurving south-westwards to Tamil Nadu coast. Moreover, its CS (7 December morning)/ SCS (7 December afternoon) and VSCS (8 December Forenoon) duration was comparatively less as it weakened into SCS in the evening of 9 December with further weakening into CS in the early hours of 11 December, into deep depression in the morning of 11 and depression in the night of 11 December. The VSCS Madi had already weakened while it was far away from the coast and turned into depression much before the landfall, which might be the reason for less number of hits to the website.

Figure 10 shows the access of the IMD website in general for the entire year 2013. It indicates that its access gets increased mainly during the bad weather period. Further it shows bimodal behaviour with primary peak in October and November and a secondary peak in May and June like the cyclonic storm over the north Indian Ocean.

6 Conclusions and Future Scope

The study indicates that the information systems of IMD have been quite efficient for collection of meteorological data and its analysis for forecast/alerts/warning etc. for the cyclones. The website of IMD (www.imd.gov.in) has been utilised optimally by public/scientists/researchers and authorities engaged in the disaster mitigation during bad weather period. People kept constant watch on the various stages of cyclones occurred during the year 2013. The watch was directly related to the intensity of the cyclone, its track and likely affected area of the country. It is clearly evident from the hits on IMD website in case of cyclone 'Phailin' vis-a-vis other

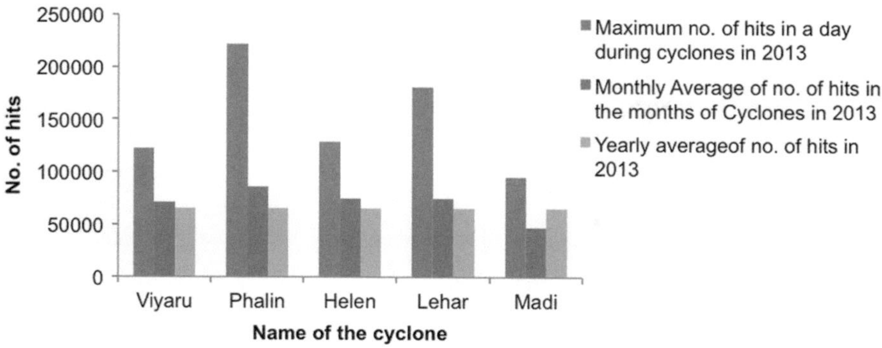

Fig. 9 IMD website access comparison for the cyclones in year 2013

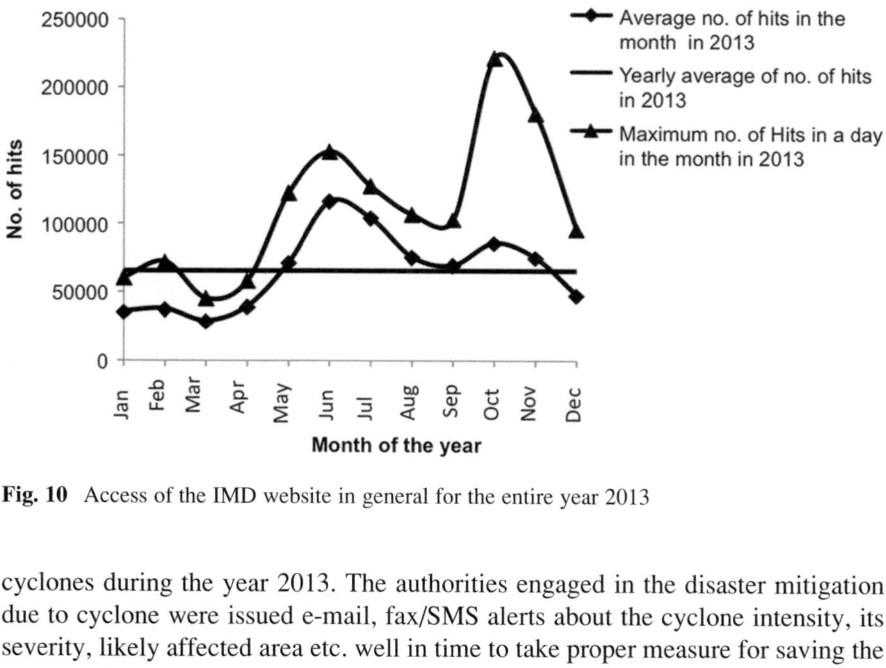

Fig. 10 Access of the IMD website in general for the entire year 2013

cyclones during the year 2013. The authorities engaged in the disaster mitigation due to cyclone were issued e-mail, fax/SMS alerts about the cyclone intensity, its severity, likely affected area etc. well in time to take proper measure for saving the loss of valuable life and property. Casualties could be minimised in case of all the cyclones for the year 2013. The study indicates the entire IMD information system must be quite efficient and run round the clock without any failure to cater such important services to the public/nation for saving the life and property.

Though IMD information system has been upgraded in 2009 during modernisation programme, still there is scope of further improvement as below:

1. All the important national links for data exchange are to be upgraded. Action for upgrading the existing national IP VPN link is in progress.
2. Internet Leased Line (ILL) bandwidth is to be upgraded for easy/faster access of IMD website by the public. Upgradation of the ILL bandwidth from existing

45–100 Mbps has been implemented. Further up gradation to 120 Mbps is in process.
3. Web server of IMD website should be capable of handling larger number of clients concurrently at a faster rate. Action for replacement of the server with higher configuration is in progress.
4. Revamping of IMD local area network (LAN) at HQ with proper security system – upgradation of IMD LAN at HQ along with security system is in the final stage. This will help smooth exchange of data over LAN and prevent any attack from outside world.
5. A new content managed GIS based website of IMD is under development, which may serve people with value added information.
6. Desktop video conferencing at all important stations are under development for instant discussion and information exchange, specially during bad weather period.

Acknowledgements Help and support by Shri Dinesh Kumar, Astt. Meteorologist, ISS Division for providing the data about the access on IMD website is gratefully acknowledged.

References

IMD (2013). Cyclone Warning in India: Standard Operation Procedure. Cyclone Warning Division, New Delhi, pp. 1–172.
RSMC, New Delhi (2014). Report on Cyclonic Disturbances over north Indian Ocean during, 2013. RSMC, New Delhi, pp. 1–236.

Management of Post-landfall Riverine Flooding

S.K. Jena

1 Introduction

Tropical Cyclones (TCs) are the most significant weather phenomena in North Indian Ocean (NIO), where these affect Indian sub-continent in general and India in particular. The inherent vulnerability features of TC in the basin lie with a long coastline of about 7516 km of flat coastal terrain, shallow continental shelf, geographical location and physiological features (Prasad and Rao 2006). Besides these, some other factors which influence vulnerability are limitations of observation networks, prediction systems, understanding of physical processes, early warning systems and disaster management, apart from socioeconomic conditions according to World Meteorological Organisation (WMO).

The character of TCs is associated with adverse weather conditions like heavy rainfall, gale wind and storm surge. Of these, the most significant impact of TCs is the resulting colossal amount of rainfall (Chen et al. 2013). Heavy rainfall associated with a storm varies from storm to storm even with the same intensity. Recorded rainfall in a cyclonic storm has been as low as trace to as high as 250 cm. It has been found that the intensity of rainfall is about 85 cm/day within a radius of 50 km and about 35 cm/day between 50 and 100 km from the centre of the storm. Precipitation of about 50 cm/day is quite common with a cyclonic storm (www.imd.gov.in). Details about the cyclone characteristics over the NIO are available in Mohapatra et al. (2013).

S.K. Jena (✉)
National Disaster Management Authority, New Delhi, India
e-mail: drm.susanta@gmail.com

© Capital Publishing Company 2017

M. Mohapatra et al. (eds.), *Tropical Cyclone Activity over the North Indian Ocean*,
DOI 10.1007/978-3-319-40576-6_6

2 Post-landfall Riverine Flooding

Landfalling TCs often bring about heavy rainfall, which typically decreases with the weakening of the TCs. However, some TCs produce even stronger rains than those at the time of their landfall (Dong et al. 2010). This represents spatial and temporal evolution of rainfall relative to the center of circulation of the storm, and the regional extent of flooding associated with landfalling TCs. During this period, it is also seen that rainfall affects a very large area (upto 500 km from the centre of the circulation) (Villarini et al. 2011).

This shows that vast areas of land experience maximum rainfall. The rainfall may be fairly widespread or with isolated heavy rains that occur in coastal regions, interior parts/upstream catchments of coastal/non-coastal state(s) starting from the genesis of a cyclone to post-landfall. This phenomenal rain causes riverine flooding in downstream areas particularly during the months of October and November. This is the most common type of flooding, in which widespread heavy rains lasting several hours to a few days over a drainage basin results in severe floods (WMO 2011).

In general, heavy rainfall may be considered as the key ingredient for flooding. However, this is not a sufficient or necessary condition as the flood extent and magnitude depend also on land use/land cover properties and soil saturation. It also depends upon its history, its projected movement, and its size (Jiang et al. 2008) and the local factors (e.g. pervious versus impervious areas, significant orography; Villarini et al. 2011).

Widespread riverine flooding from landfalling TCs is a principal threat to human life and is accompanied by major economic impacts (Chen et al. 2013). So, management of flooding associated with post-landfall TC is of paramount importance. The general approach about this has been in the form of structural and non-structural measures to prevent the flood waters from reaching potential damage centres. The same approach is also followed while flood protection measures are undertaken in India. Of all the non-structural measures, flood early warning system (EWS) is one of the measures which are gaining increased/sustained attention of the planners and acceptance of the public.

3 Present Status of Observational Network

Prediction of rainfall and flooding with landfalling TCs is a major operational challenge. Forecasting of rainfall and riverine flooding in real-time is most crucial. Early warning includes information on time of commencement and period of rainfall, its intensity and its impact in river catchment or its flood plain. This reliable advance information/warning needs the establishment of observatory network. For real-time forecast, the density of network is required to be strengthened by National Meteorology and Hydrology Services (NMHSs). In Indian context,

India Meteorological Department (IMD) and Central Water Commission (CWC) look after meteorology and hydrology services, respectively. The observatory network for cyclone monitoring by IMD has been discussed by Mohapatra et al. (2013).

Besides the NMHSs, there are several other agencies in India, who are having observatory network. Once these agencies share information with each other then it will provide not only early information/warning on rainfall but also impending floods with water level (stage forecast), discharge (flow forecast) and area likely to be submerged (inundation forecast) at various points/particular stations at a specific time.

IMD is the custodian of rainfall data, who also process the same. Central Water Commission (CWC), the nodal department for flood forecasting, develops/customises rainfall – runoff models after accessing rainfall data from IMD and its own network in river catchments. Details about the rainfall data of IMD are available in http://www.imd.gov.in/services/others/hydrometeorology-flood.htm.

The modern technique of rainfall forecast requires denser network. At present, IMD is having 675 Automated Weather Stations (AWSs) and about 1300 Automated Rain Gauges (ARGs) in addition to 552 synoptic observatories (Bandyopadhyay et al. 2014). Besides satellite image of rainfall, IMD is having the capability of Radar network for monitoring rainfall phenomena. All these real time inputs are assimilated in the numerical models. This facilitates forecasting adverse weather like severe thunderstorms resulting in high intensity rainfall, cyclones etc. In doing so, IMD is assisting CWC in flood forecasting by providing heavy rainfall warnings and quantitative precipitation forecast (QPF) through 10 Flood Meteorological Offices (http://www.imd.gov.in).

Besides IMD's input, CWC collects hydro-meteorological data at 878 sites and provides flood forecasting services at 175 stations including level forecast at 147 stations and inflow forecast to 28 reservoirs/barrages. To achieve these objectives, CWC has installed satellite based telemetry system for real time data collection, transmission, automatic flood forecasts formulation and dissemination thereof. In addition, CWC has also bilateral co-operations on flood management with Bangladesh, Bhutan, China, Nepal and Pakistan on the basis of real-time hydro-meteorological data exchange with neighbouring countries (42 stations from Nepal, 35 stations from Bhutan, 11 stations from Bangladesh and 4 stations from China).

For real-time rainfall data and flood forecasting, both the agencies are in continuous process of expanding their network to cover more areas and modernising their networks. The strategy is to make weather forecasting services at par with those of other well-developed countries of the world and to make it more efficient and reliable. To achieve this, IMD in its modernisation plan phase-I during the 12th Five Year Plan period (Mohapatra et al. 2013), installed many observation equipments but short of its target. Similarly, proposed hydro-meteorological (HM) network during the 12th Five Year Plan by CWC is still under consideration.

4 Issues Identified to Address Post-landfall Riverine Flooding

Management of riverine flooding associated with landfalling of TCs is of paramount importance. Flood early warning system is one of the non-structural measures gaining increased/sustained attention of the planners and acceptance of the public in disaster management literature. While discussing this, it is pertinent to focus (1) current gaps of density of observation network for rainfall prediction, (2) coordination of NMHSs with other agencies and (3) addressing these gaps to strengthen state planning for management of riverine flooding.

While addressing the issues of post-landfall riverine flooding, it is pertinent to discuss current gaps in density of observation network for rainfall prediction, coordination of NMHSs with other agencies and state planning for management of riverine flooding due to TCs.

4.1 Density of Observatory Network for Rainfall Prediction

The present status of observational network by NMHSs shows that national network stations are situated according to their hydrological and meteorological needs.

WMO in its manual explains that the density of meteorological stations and their reporting arrangements for resource management are insufficient in most cases for them to operate as part of a flood forecasting and warning network except for major river basins (WMO 2011). It is also found that the flood related data collection and transmission networks need to be strengthened and augmented for generation of more effective flood warnings (Mishra et al. 2012). Goyen and O'Loughlin (1999) have shown the importance of rainfall gauges and for accurate calibration for which the density of rain gauges should be at least 1 per km^2. Though this is with reference to urban run-off, the discussion shows importance of dense rain gauge network to estimate accurate precipitation over varying space and time.

Numerous studies have examined the distribution of rainfall associated with over-land and over-ocean TCs using different precipitation data sets (Kimball 2008). Satellite-based rainfall estimates represent a valuable source of information in areas of the globe where no information from ground-based stations is available, in particular over oceans. Ground-based weather radar data represent a viable alternative to satellite-based rainfall estimates. They are characterised by higher spatial and temporal resolutions (5 min and about 1 km horizontal resolution) and have already been used to study rainfall and flooding in landfalling TCs. Rain gauges represent the most direct and accurate way of measuring rainfall (Villarini et al. 2011). All these networks have certain advantages with relationship to TCs and rainfall. So, a combination of rain gauge observation, ground-based weather radar data and satellite estimates is highly recommended to take advantage of the strengths of rainfall observations. Besides, other networks are also used for real-time forecast of rainfall and flooding.

Planning Commission of India (2011) in its report made a recommendation on modernisation of flood forecasting network and its extension to other areas and reservoirs in the 12th Five Year Plan. This explains that there is scope for further enhancement of system required at national as well as regional level. This concludes that information from national monitoring networks is not sufficient to evaluate flood risk at the local scale. To validate the output of the techniques, rainfall measuring stations have to be situated locally and densely, based on land development, terrain, watershed, catchment geometry, etc. For these reasons, local monitoring networks in addition to their national networks in coordination with NMHSs may be strengthened.

4.2 Coordination of NMHSs with Other Agencies

The successful development and operation of flood forecasting depends upon cooperation/coordination among NMHSs. So, strengthening of institutional coordination, cooperation and integration between NMHSs is the fundamental requirement that uses available meteorological and hydrological real-time data. WMO (2006) outlined the 'strategy and action plan for the enhancement of cooperation amongst NMHSs for improved flood forecasting'. This emphasises strengthening of conventional procedures for riverine forecasting, which would provide useful information for inputs for the section on forecasting of riverine floods.

Besides NMHSs, other agencies like the state government of Andhra Pradesh, Karnataka and Municipal Corporation of Greater Mumbai (MCGM) have established observational network for forecasting of real-time rainfall data.

The state government of Andhra Pradesh has customised the earth observation network for river systems including major rivers covering an area of 6.85 lakh km^2. This network estimates the run off contributing to the river systems based on estimated rainfall. For which, the state government has already established 2050 Automated Weather Stations, 116 river gauges on all 40 river systems of AP, 100 soil moisture meters and 150 sunshine meters. Besides, the observation system is operated through the hardware of 50 Telemetry Rain gauges, 116 River Gauges, 4 Reservoir Level Recorders and 2 Earth Receiving Stations.

Similarly, the state government of Karnataka has installed 758 Standard Rain Gauge Stations (SRG), 166 Autographic Rain Gauge Stations and 86 Hydro-meteorological Stations (HMS) spread throughout the state for real-time monitoring and warning of weather events. Also, MCGM has installed 60 ARGs to give rainfall intensity in real-time basis of rainfall warning in every 15 min.

This shows that National Hydrological Services (NHS) should collect data from other agencies and do the modelling to provide reliable and timely forecasts for catchments. Also, the warnings should be made in real-time, duplication of work by various agencies should be avoided and special provision should be made for redundancy. Such additional data will be immensely useful for much improved early warning. So, coordination and cooperation between meteorologists and hydrologists is called for, since it constitutes an important corner stone in the development of new, integrated forecasting products with the overall goal to assist in the process of disaster management.

4.3 State Planning for Management of Riverine Flooding

The subject of disaster management has received a new orientation with the passage of the Disaster Management Act, 2005. The orientation for handling disaster situations has been changed from a relief-centric to a holistic, multi-dimensional and multi-disciplinary approach. The new approach encompasses the entire gamut of disaster management activities, i.e. prevention, mitigation, preparedness and response.

Here the discussion of post-landfall riverine flooding is a multi-disaster phenomenon and hence requires a multipronged strategy to deal with. In order to deal with the riverine flooding, the state government also has to improve the density of observatory network to provide additional local inputs to NMHSs. Such additional data will be immensely useful for much improved early warning and will result in better response and management of disasters.

In such a multi-hazard scenario of TCs and riverine flooding, the multi-actor agencies should work together on a common platform. Coordinated effort by various government departments/ agencies can give better results. There are numerous actors (such as line departments of state government, urban/rural local governments, townships, and local communities) involved in early warning preparedness, response and decision-making to synchronise the efforts of the state government(s).

Thus, management of riverine flooding requires a variable degree of decision-making from the national hydrological services to state level and local level authorities. Real-time flood forecasting will provide information to these users for appropriate preparedness and response measures at the most extreme level with a short span of time. Only then it could be developed as a part of holistic disaster management programme.

5 Conclusion

The establishment of an effective flood forecasting, warning and response depends on a thorough analysis of existing forecasting capabilities and identification of key users and their information needs. With regard to post-landfall riverine flooding, effective/appropriate actions by the concerned state government(s) require dense network of infrastructure, appropriation of monitoring rainfall and flooding, where strengthening of meteorological and hydrological networks are necessary.

The adoption of a multi-agency approach in all flood forecasting related activities, especially by inclusion of inputs from different agencies in forecasting system development, offers an opportunity to increase flood forecasting effectiveness. Thus, close coordination amongst national agencies of meteorology and hydrology with other agencies will strengthen the system and will provide real-time early warning to end-users. Once accurate warnings are provided on the basis of real time data, atleast the state government(s) will be able to take appropriate multi-hazard measures for management of riverine flooding other than landfall of a TC at a particular time.

Apart from this, there is a need to strengthen the observatory network and application of robust technologies to reach all the stakeholders. For this, integrated flood management approach should be adopted at the national and regional level.

References

Bandyopadhyay, B.K, Mohapatra, M. and Ray, K. (2014). Monitoring and prediction of heavy rainfall due to landfalling tropical cyclones over the North Indian Ocean, National workshop on enhanced and unique cyclonic activity during 2013, (24–25 July 2014), IMD, New Delhi.

Chen, Y, Ebert, E.E, Walsh, K.J.E. and Davidson, N.E. (2013). Evaluation of TMPA 3B42 daily precipitation estimates of tropical cyclone rainfall over Australia. *Journal of Geophysical Research: Atmospheres*, **118**, 1–13.

Dong, M, Chen, L, Li, Y. and Lu, C. (2010). Rainfall reinforcement associated with landfalling tropical cyclones, *American Meteorological Society*, **67**, 3541–3558.

Jiang, H, Halverson, J.B, Simpson, J. and Zipser, E.J. (2008). Hurricane "rainfall potential" derived from satellite observations aids overland rainfall prediction, *Journal of Applied Meteorology and Climatology*, **47**, 944–959.

Kimball, S.K. (2008). Structure and evolution of rainfall in numerically simulated landfalling hurricanes, *Monthly Weather Review*, **136**, 3822–3847.

Goyen, A.G. and O'Loughlin, G.G. (1999). The effects of infiltration spatial and temporal patterns on urban run-off, the 25th hydrology and water resources symposium, Brisbane.

Mishra, O.P, Ghatak, M. and Ahmed, K. (2012). SAARC workshop on flood risk management in south Asia, 9–10 October 2012, SDMC, New Delhi.

Mohapatra, M, Sikka, D.R, Bandyopadhyay, B.K. and Tyagi, A. (2013). Outcomes and challenges of Forecast Demonstration Project (FDP) on landfalling cyclones over Bay of Bengal. *Mausam*, **61**, 1–15.

Planning Commission (2011). Report of working group on flood management and region specific issues for XII Plan, Planning Commission, Govt. of India, New Delhi.

Prasad, K. and Rao, Y.V.R. (2006). Simulation studies on cyclone track prediction by quasi-lagrangian model (QLM) in some historical and recent cases in the Bay of Bengal, using global re-analysis and forecast grid point data sets, SMRC – No. 15, SAARC Meteorological Research Centre, Dhaka.

Villarini, G, Smith, J.A, Baeck, M.L, Marchok, T. and Vecchi, G.A. (2011). Characterization of rainfall distribution and flooding associated with U.S. landfalling tropical cyclones: Analyses of hurricanes Frances, Ivan, and Jeanne (2004). *Journal of Geophysical Research: Atmospheres*, **116**, 1–19.

World Meteorological Organisation (2006). Report of the international workshop on Strategy and action plan for the enhancement of cooperation between national meteorological and hydrological services for improved flood forecasting, WMO, Geneva.

World Meteorological Organisation (2011). Manual on flood forecasting and warning (WMO 1072), WMO, Geneva.

Part II
Climatological Aspects and Rapid Changes in Tropical Cyclones

Tropical Cyclone Track, Structure and Intensity Changes at Landfall

M. Mohapatra

1 Introduction

Cyclone Warning Division of India Meteorological Department (IMD), New Delhi acts as a Regional Specialised Meteorological Centre (RSMC) for north Indian Ocean (NIO) and provides tropical weather outlook and tropical cyclone (TC) advisories to the World Meteorological Organisation (WMO)/Economic and Social Cooperation for Asia and the Pacific (ESCAP) Panel member countries, viz., Bangladesh, Myanmar, Thailand, Srilanka, Maldives, Pakistan and Oman. The low pressure systems over the NIO are classified (Table 1) based on the associated maximum sustained surface wind (MSW) at the surface level (IMD 2013a). The entire process of TC monitoring and forecasting (Mohapatra et al. 2013a) is shown in a schematic diagram (Fig. 1). Like other Ocean basins, the TCs over the NIO also show sometimes significant changes in track, structure and intensity prior to, during and after the landfall. Out of 5–6 TCs developing over the NIO, about 3–4 make landfall (Tyagi et al. 2010; Mohapatra et al. 2014). The strategies adopted by RSMC, New Delhi to predict the changes in track, structure and intensity of landfalling TCs along with difficult situations and future plans are presented in following sections.

M. Mohapatra (✉)
India Meteorological Department, Mausam Bhawan, Lodi Road, 110003 New Delhi, India
e-mail: mohapatraimd@gmail.com

© Capital Publishing Company 2017

M. Mohapatra et al. (eds.), *Tropical Cyclone Activity over the North Indian Ocean*,
DOI 10.1007/978-3-319-40576-6_7

Table 1 Classification of cyclonic disturbances over the NIO (since 2015)

Low pressure system	Maximum sustained surface winds (knots)	Number of closed isobars at interval of 2 hPa within 5° latitude/longitude square
Low pressure area	<17	1
Depression	17–27	2
Deep depression	28–33	3
Cyclonic storm	34–47	4–7
Severe cyclonic storm (SCS)	48–63	8–10
Very severe cyclonic storm (VSCS)	64–89	11–25
Extremely severe cyclonic storm (ESCS)	90–119	26–39
Super cyclonic storm (SuCS)	120 and above	40 or more

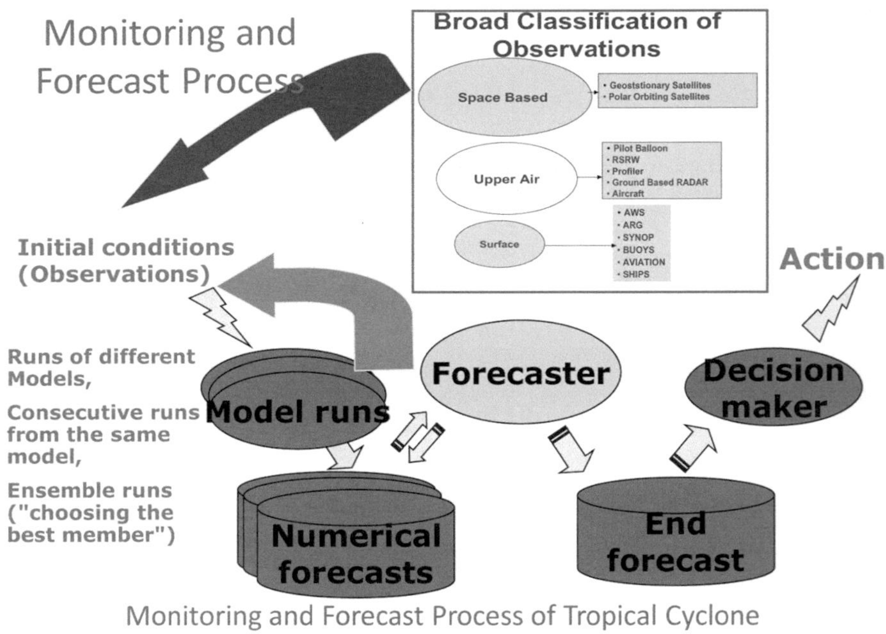

Fig. 1 Monitoring and forecasting process of tropical cyclone (after Mohapatra et al. 2013a)

2 Current Strategy of Monitoring and Prediction of Track, Structure and Intensity

Various kinds of analytical procedure are described in Cyclone Manual (IMD 2003, 2013a). A systematic check list is prepared for identification of location and intensity of TC and also for the prediction of track and intensity. The procedure necessarily deals with determination of location and intensity along with other characteristics of the TC like associated MSW, estimated central pressure and pressure drop at the centre, shape and size, radius of outermost closed isobar, point and time of landfall, if any or area of dissipation etc. with the available observations in the storm region.

To ensure the availability of the data and forecast products from various national and international sources at Cyclone Warning Division, IMD, New Delhi, an institutional mechanism has been developed in consultation with all the stake holders. A standard operation procedure (SOP) has been prepared for monitoring and prediction of cyclonic disturbances and issue of warning. It includes the road map and check lists for this purpose.

The TC analysis, prediction and decision-making process is made by blending scientifically based conceptual models, dynamical and statistical models, meteorological datasets, technology and expertise. A weather analysis and forecasting system in a digital environment is used to plot and analyse different weather parameters, satellite, Radar and numerical weather prediction (NWP) model products. In this hybrid system, synoptic method could be overlaid on NWP models supported by modern graphical and GIS applications to produce high quality analyses and forecast products. The automation of the process has increased the efficiency of system, forecast accuracy, visibility of IMD and utility of warning products (Mohapatra et al. 2013a). There has been improvement in recent years due to modernisation programme of IMD resulting in improved observational system like DWR and AWS along the coast, improved modelling technique and ongoing forecast demonstration project (FDP) on landfalling TCs over the Bay of Bengal.

3 TC Track Changes During Landfal

3.1 Current Strategy for TC Tracking

The location of the centre of the TC is determined based on consensus arising from (a) Synoptic position, (b) Satellite position (INSAT/METSAT/microwave position), (c) Radar position and (d) Centre determined by other satellite centres. Details are discussed in Manual on standard operation procedure published by IMD (2013a). The average confidence level of locating the centre of the system over the NIO is about 50 km (Mohapatra et al. 2012a; Goyal et al. 2013). The location error of a depression is more than that of a TC over the open sea area due to similar error in estimation by the satellite. The induction of DWR reduces the error in fixing the centre of the TCs in radar range. The landfall point estimation error has been reduced to about 25 km by 2010 mainly due to installation of coastal AWS during late 2000s (Mohapatra et al. 2012a). The landfall time estimation error may be about 30 min since 1974 with the introduction of coastal hourly observations and satellite and radar observations.

3.2 Current Strategy of TC Track Forecasting

The challenge for the Indian researchers in the last couple of years has been to reduce the track prediction error. Several groups in India have applied high resolution meso-scale models of the US origin with a variety of physical parameterisation in isolation and also in combination. Several papers on recent advances on TC prediction have been published in the special issue of Journal, *Mausam* (IMD 2013b), book on ‘Monotoring and Prediction of Indian Ocean Tropical Cyclones and Climate Change’ edited by Mohanty et al. (2014) and another book on ‘High Impact Weather Events over SAARC Region’ edited by Ray et al. (2014). Meso-scale WRF models were introduced in IMD and other research agencies during 2009 with spatial resolution of 27 km (Mohanty et al. 2014). It was further improved with local data assimilation and increase in resolution to 9 km in 2010 and 2011. Currently, an effort is underway in which high resolution hurricane WRF model with the support from NCEP, USA is being used in track and intensity predictions. Currently, following methods are used by IMD for TC track forecasting. Synthetic vortex creation in global models run at NCMRWF, India has helped in improving the model forecast also (Chourasia et al. 2013).

1. Statistical Techniques
    (a) Analogue (digitised tracks of TCs during 1891–2013 available for selecting analogue (IMD, 2008)).
    (b) Persistence, Climatology and Climatology and persistence (CLIPER).
2. Synoptic, satellite and radar Techniques – Empirical techniques.
3. NWP Models.
    (a) Individual models (global and regional).
    (b) IMDGFS (574), Global Tropical Model (Meteo-France), ECMWF, JMA, UKMO, NCEP, WRF (IMD, IITD, IAF), HWRF (IMD), Quasi-Lagrangian model (QLM).
    (c) MME (IMD) and MME based on TC Module.
    (d) Ensemble prediction system (EPS) for Strike probability and location specific probability.

Although, the synoptic, statistical and satellite/radar guidances help in short range forecast (upto 12/24 h), the NWP guidance is mainly used for 24–120 h forecasts. Hence, the RSMC forecast tracks result from a manually analysed forecasting process, which relies on output from several NWP models as discussed above. Consensus forecasts that gather all or part of the numerical forecast tracks and uses synoptic and statistical guidance are utilised to issue official forecast (Mohapatra et al. 2014).

All the NWP models as mentioned above except QLM and CLIPER are late models as the model output is available to forecasters as late as 6–12 h. Similarly, all the NWP models as mentioned above do not use synthetic vortex and vortex relocation. Although QLM, NCEP use synthetic vortex, global model like ECMWF does not use these processes. MME developed by IMD (Kotal and Roy Bhowmik 2011) provides better guidance compared to deterministic model. This has been illustrated with an example of very severe cyclonic storm, Phailin over the Bay of Bengal during 8–14 October 2013. The MME and other NWP models forecast tracks based on different initial conditions along with the observed track is depicted in Fig. 2. The figure shows that from the day 1 (00 UTC 8 October to 10 UTC 10 October 2013), MME could be able to predict correctly and consistently the landfall at Gopalpur (Odisha). The mean direct position error (DPE) of MME was about 65 km at 12 h to 150 km at 120 h, which is less than that of most of the deterministic models (RSMC, New Delhi 2014).

3.3 Operational Track Forecast Products to Maximise the Utility of Track Forecasts

Considering recent development in prediction capability, IMD extended the objective TC track forecast valid upto 72 h in 2009 (Mohapatra et al. 2013b). The track forecast is issued by RSMC, New Delhi from deep depression stage onwards considering the fact that the Bay of Bengal and Arabian Sea are small basins and the life period of TCs over these basins are small (3–5 days). It introduced 96 and 120 h forecasts in 2013. The TC forecast is issued 6 times a day at the interval of 3 h, i.e. based on 00, 03, 06, 09, 12, 15, 18 and 21 UTC observations. The forecasts are issued about 3 h after the above mentioned observation time. An example of the product during TC, Phailin is shown in Fig. 3.

To take care of uncertainty in track forecast, the cone of uncertainty (COU) in the forecast of IMD has been introduced with effect from the TC, 'WARD' during December 2009 valid upto 72 h and upto 120 h since 2013. The radii of circles used to construct the COU are 75, 150, 200, 250, 300 and 350 km, respectively for 12, 24, 36, 48, 60 and 72 h forecasts based on past average errors of official forecasts. A typical example of COU forecast showing the uncertainty circles for different forecast periods are shown in Fig. 3 following Mohapatra et al. (2012b). It is

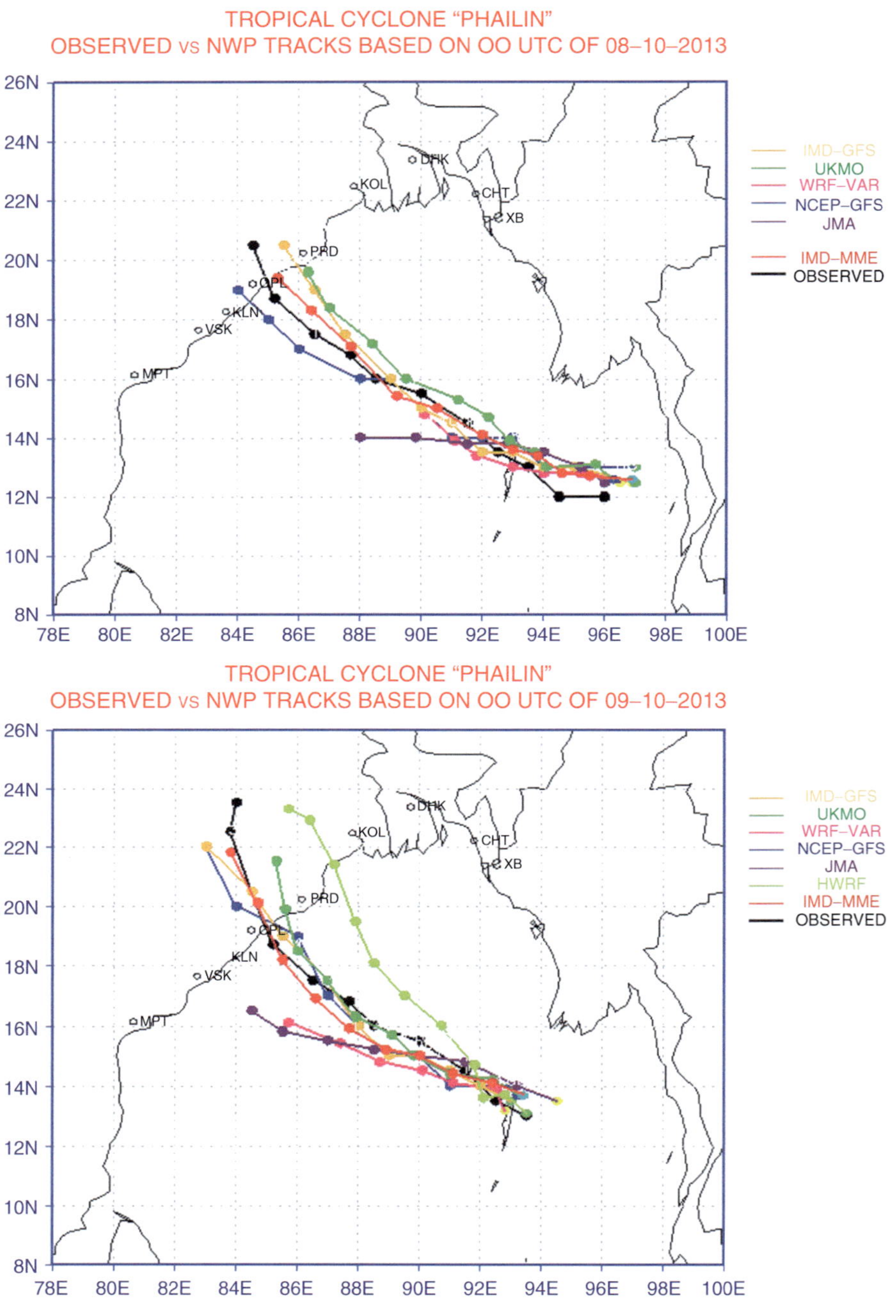

Fig. 2 Track prediction of very severe cyclonic storm, Phailin (8–14 October 2013) over Bay of Bengal

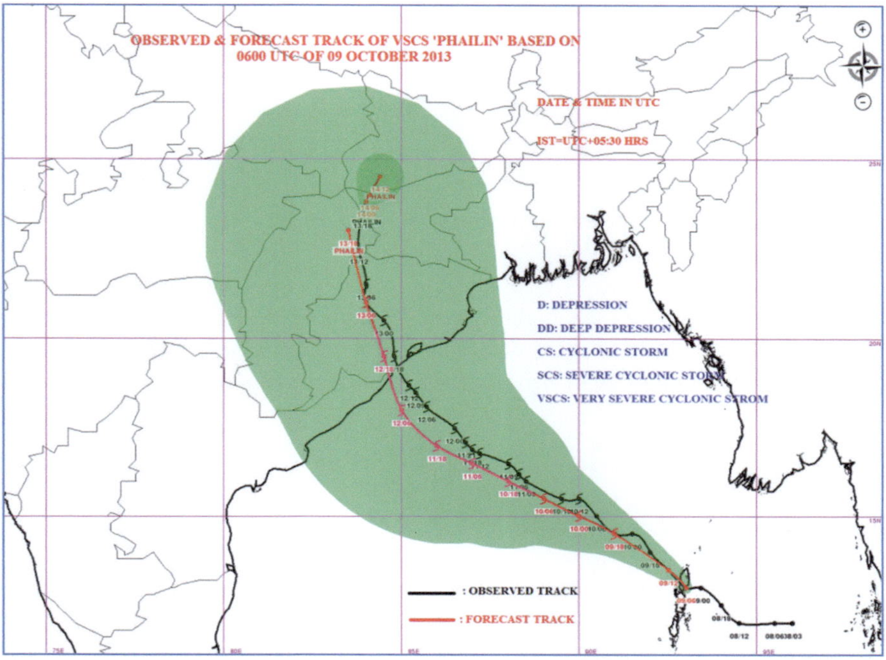

Fig. 3 A typical example of observed and forecast track of TC Phailin during 8–14 October 2013

found that the observed track lies within the forecast COU in about 60 % of the cases.

3.4 Difficult Situations in Track Forecasting for a Landfalling Cyclone

3.4.1 Average Track Forecast Errors over the NIO

Mohapatra et al. (2013b) evaluated the TC track forecast issued by IMD during 2003–2011 (9 years) by calculating the direct position error (DPE) and skill in track forecast. The average DPE is about 124, 202 and 268 km and skill is about 36 %, 53 % and 62 %, respectively for 24, 48 and 72 h forecasts over the NIO as a whole during 2009–2013 (RSMC New Delhi 2014). Scarcity of observational data leads to poor initial condition in NWP models. It further leads to increase in forecast error of NWP models, which are the backbone for official track forecast of TCs, especially in 36–120 h forecast range (Osuri et al. 2012; Mohanty et al. 2010, 2013). The error is higher than that of National Hurricane Centre (NHC), USA by about 40–60 km. To overcome the lag, various initiatives like forecast demonstration projects, aircraft probing, augmentation in observational network, introduction of HWRF model, data assimilation, synthetic vortex generation, MME and operationalisation of EPS products are being attempted. Also frequent update of bulletins along with cone of uncertainty and liaison with disaster managers help in maximising the track forecast accuracy and addressing the uncertainty in forecasts.

3.4.2 Track Forecast over Data Sparse Region and for Recurving/ Looping TCs

The track forecast errors are higher over the Arabian Sea than over the Bay of Bengal, as Arabian Sea is more data sparse. The error is also higher in case of recurving and looping TCs than in case of straight moving TCs (Mohapatra et al. 2013b).

3.4.3 Rapid Changes in Tracks

Track forecasting is more difficult when there is a rapid change in track near the landfall point. Such difficult situations include the (i) recurving TCs and (ii) rapid movement of TCs during landfall or slow movement/stationarity of TCs near the coast. The average track forecast errors of the cyclones with rapid changes in tracks are shown in Table 2. It is found that the error is higher by about 5–20 % for 12–72 h lead period of forecasts in case of TCs with rapid track changes as compared to the mean track forecast errors based on the data of 2003–2013. Comparing the track forecast errors of cyclones with sudden changes in track direction, rapid movement and slow movement, the error is maximum in case of sudden change in direction followed by rapidly moving TCs. The landfall forecast errors of TC, Viyaru over the Bay of Bengal during 10–16 May 2013 (Fig. 4) are shown in Table 3 to illustrate difficulty in predicting the track of recurving as well as rapidly moving TCs (RSMC, New Delhi 2014). It may be mentioned that the TC, Viyaru moved with a speed of 40–50 kmph during 12 h prior to landfall. The cyclonic storm, Viyaru crossed Bangladesh coast near latitude 22.8° N and longitude 91.4° E, about 30 km south of Feni around 0800 UTC of 16 May 2013 with MSW of about 85–95 kmph. Due to the faster movement, the adverse weather due to the cyclonic storm was relatively less and was over-predicted. The landfall error was higher especially in terms of time of landfall (Table 3). The average landfall point forecast error based on 2009–2013 is 75, 100 and 124 km for 24, 48 and 72 h forecasts, respectively.

Table 2 Comparison of track forecast errors of cyclones with rapid changes in tracks with that of all the cyclones during 2003–2013

Lead period (h)	Track forecast error (km) for all cyclones (A)	Track forecast error (km) for cyclones with rapid changes of tracks (B)	Increase in error (in %) in case of cyclones with rapid changes of tracks ((B – A)/A × 100)
12	79.9	83.6	4.6
24	134.6	148.7	10.5
36	163.7	187.0	14.2
48	202.1	230.3	14.0
60	233.7	272.0	16.4
72	268.2	319.7	19.2

4 Intensity Changes During Landfall

4.1 Current Strategy in TC Intensity Monitoring

The consensus-based intensity estimation takes into consideration (a) satellite, (b) Radar and (c) synoptic analysis (IMD 2013a; Mohapatra et al. 2014). As operational practice, a check list is prepared for this purpose to take care of all the available information, which in turn helps in decision-making process to determine the above characteristics of TC. Operationally, the value of MSW is generally inferred from satellite using the Dvorak technique (Dvorak 1984). As satellite and radar based wind are not available at surface level, the wind observations from these techniques are converted to 10 m wind using the suitable conversion technique like those used in case of aircraft reconnaissance technique in Atlantic. The empirical relationship between *T* number and the maximum wind speeds (Dvorak 1984) gives the pressure depths (peripheral pressure minus central pressure in hPa) as applicable for Indian Sea area using the relation $V_{max} = 14.2$ X SQRT $(P_n - P_o)$ following Mishra and Gupta (1976). As there is no aircraft reconnaissance in the NIO, Dvorak's technique, which is based on 1 min averaging has not been verified. However, coastal stations, equipped with cup anemometer and Dynes P.T. anemograph use 3-min averaging in Indian region. The brightness temperature from the polar orbiting satellites is utilised to estimate the central pressure and MSW. Though this technique has not been validated over the NIO, the algorithm used in north Atlantic basin is being used for the NIO (Jha et al. 2013). Comparison of satellite-based intensity and the best track estimates of IMD indicate a difference of about 0.5 T (Goyal et al. 2013). The average error in MSW estimation could have been (05–20 knots or 3–10 mps) with the introduction of Dvorak's classification of intensity since 1974 (Mohapatra et al. 2012a).

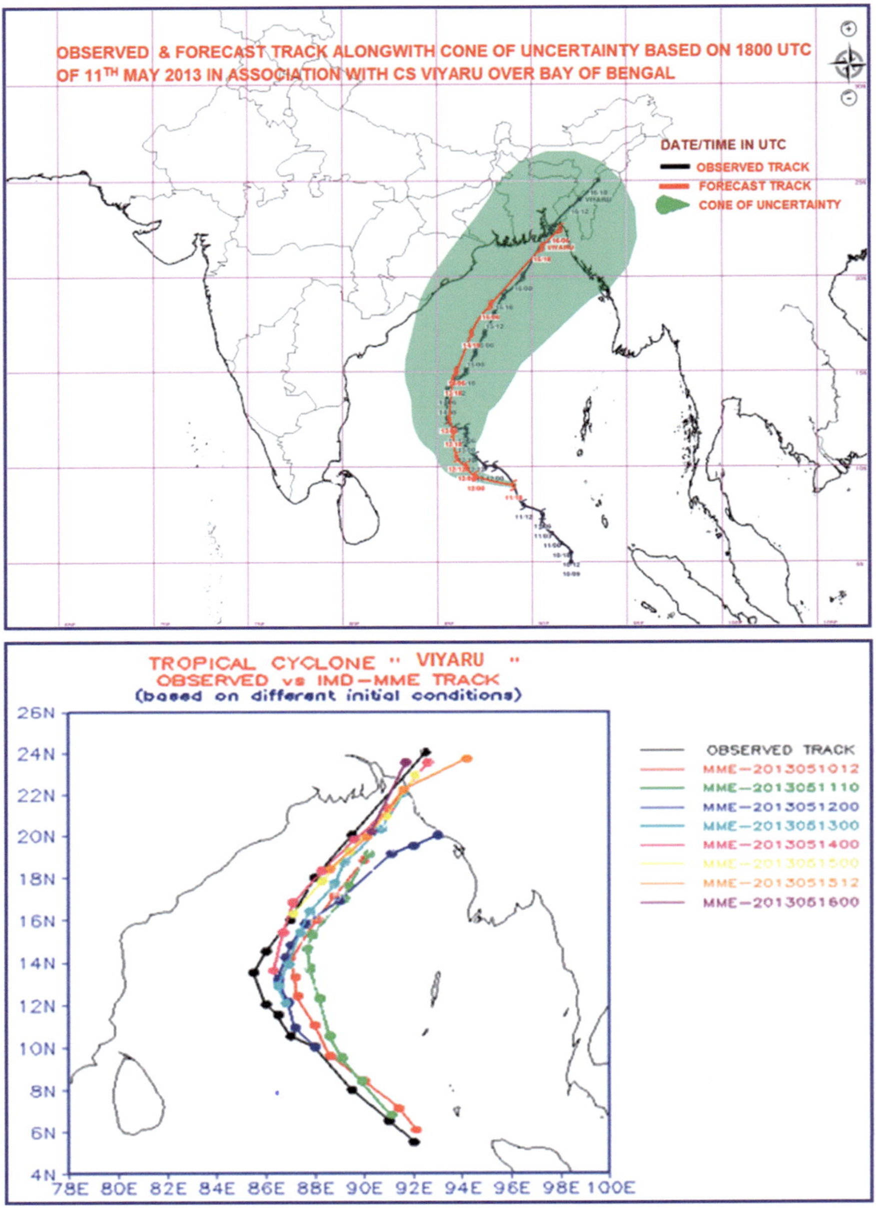

Fig. 4 Observed and official forecast tracks of TC, Viyaru over Bay of Bengal alongwith MME based track forecast

Table 3 Landfall point (km) and time (h) forecast errors of TC, Viyaru over the Bay of Bengal during 10–16 May 2013

	FC based on 00 UTC/14 May 2013	FC based on 00 UTC/15 May 2013
Model	Lead time: 56 h	Lead time: 32 h
IMD-GFS	NO LF	NO LF
IMD-WRF	NO LF	147/+5
IMD-QLM	NO LF	63/+15
JMA	137/+10	63/+12
NCEP-GFS	289/−4	169/+4
ECMWF	259/+4	274/+12
IMD-MME	63/+10	63/+13
IMD-HWRF	84/+10	174/+4
Official	89/+10	94/+13

+: Model prediction lagged behind the actual, −: Model prediction led the actual.

4.2 Current Strategy for Intensity Forecasting

The intensity forecast has been issued by RSMC, New Delhi from deep depression stage onwards since 2009 for 12, 24, 36, 48, 60 and 72 h forecast periods (Mohapatra et al. 2013c). The TC intensity forecast is issued 4 times a day at the interval of 6 h, i.e. based on 00, 06, 12 and 18 UTC observations with every 3-h updates and validity period extended upto 120 h since 2013. The forecasts are issued about 3 h after the above-mentioned observation time. The tools and methods used by IMD for intensity forecasting of TCs over the NIO are same as that for track forecasting (Mohapatra et al. 2014). In addition, there is a Dynamical Statistical Model (statistical cyclone intensity prediction – SCIP, Kotal et al. 2008; Roy Bhowmik et al. 2007). Consensus forecasts that gather all or part of the numerical forecast intensities and uses synoptic and statistical guidance are utilised to issue official forecast. An example of this product is shown in Fig. 5.

4.3 Difficult Situations

4.3.1 TC Intensity Prediction Error

Mohapatra et al. (2013b) evaluated the TC intensity forecast issued by IMD during 2005–2011 (7 years) by calculating the absolute error (AE), root mean square error (RMSE) and skill in intensity forecast in terms of MSW. The study shows that the average AE (RMSE) in intensity forecast is about 11(14), 14(19) and 20(26) knots, respectively, for 24-, 48- and 72-h forecasts over the NIO as a whole during 2009–2013. The skill of intensity forecast is about 44 %(48 %), 60 %(58 %) and 60 %(65 %) for 24-, 48- and 72-h forecasts during 2009–2013 with respect to AE

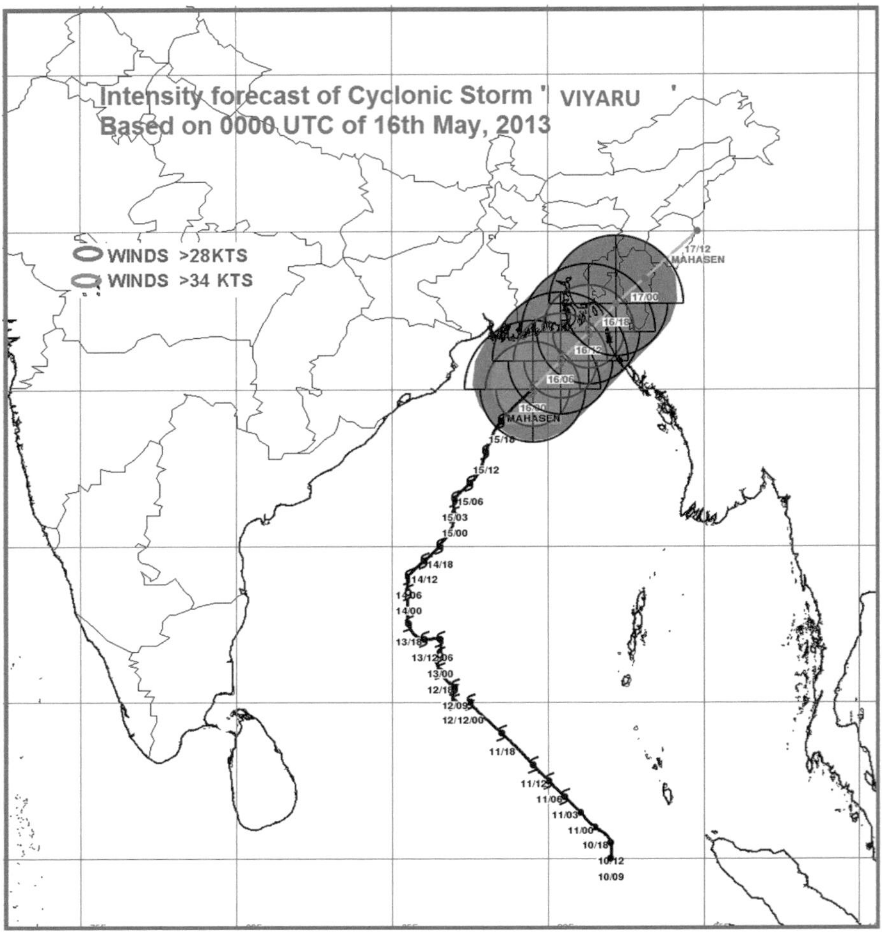

Fig. 5 A typical graphical presentation of intensity and quadrant wind forecast during TC, Viyaru (10–17 May 2013)

(RMSE). There is no significant improvement in terms of reduction in AE and RMSE of MSW forecast over the NIO during 2005–2013.

4.3.2 Intensity Prediction of Intense TCs and over Data Sparse Region

The intensity forecast error is higher for intense TCs than in weaker TCs. Intensity prediction is more difficult over Arabian Sea than over Bay of Bengal as it is more data sparse leading to poor initial conditions in NWP models. There is no DWR in Arabian Sea rim countries. Number of buoys are also very less.

4.3.3 Prediction of Rapid Intensification

In some cases, the rapid intensification (change in central pressure by 30 hPa in 24 h) takes place prior to landfall putting a challenge to forecasters. For this purpose, a rapid intensification index (RII) has been developed by IMD (Kotal and Roy Bhowmik 2013) based on dynamical statistical model. During the period, 2011–2012, for 29 forecast events, the Brier score (BS) on prediction of rapid intensification based on RII is found to be 0.005, which shows RII achieved a good score for RI forecasting during the period. The probability forecasts of Rapid intensification for cyclone Phailin over Bay of Bengal (RSMC, New Delhi 2014) during 08–14 October 2013 is given in Table 4 as an example. The table shows that the RII could be able to predict OCCURENCE as well as NON-OCCURENCE of Rapid Intensification of cyclone PHAILIN during its lifetime except forecast for 12 UTC of 09 October 2013 and 00 UTC of 11 October 2013 (Kotal and Roy Bhowmik 2013). Of course, there was some over-forecasting in the same case as can be seen from Table 4. Hence, further work is needed in this regard.

4.3.4 Intensity Prediction After Landfall

There are cases, when, the TC maintains its intensity with very slow decaying process after landfall. It specially occurs, when the TC moved over a plain surface, especially over deltaic region and during post-monsoon season (October–December). It occurs in post-monsoon season due to availability of moisture over the land surface and atmosphere due to monsoon circulation which prevailed during June–September and may extend to October during its withdrawal phase. A statistics based DECAY model for intensity prediction after landfall using the intensity at the time of landfall has been developed by IMD (Roy Bhowmik et al. 2005). The Average absolute error (AAE) of this model is ranged from 1 to 5 knots for forecasts up to 24 h. The error statistics shows that the model forecasts are reasonably good for predicting decay in intensity after landfall. For the cyclone Phailin as an example, decay (after landfall) prediction curve (6-h up to 30 h; Fig. 6) shows fast decay compared to observed decay. As the model is statistical in nature, it does not take into consideration various dynamical and hydro-dynamical processes governing the decay of a TC after landfall. Hence prediction of decay of intensity of TC after landfall is still a challenge.

4.3.5 Rapid Weakening of TC over the Sea Before Landfall

Another interesting part is the dissipation/rapid weakening of systems over NIO. It poses a great challenge to a forecaster as it very often leads to wrong/over-warning. The dissipation/rapid weakening over the sea may happen due to various reasons including colder ocean thermal energy, entrainment of dry and cold continental air

Table 4 Probability of rapid intensification for cyclone Phailin over Bay of Bengal during 8–14 October 2013

Forecast based on	Probability of RI predicted (%)	Chances of occurrence predicted	Intensity changes (knot) in 24 h	Occurrence
00 UTC/08 October 2013	9.4	Very low	5	NO
00 UTC/09 October 2013	9.4	Very low	15	NO
12 UTC/09 October 2013	9.4	Very low	40	YES
00 UTC/10 October 2013	72.7	High	65	YES
12 UTC/10 October 2013	72.7	High	40	YES
00 UTC/11 October 2013	72.7	High	5	NO
12 UTC/11 October 2013	32.0	Moderate	0	NO

into the core of TC and increase in vertical wind shear in the horizontal wind. An example of very severe cyclonic storm, Lehar (23–28 November 2013) over Bay of Bengal (RSMC, New Delhi 2014), which rapidly weakened before the landfall (Fig. 7) is presented herewith. It rapidly weakened over the sea from the stage of very severe cyclonic storm (75 knots) to depression (25 knots) in 18 h. It had landfall near Machilipatnam (Andhra Pradesh) as a depression. It did not cause any significantly heavy rainfall over Andhra Pradesh. The intensity forecast errors were very high as can be seen from Table 5. The error in intensity prediction led to large error in prediction of rainfall, wind and storm surge. However, the situation was managed by providing frequent update and immediate revision of forecasts with the sign of weakening envisioned through synoptic analyses. There is a need for development of dynamical statistical model for rapid weakening of TCs over the sea.

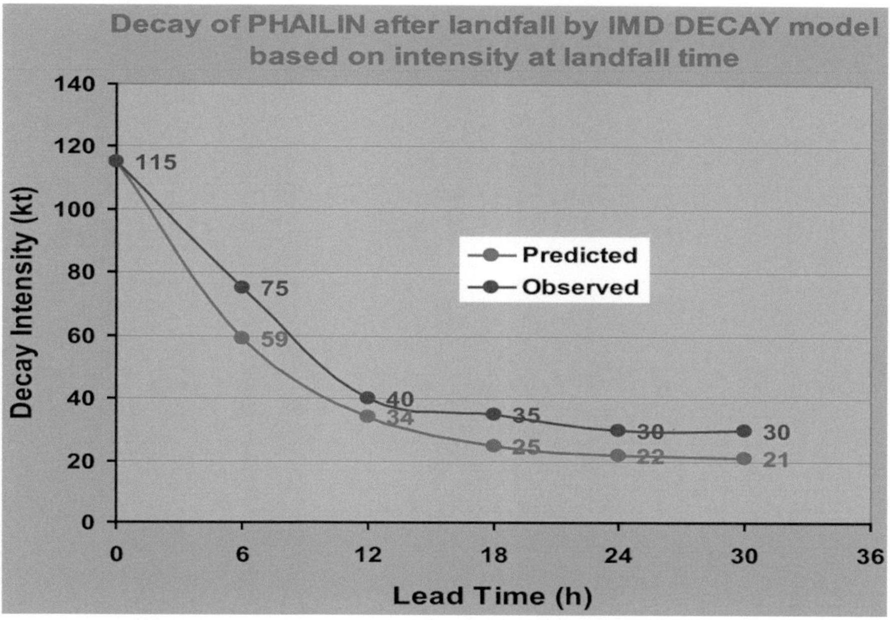

Fig. 6 IMD's decay model based forecast and observed MSW after the landfall of TC Phailin on 12 October 2013

5 Structure Changes During Landfall

Due to absence of aircraft reconnaissance, detailed structural characteristics like wind and temperature distribution along the vertical and horizontal structure of eye and eye wall etc. over the NIO are not yet known. Studies made so far in this respect are mainly based on satellite and radar observations (Raghavan 2013; Bhatia and Sharma 2013). In recent years, the microwave imageries from the polar orbiting satellites have helped further to understand the structure of TC as it can provide the imageries in different levels of the TC (Jha et al. 2013). During pre-monsoon and post-monsoon seasons, the inter-tropical convergence zone (ITCZ) and convection play pre-dominant role in determining the structure of TCs at landfall in addition to the large scale environment like Madden Julian Oscillation (MJO). According to Mohapatra and Adhikary (2011), 60 % of total genesis and intensification over the NIO are associated with the MJO conditions. The structure of TC varies with respect to area of genesis, viz., Bay of Bengal and Arabian Sea, season of formation (pre-monsoon and post-monsoon seasons), intensity of TCs (Mohapatra and Sharma 2015).

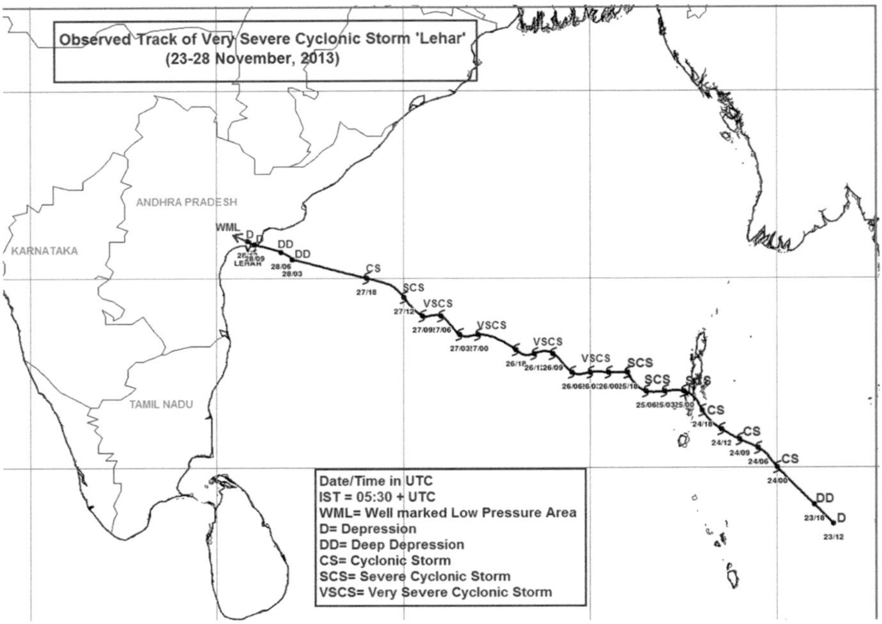

Fig. 7 Track of very severe cyclonic storm, Lehar (23–28 November 2013) over the Bay of Bengal

Table 5 Intensity forecast errors of RSMC, New Delhi in case of TC, Lehar over Bay of Bengal during 23–28 November 2013

Lead period	Intensity forecast error (knots)		Long period average (2008–2012): Absolute error (knots)	Long period average (2008–2012): RMS error (knots)
	Absolute error	Root mean square error		
12	10.2	13.4	7.3	9.9
24	18.4	24.5	10.4	13.5
36	25.5	33.1	12.7	16.1
48	29.2	38.0	13.4	17.8
60	32.4	39.3	13.4	15.3
72	41.0	47.3	19.0	24.0
84	47.0	49.6	–	–
96	48.5	54.3	–	–
108	34.5	36.9	–	–
120	–	–	–	–

5.1 *Current Strategy of Structure Monitoring and Forecasting*

In addition to routinely analysing position (and uncertainty), MSW (3-min mean) and the mean sea level pressure (MSLP), radii of 28, 34, 50 and 64 knot 3-min winds are analysed in quadrants. The radius of outermost closed isobar (ROCI), pressure of outermost closed isobar (POCI), radius of maximum wind (RMW) and the vertical depth/extent of the TC (deep/medium/shallow) are estimated every 6 h based on 00, 06, 12 and 18 UTC from the stage of deep depression (MSW of 28 knots or more) onwards, if the system is expected to intensify into a CS. This has been introduced since TC, Giri over the Bay of Bengal in 2010 in both text and graphics form (RSMC, New Delhi 2014). This product is also provided in coded format to the NWP and storm surge modelling groups. In addition, the analysis and forecast wind radii of 28, 34, 50 and 64 knots threshold are generated alongwith the MSW and location every 6 h based on 00, 06, 12 and 18 UTC valid upto 120 h in the interval of 6 h for first 24 h forecast period and in the interval of 12 h for the subsequent forecast periods. The primary methods for TC wind field estimation by IMD involves satellite-based scatterometer estimates, cloud motion vectors, water vapour based wind vectors, wind estimates from brightness temperatures, estimates from RADAR products and NWP model analyses, persistence and climatology products. The IMD's NWP model system guidance includes IMD-Global Forecast system (GFS), UK Meteorological Office, Japan Meteorological Agency (JMA), ARP-Meteo-France, National Centre for Environmental Prediction (NCEP) – GFS, Weather Research and Forecast (WRF) and Hurricane WRF (HWRF) models. The multi-platform satellite based surface wind analysis as developed by CIRA for the TCs on real time basis based on 00, 06, 12 and 18 UTC observations are also utilised by IMD. Consensus analysis and forecasts that gather all or part of the numerical forecast and uses synoptic and climatological guidance are utilised to issue analysis and forecast surface wind radii in four geographical quadrants. A typical graphical product is shown in Fig. 5.

5.2 *Difficult Situations*

5.2.1 Error in Quadrant Wind Forecast due to Error in MSW Forecast

Over the past several years, there have been large improvements in track forecast skill (Mohapatra et al. 2013b) and modest improvements in the intensity skill (Mohapatra et al. 2013c) like other ocean basins. These errors, particularly the intensity errors negatively affect wind radii forecasts. The poor intensity forecast is particularly pronounced when intensity forecast fail to or falsely forecast winds that exceed the 34, 50 and 64 knots thresholds.

5.2.2 Shifting of Convection and Wind Maxima During Landfall

There is shifting of convection and wind maxima during landfall sometimes leading to error in predicting heavy rainfall and wind over the land regions. To overcome this problem, R&D activities have been taken up to develop R-CLIPER for prediction of rainfall, QPE estimates based on Radar and INSAT 3D, introduction of HWRF etc.

6 Conclusions

India Meteorological Department is serving as the Regional Specialised Meteorological Centre for Tropical Cyclones for the North Indian Ocean basin and provides tropical cyclone advisories to the member countries of WMO/ESCAP panel, viz., Bangladesh, Maldives, Myanmar, Oman, Pakistan, Sri Lanka and Thailand. TC analysis, prediction and decision-making process adopted at RSMC New Delhi involves a judicious blend of scientifically based conceptual models, dynamical and statistical model products, use of multi-platform high resolution data and products along with human experience and expertise. There has been vast improvements in TC forecasting over the North Indian Ocean due to the modernisation programme of IMD during the recent decade. However, there are some difficult situations wherein TC forecasting is still tricky and more challenging (such as recurving and rapidly intensifying/weakening TCs near the landfall point) than in other situations. IMD is regularly upgrading its observational and analytical capabilities to meet these challenges and to improve its forecasting skills further in the near future.

7 Future Scope

1. The FDP on landfalling TCs over the Bay of Bengal has been taken up. It will help us in minimising the error in prediction of TC track and intensity forecasts and hence adverse weather. During pilot phase (15 October to 30 November 2008–2013), several national institutions participated for joint observational, communicational and NWP activities resulting in improved forecast and delivery of services (Mohapatra et al. 2013a). With possible manned and unmanned aircraft reconnaissance, TC track and intensity forecasting will improve and hence the adverse weather warning over the NIO region, as demonstrated in Atlantic and Pacific Ocean basins. With the completion of ongoing modernisation programme, the error is likely to reduce by about 20 % by 2015 and by 40 % by 2020 from the base year of 2010 (Rao and Mohapatra 2010).

2. IMD continuously expands and strengthens its activities in relation to observing strategies, forecasting techniques, disseminating methods and research relating to different aspects of TCs to ensure most critical meteorological support through observations, analysis, predictions and warnings to disaster managers and decision makers not only in India but also to the NIO rim countries. Some of the planned activities include (i) augmentation of observational network, (ii) implementation of HWRF model, (iii) introduction of probabilistic wind forecast, (iv) location of specific heavy rainfall forecast, (v) operationalisation of coastal inundation model, which is running experimentally since 2013, (vi) severe weather forecast demonstration project (SWFDP) over Bay of Bengal, (vii) coastal inundation forecast demonstration project (CIFDP)-Bangladesh and (viii) observed and satellite based merged data set for TC rainfall etc.

References

Bhatia, R.C. and Sharma, A.K. (2013). Recent advances in observational support from space-based systems for tropical cyclones, *Mausam*, 64, 97–104.

Chourasia, M, Ashrit, R.G. and George, J.P. (2013). Impact of cyclone bogusing and regional assimilation on tropical cyclone, track and intensity predictions, *Mausam*, **64**, 135–148.

Dvorak, V.F. (1984). Tropical cyclone intensity analysis using satellite data. Technical Report (NOAA TR NESDIS 11), National Oceanic and Atmospheric Administration, National Environmental Satellite, Data, and Information Service, 47 pp.

Goyal, S, Mohapatra, M. and Sharma, A.K. (2013). Comparison of best track parameters of RSMC, New Delhi with satellite estimates over north Indian Ocean, *Mausam*, **64**, 25–34.

IMD (2003). Cyclone Manual, IMD New Delhi.

IMD (2008). Track of storm and depressions over the Indian Seas during 1891–2007, Cyclone e-Atlas of IMD. IMD, Chennai, India.

IMD (2013a). Cyclone Warning Services: Standard Operation Procedure. Cyclone Warning Division. IMD: New Delhi.

IMD (2013b). Special Issue on Proceedings of National Conference on Bay of Bengal Tropical Cyclone Experiment (BOBTEX 2011), *Mausam*, **64**, 1, pp. 1–210.

Jha, T.N, Mohapatra, M. and Bandyopadhyay, B.K. (2013). Estimation of intensity of tropical cyclone over Bay of Bengal using microwave imagery, *Mausam*, **64**, 105–116.

Kotal, S.D. and Roy Bhowmik, S.K. (2011). A multimodel ensemble technique for cyclone track prediction over north Indian Sea, *Geofizika*, **28**, 275–291.

Kotal, S.D. and Roy Bhowmik, S.K. (2013). Large-scale characteristics of rapidly intensifying tropical cyclones over the Bay of Bengal and a Rapid Intensification (RI) Index. *Mausam*, **64**, 13–24.

Kotal, S.D, Roy Bhowmik, S.K, Kundu, P.K. and Das, A.K. (2008). A statistical cyclone intensity prediction model for Bay of Bengal. *Journal of Earth Systems Science*, **117**, 157–168.

Mishra, D.K. and Gupta, G.R. (1976). Estimates of maximum wind speed in tropical cyclones occurring in the Indian Seas. *Indian Journal of Metereology and Geophysics*, **27**, 285–290.

Mohanty, U.C, Osuri, K.K, Routray, A, Mohapatra, M. and Pattanayak, S. (2010). Simulation of Bay of Bengal Tropical Cyclones with WRF model: Impact of initial and boundary condition. *Marine Geodesy*, **33**, 294–314.

Mohanty, U.C, Osuri, K.K. and Pattanayak, S. (2013). A study on high resolution mesoscale modeling systems for simulation of tropical cyclones over the Bay of Bengal, *Mausam*, **64**, 117–134.

Mohanty, U.C, Mohapatra, M, Singh, O.P, Bandyopadhyay, B.K. and Rathore, L.S. (2014). *In:* Mohanty, U.C, Mohapatra, M, Singh, O.P, Bandyapadhyay, B.K. and Rathore, L.S. (ed.), Monitoring and prediction of tropical cyclones over the Indian Ocean and climate change. Springers and Capital publishers, New Delhi, India, pp. 1–428.

Mohapatra, M. and Adhikary, S. (2011). Modulation of cyclonic disturbances over the north Indian Ocean by Madden – Julian oscillation, *Mausam*, **62**, 375–390.

Mohapatra, M, Bandyopadhyay, B.K. and Tyagi, A. (2012a). Best track parameters of tropical cyclones over the North Indian Ocean: A review; *Nature Hazards*, **63**, 1285–1317, doi: 10. 1007/s11069-011-9935-0.

Mohapatra, M, Nayak, D.P. and Bandyopadhyay, B.K. (2012b). Evaluation of cone of uncertainty in tropical cyclone track forecast over north Indian Ocean issued by India Meteorological Department. *Tropical Cyclone Research and Review*, **2**, 331–339.

Mohapatra, M, Sikka, D.R, Bandyopadhyay, B.K. and Tyagi, A. (2013a). Outcomes and challenges of Forecast Demonstration Project (FDP) on landfalling cyclones over the Bay of Bengal, *Mausam*, **64**, 1–12.

Mohapatra, M, Nayak, D.P, Sharma, R.P. and Bandyopadhyay, B.K. (2013b). Evaluation of official tropical cyclone track forecast over north Indian Ocean issued by India Meteorological Department. *Journal of Earth System Sciences*, **122**, 589–601.

Mohapatra, M, Bandyopadhyay, B.K. and Nayak, D.P. (2013c). Evaluation of operational tropical cyclone intensity forecasts over north Indian Ocean issued by India Meteorological Department. *Nature Hazards*, **68**, 433–451, doi: 10.1007/s11069-013-0624-z

Mohapatra, M, Bandyopadhyay, B.K. and Tyagi, A. (2014). Construction and Quality of best tracks parameters for study of climate change impact on Tropical Cyclones over the North Indian Ocean during satellite era. *In*: Mohanty, U.C, Mohapatra, M, Singh, O.P, Bandyopadhyay, B.K. and Rathore, L.S. (ed.) Monitoring and Prediction of Tropical Cyclones over the Indian Ocean and Climate Change. Springers and Capital publishers, New Delhi, India, pp. 1–17.

Mohapatra, M. and Sharma, M. (2015). Characteristics of surface wind structure of tropical cyclones over the north Indian Ocean, *J. Earth Sys. Sci.,* **124 (7)**, 1573–1598.

Osuri, K.K, Mohanty, U.C, Routray, A. and Mohapatra, M. (2012). The impact of satellite-derived wind data assimilation on track, intensity and structure of tropical cyclones over the North Indian Ocean. *International Journal of Remote Sensing*, **33**, 1627–1652, doi: 10.1080/ 01431161.2011.596849.

Raghavan, S. (2013). Observational aspects including weather radar for tropical cyclone monitoring. *Mausam*, **64**, 89–96.

Rao, A.D. and Mohapatra M. (2010). Cyclone and storm surge, *In*: Vision Document on Atmosphere Sciences, Ministry of earth sciences, Government of India, New Delhi, pp. 55–58.

Ray, K, Mohapatra, M, Bandyopadhyay, B.K. and Rathore, L.S. (2014). *In:* Kamaljit Ray, Mohapatra, M, Bandyapadhyay, B.K. and Rathore, L.S. (ed.), High Impact Weather Events over the SAARC Region. Springers and Capital publishers, New Delhi, India, pp. 1–392.

Roy Bhowmik, S.K, Kotal, S.D. and Kalsi, S.R. (2005). An empirical model for predicting decaying rate of tropical cyclone wind speed after landfall over Indian region. *Journal of Applied Meteorology*, **44**, 179–185.

Roy Bhowmik, S.K. Kotal, S.D. and Kalsi, S.R. (2007). Operational tropical cyclone intensity prediction – An empirical technique. *Nature Hazards*, **41**, 447–455.

RSMC, New Delhi (2014). Report on Cyclonic Disturbances Over the north Indian Ocean during 2013. RSMC, IMD, New Delhi, pp. 1–273.

Tyagi, A, Mohapatra, M, Bandyopadhyay, B.K. and Kumar, N. (2010). Inter-annual variation of frequency of cyclonic disturbances landfalling over WMO/ESCAP Panel Member Countries, WMO/TD-No. **1541** on Ist WMO International conference on Indian Ocean Tropical Cyclones and climate change, Muscat, Sultanate of Oman, 08–11 March 2009, WWRP-**2010/2**, pp. 1–7.

Very Severe Cyclonic Storm MADI over Bay of Bengal, 6–13 December 2013: A Diagnostic Study

B. Sabade and M. Mohapatra

1 Introduction

A tropical cyclone (TC) is a warm core large scale low pressure system with maximum sustained wind (MSW) of 34 knots or more around the centre which rotates in anti-clockwise direction in Northern Hemisphere and clockwise in Southern Hemisphere. The TCs are most destructive due to the associated storm surge, heavy rain and wind. Indian coast is more prone to TC though the proneness varies from districts to districts (Mohapatra et al. 2012a, b, Mohapatra 2015a, b). Though there have been significant improvement in the track and intensity forecast of the TCs over North Indian Ocean (NIO) (Mohapatra et al. 2012a, b, 2013a, b, c; Mohapatra 2015a, b), still there are challenges in the unique cases like rapid track change, rapid intensity change etc. Hence there is need for further understanding of physical processes associated with such unique cases of TCs developing over the NIO.

From the India Meteorological Department (IMD) data of 1891–2013 (IMD 2011), it is seen that the month of December accounts for about 19 % of the intense systems that form over the Bay of Bengal during post-monsoon season. In all the months of post-monsoon season, when the storms moved southwestward they were definitely subjected to weakening (Muthuchami and Sridharan 2008). Their study also concludes that by December, the Sea Surface Temperatures (SST) over North Bay of Bengal becomes cooler and upper level sub-tropical westerly jet shifts southward and so the storms moving into this area have a tendency to weaken. Muthuchami (2000) concluded that during post-monsoon season in the Bay of Bengal, cyclonic storms have a particular annual behaviour in direction of motion in each year, i.e. the storms in a year tend to move in a particular direction. Though

B. Sabade (✉) • M. Mohapatra
India Meteorological Department, Mausam Bhawan, Lodi Road, 110003 New Delhi, India
e-mail: bsabade@yahoo.co.in

© Capital Publishing Company 2017

M. Mohapatra et al. (eds.), *Tropical Cyclone Activity over the North Indian Ocean*,
DOI 10.1007/978-3-319-40576-6_8

in the year 2013, four storms formed, three had re-curving tracks. But out of the three storms, two cyclonic storms moved in southwest direction and one in the northeast direction.

A cyclonic storm (CS) 'Madi' formed over southwest Bay of Bengal on 7 December 2015. It initially moved northwards and intensified upto very severe cyclonic storm (VSCS) with MSW of 65 knots. The system intensified to VSCS while moving in northeastward direction and hence became an exception defying previous studies which indicated higher rate of weakening when the system moves northeastward due to wind shear associated with westerlies. After crossing latitude 15° N it weakened due to unfavourable conditions and re-curved southwestwards. Generally, the system moving southwestwards dissipate over the sea, but in this case, though the system weakened during its southwestward course, it intensified for a temporary period and could retain the intensity of depression till landfall.

VSCS Madi depicted unique intensification and movement characteristics. Comparing the tracks, the track of Madi was most unique in nature and had a rare analogue with past records.

To understand these salient synoptic features the role of physical para-meters in, intensification/weakening of the storm such as TC Heat Potential (TCHP) and SST, relative humidity (RH), total precipitable water (TPW) and vertical wind shear in the Bay of Bengal during VSCS Madi is studied and the results are presented. The possible reason for storm moving in southwestwards direction is also presented here.

2 Brief Life History of VSCS Madi

A depression (D) formed over southwest Bay of Bengal in the morning of 6 December 2013 and became deep depression (DD) in the same midnight. It moved very slowly northward and intensified into a cyclonic storm, 'Madi' in the morning of 7 December. It continued to move slowly and intensified into a severe cyclonic storm (SCS) in the afternoon of 7 December and into a very severe cyclonic storm in the forenoon of 8 December 2013 with a maximum sustained wind speed of 65 knots. However, due to entrainment of cold air, colder sea and increase in vertical wind shear the VSCS weakened into severe cyclonic storm in the evening of 9 December. Due to weakening, the system moved southwestward after reaching the latitude of 15.7° N under the influence of lower and middle tropospheric steering ridge. It further weakened into cyclonic storm in the early hours, into deep depression in the morning, depression in night of 11 December. It crossed Tamil Nadu coast near Vedaranniyam around 1330 UTC of 12 December, emerged into Palk strait around 1500 UTC and again crossed Tamil Nadu coast near Tondi around 1700 UTC. It then emerged into southeast Arabian Sea as a well marked low pressure area in the early morning of 13 December 2013.

The track of the system is shown in Fig. 1.

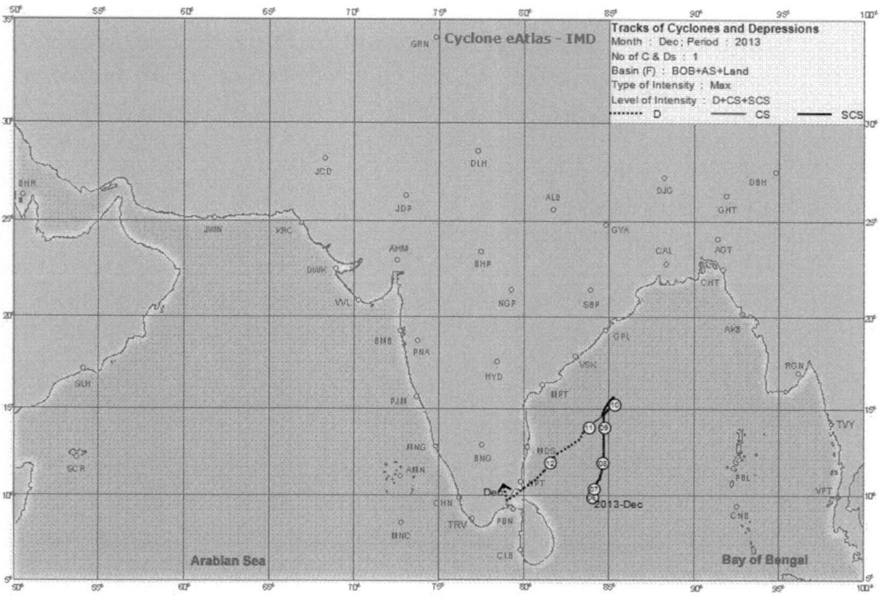

Fig. 1 Track of VSCS Madi (www.rmcchennaieatlas.tn.nic.in)

3 Data and Methodology

Regional Specialised Meteorological Centre (RSMC) report 2014 (RSMC 2014) prepared by Cyclone Warning Division New Delhi, formed the main basis for the study. To analyse the various characteristics of Madi, the best track data of RSMC, New Delhi (RSMC 2014) were considered. The VSCS, MADI was monitored mainly with satellite supported by meteorological buoys, coastal and Island observations and Doppler Weather Radars (DWRs). The synoptic and NWP models analyses have also been considered. The intensification, movement and dissipation of the VSCS Madi are analysed using the synoptic, thermodynamic and dynamic parameters.

Gray (1968, 1975, 1979) demonstrated that the genesis and intensification may be related to environmental factors such as large value of low-level relative vorticity, location of at least few degrees poleward of equator giving a significant value of planetary vorticity, weak vertical shear of horizontal winds, SST exceeding 26 °C and a deep thermocline, conditional instability through deep atmospheric layer and large values of relative humidity in the lower and middle troposphere. The parameters for intensification (Gray 1981) of the TC have been analysed. SST and mid-level moisture effects are included in the thermodynamic term and vertical wind shear and low-level vorticity effects are included in the dynamical term.

Generally TCHP and SST, humidity and vertical wind shear influence the intensification. In order to study the influence of these factors in the movement and weakening of the VSCS Madi over the Bay of Bengal during post-monsoon season, products from the web sites of US, National Oceanic and Atmospheric Administration (NOAA), Atlantic Oceanographic and Meteorological Laboratory (AOML) (http://www.aoml.noaa.gov/), Cooperative Institute for Meteorological Satellite Studies (CIMSS), University of Wisconcin (http://tropic.ssec.wisc.edu/) and NOAA, National Environmental Satellite, Data and Information Service (NESDIS), Coperative Institute for Research in the Atmosphere (CIRA) (http://rammb.cira.colostate.edu/).

To study the movement, day to day position of the ridge at upper level along with mid-tropospheric level circulation and northeast monsoon circulation at the lower level over the Indian region has been studied to understand the initial north to northeastward movement and then finally the re-curvature to southwestwards.

4 Results and Discussions

The intensification characteristics are discussed in Sect. 4.1. The characteri-stic of unique track traversed by VSCS Madi is discussed in Sect. 4.2.

4.1 Intensity Characteristics of VSCS, Mai

The system intensified while moving north/northeastwards when the SST, TCHP and low vertical wind shear near the storm region favoured its intensification and weakened during its southwestward course.

The salient features of VSCS Madi are:

- As the very severe cyclonic storm moved to the north of 13.0° N i.e. to west central Bay of Bengal, it experienced colder sea surface temperature, low Ocean thermal energy (<50 KJ/cm^2) and high vertical wind shear (20–30 knots), it weakened into a severe cyclonic storm.
- The system had temporary intensification also during its weakening phase. But thereafter with the combined impact of colder sea surface temperature, low Ocean thermal energy, high vertical wind shear and incursion of cold and dry air into the core of the cyclone, it gradually weakened into a Depression.
- The system re-entered into the warmer SST region and could maintain the intensity of Depression inspite of the core being isolated from the warm and moist feed.

The above salient features are discussed in the following sub-sections:

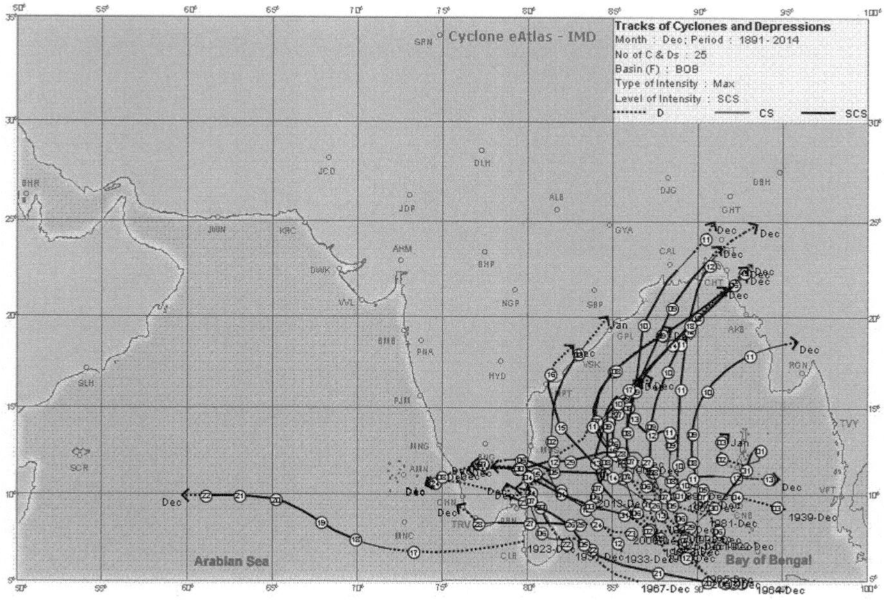

Fig. 2 Tracks of SCS over BOB, in the month of December during 1891–2014 (www.rmcchennaieatlas.tn.nic.in)

4.1.1 Climatological Aspect of Weakening of Systems in the Bay of Bengal

Out of the 50 cyclonic storms/severe cyclonic storms that formed in December, 50 % weakened into depressions before they crossed the coast or dissipated over the sea while moving in different direction. About 50 % of the storms that weakened moved in northeast direction. The higher rate of weakening when they move towards northeastward is due to shear associated with westerlies. Also the percentage of the severe cyclonic storms that weakened before crossing was less than the cyclonic storms that weakened. The tracks of SCS over Bay of Bengal (BOB) in the month of December are shown in Fig. 2.

4.1.2 Low Level Relative Vorticity

Tropical cyclones form only in regions of large positive low level (~850 hPa) vorticity. The larger this low level vorticity, greater the potential for cyclone genesis. In case of VSCS Madi, the positive vorticity was present in the form of a pre-existing low pressure area over south China Sea which moved across Malay peninsula and emerged into south Andaman Sea on 01 December. Moving westwards it lay as well-marked low pressure area over southwest Bay of Bengal off Sri Lanka coast on 4 December 2013.

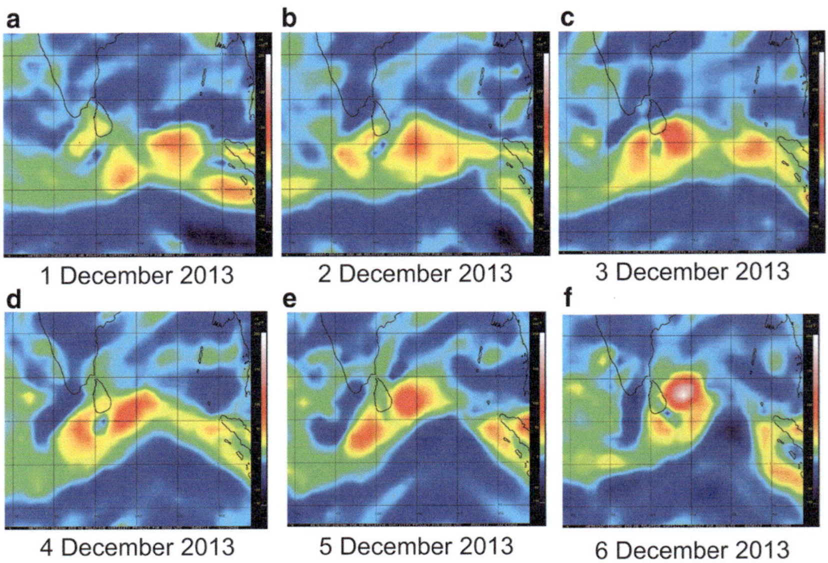

Fig. 3 (**a–f**) Relative vorticity at 850 hPa during 1–6 December 2013 (Source: http://tropic.ssec.wisc.edu/)

Figure 3a–e shows the positive vorticity at 850 hPa from 1 December 2013 to 5 December 2013. It is clearly seen that the lower level convergence and relative vorticity increased from 5 to 6 December 2013.

4.1.3 Low Level Wind Shear

A vertical wind shear of less than 10 m/s (20 knots, 22 mph) between the 200 and 850 hPa level, is favoured for tropical cyclone development. The vertical wind shear was moderate (10–20 knots) upto 9 December 2013, when the system intensified from CS to VSCS as seen from the figures (Fig. 4a–g). From 0900 UTC of 9 December till the landfall the wind shear increased to 20–30 knots and the system weakened from VSCS to CS except for a brief period from 0300 to 0900 UTC of 10 December when it had re-intensified.

In spite of the high wind shear, the system had re-intensified from SCS to VSCS from 0300 to 0900 UTC of 10th while moving in northeastward direction because the system came in the forward sector of a mid and upper tropopsheric trough (Fig. 5) west of it and the trough enhanced the poleward outflow from the system. And at 1200 UTC of 10th the trough was seen east of the system and also around the same time, the anti-cyclone to the east of the system ceased to affect the storm motion.

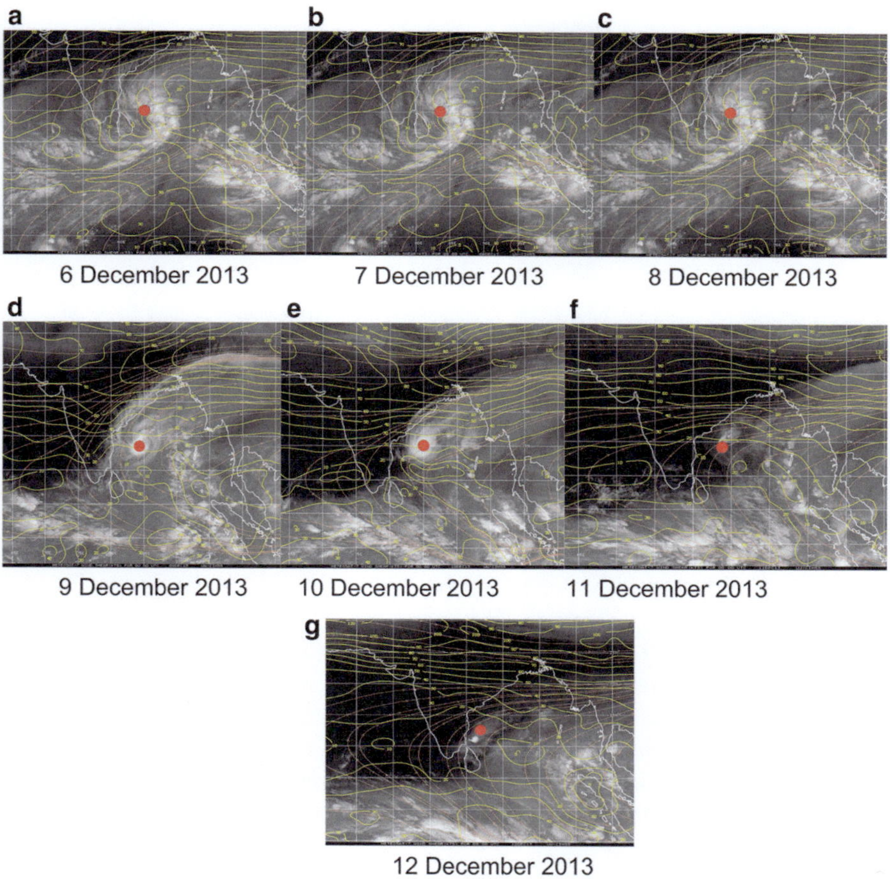

Fig. 4 (**a–g**) Wind shear from 6 to 12 December (Source: http://tropic.ssec.wisc.edu/)

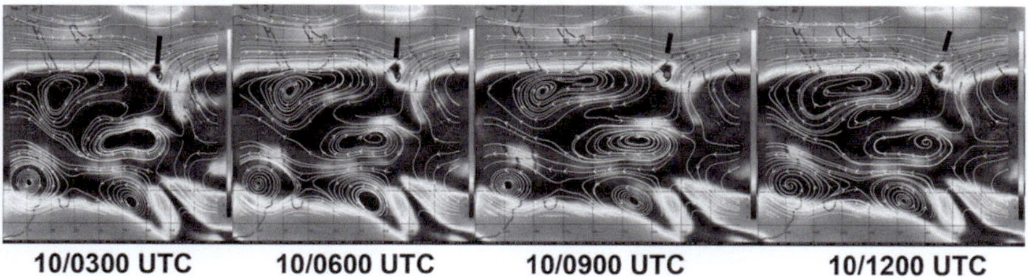

Fig. 5 Environmental wind (200–700 hPa layer mean) on 10th December 2013 (Source: http://tropic.ssec.wisc.edu/). Positon of westerly trough w.r.t the centre of the system is indicated by a *black line*

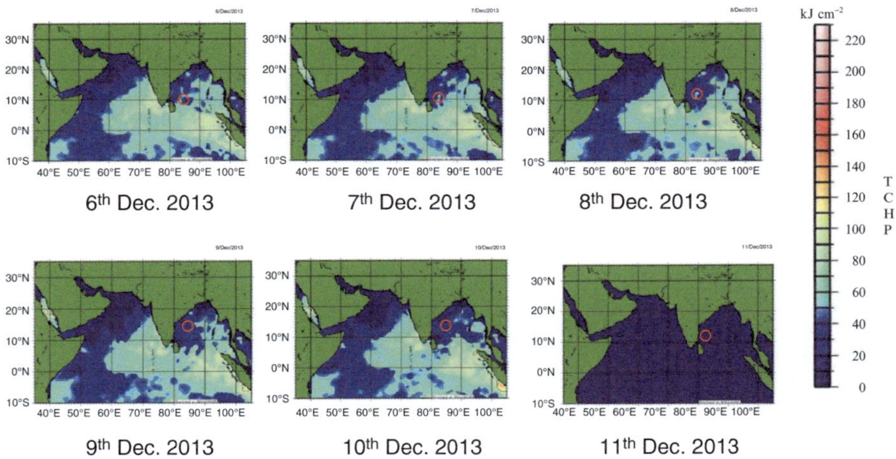

Fig. 6 TCHP during 6th–11th Dec. 2013 alongwith storm centresongwith the storm centres

4.1.4 SST and Ocean Thermal Energy

A warm ocean waters (of at least 26.5 °C) throughout a sufficient depth (at least on the order of 60 m) are necessary to fuel the tropical cyclone heat engine. An ocean thermal parameter related to the sea surface temperature and the depth of the sea temperature is given by the TCHP. The TCHP was more than 50 kJ/cm^2 during 6–8 December 2013, when the system intensified from Depression to VSCS and less than 50 kJ/cm^2 from 9 to 11 when the system weakened from VSCS to Depression (Fig. 6).

The SST (Fig. 7) during genesis was about 26–28 °C but when the system moved to the north of 13.0° N i.e. to westcentral Bay of Bengal, it experienced colder SST and low Ocean thermal energy (<50 KJ/cm^2). Also the vertical wind shear of horizontal wind gradually increased and became high (20–30 knots). As a result, the very severe cyclonic storm weakened into a severe cyclonic storm at 1200 UTC of 09. It then again moved to warmer SST over southwest BOB and therefore the system could retain its intensity of Depression on 11 and 12 till the time of landfall.

4.1.5 Moist Static Stability

An atmosphere in which temperatures decrease fast enough with height such that it is potentially unstable to moist convection is conducive for tropical genesis. It is the precipitating convection typically in the form of convective complexes that allows the heat stored in the ocean waters to be liberated for the tropical cyclone development. The Total Precipitated Water (TPW) imageries during 6–12 December (Fig. 8) indicate that the dry and cold air penetrated into the southwestern periphery of the cyclone from 10 December. It gradually penetrated further towards the centre

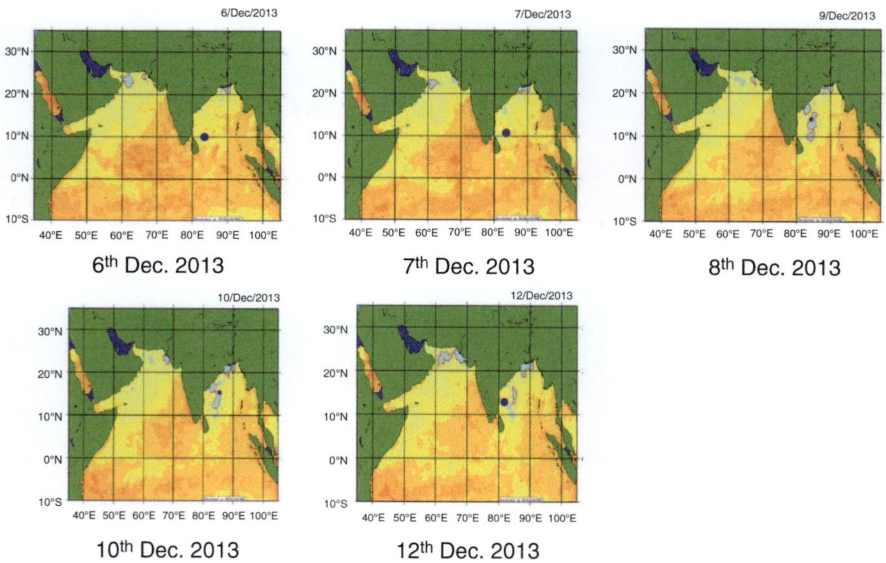

Fig. 7 SST near the (storm centre as shown by the *blue dot*)

of the cyclone from the southern side. As a result, it isolated the core of the cyclone from the warm and moist air from the southeast sector at 0500 UTC of 11 December.

4.1.6 Middle Tropospheric Relative Humidity

Dry middle levels are not conducive for allowing the continuing development of widespread thunderstorm activity. According to Gray (1968, 1975, 1979, 1981), a minimum threshold of 40 % mid-tropospheric RH is necessary for tropical cyclogenesis. Entrainment of relatively dry environmental air near the system centre would lead to decrease of low and mid tropospheric humidity and hence weaken the system.

As can be seen from the Fig. 9, the relative humidity at 700 and 500 hPa on 8th was more than 70 % while on 11th near the storm centre, it was less than 40 % at 500 hPa.

Hence, it can be seen that the SST > 26 °C, moderate wind shear (10–20 knots), TCHP > 50 kJ/cm^2 led to initial intensification while SST < 26 °C, high wind shear (>20 knots), TCHP < 50 kJ/cm^2, entrainment of dry and cold air from Indian landmass lead to subsequent weakening. The entrainment of dry and cold air isolated the core from the warm and moist feed from southeast sector and there was a reduction in up-draft parcel buoyancy. The system re-intensified into VSCS with high vertical wind shear on 10 December while moving in northeastward direction because of the increase in poleward outflow which further deepened the system. During its weakening stage, the system again entered into the area with

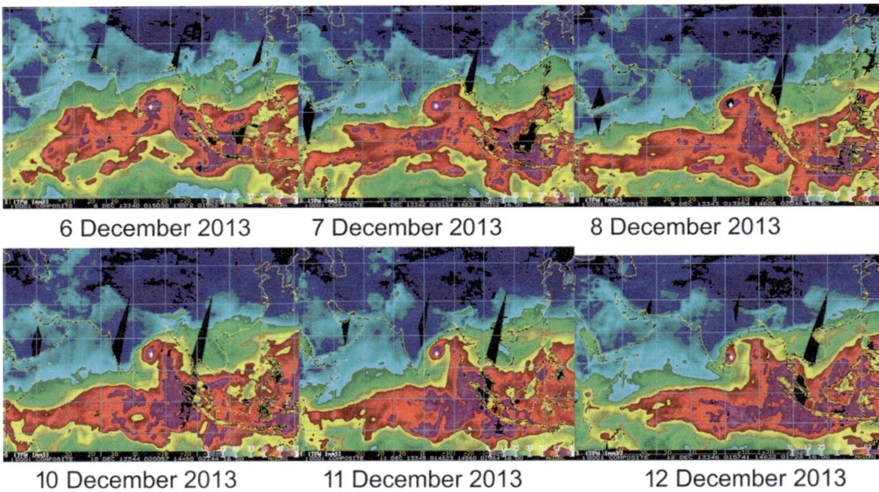

Fig. 8 Total Precipitable Water imageries during 6–12 December 2013 (Source: http://rammb.cira.colostate.edu/)

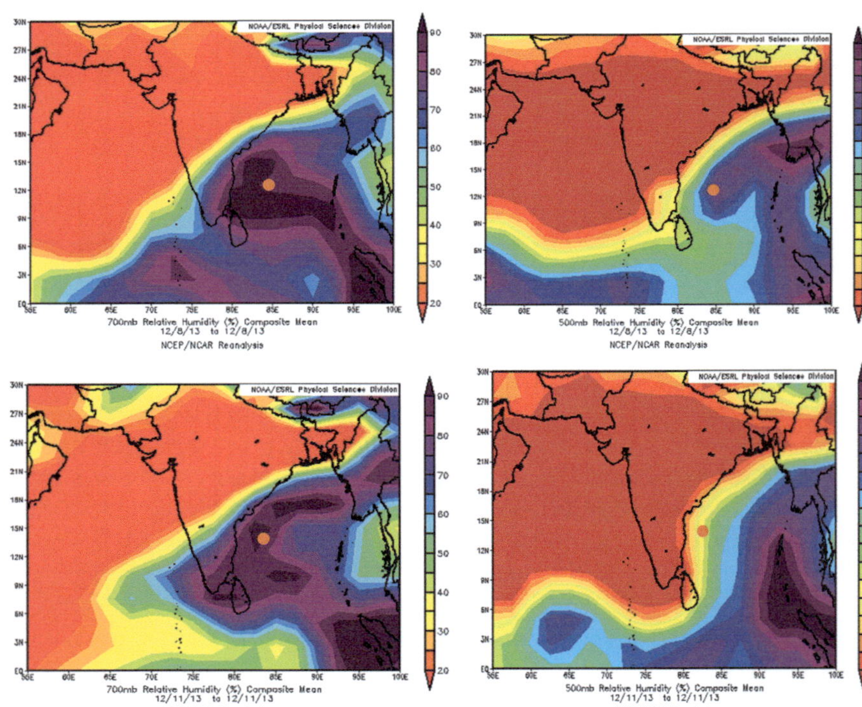

Fig. 9 Mean relative humidity distribution at 700 hPa level and 500 hPa level on 8th Dec. (day of Intensification, mean position of the storm 12.5 °N/84.7 °E) and on 11th Dec. (day of weakening, mean position of the storm 13.8 °N/83.4 °E)

SST > 26 °C and hence could maintain the intensity of depression till the time of landfall in spite of the fact that low relative humidity (RH < 40 %) prevailed at mid-tropospheric level (500 hPa) due to the entrainment of relatively dry and cold air into the core of the cyclone.

4.2 *Movement of the Storm*

The salient features with respect to the track of VSCS Madi are:

- It had a unique track with near northward movement till 15.7° N and then it recurved southwestwards to Tamil Nadu coast.
- It moved very slowly during its northward journey and speed peaked up gradually after the recurvature to southwest.

It is attempted to explain the above salient features in the following sub-section:

4.2.1 Track in Relation to Steering Wind

Figure 10 indicates the position of the ridge and the storm centre from 6 to 12 December 2013. Ridge at 200 hPa ran about 10° N during the cyclone's life period (Fig. 10). The system lay to the south but close to the ridge during 6–9 December and hence moved northwards slowly. According to Srinivasan and Ramamurthy (1973), there is a pre-pondered northerly motion when the centre is close to the ridge. The upper tropospheric ridge moved northwards with the northward movement of the system. As it lay north of the ridge on 10th, it moved north-northeastwards under influence of upper tropospheric anti-cyclonic circulation lying to the east of the system. However, due to gradual weakening of system, the steering level changed from upper troposphere to lower and middle troposphere (Fig. 11). The influence of the upper tropospheric anti-cyclonic circulation to the east of system centre decreased and that of lower and middle level anti-cyclonic circulation (Fig. 12) lying to the west of the system centre (over central India) increased.

As a result, on 11th and 12th the system was steered by the anti-cyclonic circulation lying to its west along with the northeast monsoonal circulation flow.

The Fig. 11 shows deep layer steering between 200 and 700 hPa layer at 0900 and 1200 UTC of 10th. The figure shows that at 1200 UTC of 10th the anti-cyclone to the west of the system strengthened to steer the system southeastwards.

Though the ridge on 11th and 12th lay around the same latitude (Fig. 10), the system started moving in southwestward direction which clearly indicated that the anti-cyclonic circulation to the east of the system was not the steering force when the system weakened. But it was the anti-cyclonic circulation lying to the west of the system at mid-tropopsheric level that had strengthened along with the northeast monsoon circulation at the lower levels that steered the system southwestwards.

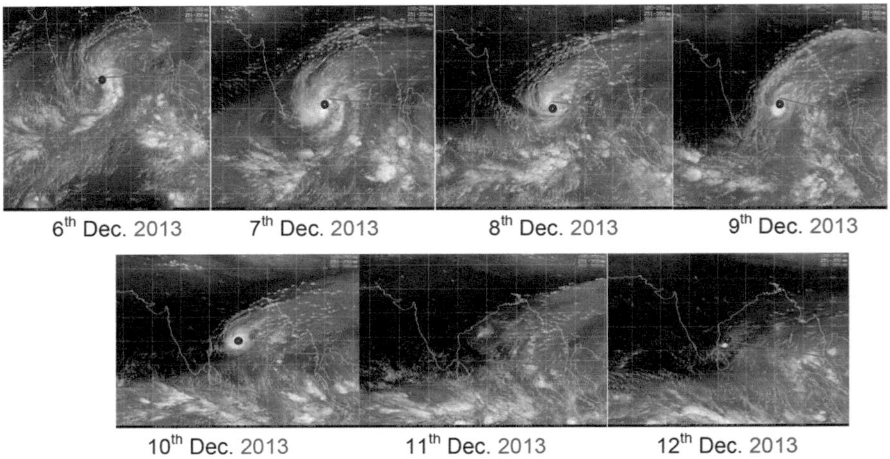

Fig. 10 METEOSAT-7 Upper level winds during 6–12 Dec. 2013 (Source: http://tropic.ssec.wisc.edu/)

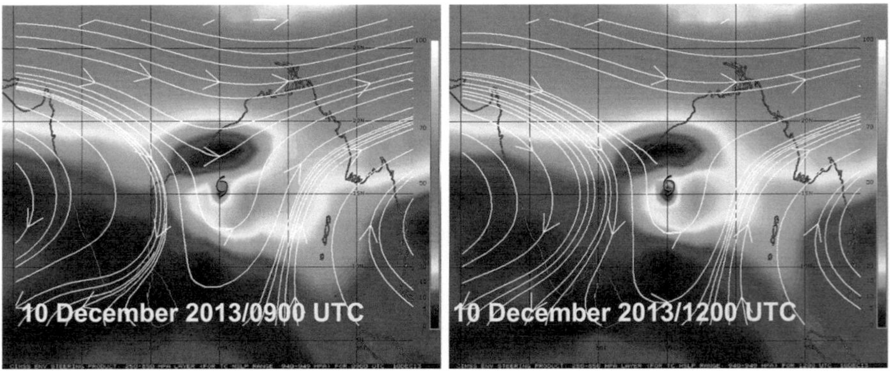

Fig. 11 Steering of TC Madi by the mid and upper tropospheric level circulation on 10th December 2013 (Source: http://tropic.ssec.wisc.edu/)

Hence the reasons which led to its unique southwestward movement during its weakening stage were the lower and mid-tropospheric steering instead of the upper-tropospheric steering. Thus VSCS Madi is a very good example to demonstrate the fact that the intensity of a cyclone also plays a dominant role in determining the steering flow and hence the track of the cyclone. Figure 12 shows the anti-cyclonic flow on 10th and 11th December 2014 at 850, 500 and 200 hPa.

The movement during its intensification stage was steered by the ridge at 200 hPa lying close to its north in association with an anti-cyclonic circulation to the northeast of the system. The system moved slowly when close to the ridge. When the system was north of the ridge, it moved in a north-northeastward direction. Though the ridge towards the east of the system remained at 10° N, when the system weakened, the

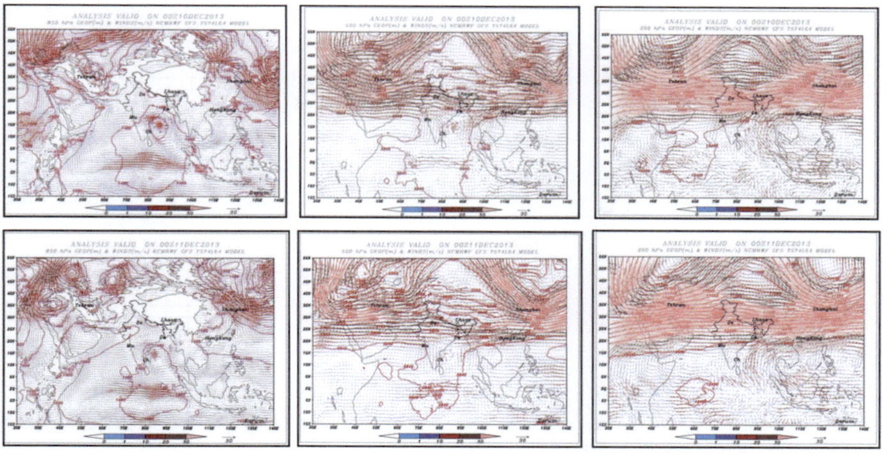

Fig. 12 Anticyclonic flow on 10th and 11th Dec. 2014 at 850, 500 hPa and 200 hPa

upper air steering ceased to affect the path of the storm. The system was embedded in the large scale NE monsoon circulation, also, the anti-cyclone to the west of the system became dominant. Hence the reasons which led to its unique southwestward movement during its weakening stage were the lower and middle tropospheric steering instead of the upper tropospheric steering.

5 Conclusions

The study of intensity and track characteristics of VSCS Madi concluded the following:

- The study restated the fact that SST (>26 °C), TCHP (>50 kJ/cm^2), vertical wind shear (low to moderate, < 20 knots), relative vorticity ($>10 \times 10^{-5}$ per s) and mid-tropospheric relative humidity ($\geq$50 %) is essential for tropical cyclogenesis and intensification.
- Ocean thermal energy, wind shear together with SST were responsible for weakening of the system.
- VSCS Madi intensified temporarily on 10th because of enhanced poleward outflow as the system lay in the forward sector of an upper-tropospheric trough and experienced warmer SST.
- Study suggests that TPW can be utilised as a precursor for predicting intensification/weakening of the system.
- The intensity of the cyclone is conjunction with vertical wind shear and steering flow plays a dominant role in determining the steering flow. When the system was of VSCS intensity, it was steered by the upper-tropospheric anti-cyclonic circulation and when it was SCS, it was steered by mid- and upper-tropospheric

anti-cyclonic circulation. The change in the track of Madi during its life period could be attributed to above impact of TC on steering flow.

- The track of the cyclone in turn also plays a dominant role in determining the intensity of the system. The initial north/north-northeastward track of Madi helped in increasing beta effect and favourable outflow leading to intensification. Its southwestward movement in the later part of the life cycle led to weakening, due to low SST, Less TCHP ($<$50 kJ/cm^2), high wind shear ($>$20 knots) and dry and cold air intrusion from the Indian sub-continent.

References

Gray, W.M. (1968). Global view of the origin of tropical disturbances and storms. *Monthly Weather Review*, **96**, 669–700.

Gray, W.M. (1975). Tropical Cyclone Genesis. Dept. of Atmos. Sci. Paper No. 234, Colorado State University, Ft. Collins, CO, 121 pp.

Gray, W.M. (1979). Hurricanes: Their formation, structure and likely role in the general circulation. *In*: Shaw, D.B. (ed.), Meteorology Over the Tropical Oceans. Royal Meteorological Society, James Glaisher House, Grenville Place, Bracknell, Berks, RG 12 1BX, 155–218.

Gray, W.M. (1981). Recent advances in tropical cyclone research from rawinsonde composite analysis. World Meteorological Organisation, 407 pp.

IMD (2011). Track of storm and depressions over the Indian Seas during 1891–2013, 'Cyclone e-Atlas of IMD'. IMD, Chennai.

Mohapatra, M, Mandal, G.S, Bandyopadhyay, B.K, Tyagi, A. and Mohanty, U.C. (2012a). Classification of cyclone hazard prone districts of India. *Natural Hazards*, **63**, 1601–1620.

Mohapatra, M, Nayak, D.P. and Bandyopadhyay, B.K. (2012b). Evaluation of cone of uncertainty in tropical cyclone track forecast over North Indian Ocean Issued by India Meteorological Department. *Tropical Cyclone Research and Review*, **01**, 331–339.

Mohapatra, M, Nayak, D.P, Sharma, R.P. and Bandyopadhyay, B.K. (2013a). Evaluation of official tropical cyclone track forecast over north Indian Ocean issued by India Meteorological Department. *Journal of Earth System Science*, **122**, 589–601.

Mohapatra, M, Sikka, D.R, Bandyopadhyay, B.K. and Tyagi, A. (2013b). Outcomes and challenges of forecast demonstration project (FDP) on landfalling cyclones over the Bay of Bengal. *Mausam*, **64**, 1–12.

Mohapatra, M, Bandyopadhyay, B.K. and Nayak, D.P. (2013c). Evaluation of operational tropical cyclone intensity forecasts over north Indian Ocean issued by India Meteorological Department. *Natural Hazards*, **68**, 433–451.

Mohapatra, M. (2015a). Cyclone hazard proneness of districts of India. *Journal of Earth System Sciences*, **124**, 515–526.

Mohapatra, M, Nayak D.P, Sharma, M, Sharma, R.P. and Bandyopadhyay, B.K. (2015b). Evaluation of official tropical cyclone landfall forecast Issued by India Meteorological Department. *Journal of Earth System Sciences*, **124**, 861–874.

Muthuchami, A. (2000). Origin and movement of cyclonic storms in the Bay of Bengal. *Mausam*, **51**, 225–230.

Muthuchami, A. and Sridharan, S. (2008). Intensification and movement of cyclonic storm in the Bay of Bengal during post monsoon season. *Mausam*, **59**, 51–68.

RSMC New Delhi (2014). Report on Cyclonic Disturbances over North Indian Ocean during 2013, Met. Monograph no. ESSO/IMD/RSMC-Tropical Cyclones Report No. 01 (2014)/5

Srinivasan, V. and Ramamurthy, K. (1973). Northeast monsoon FMU Report No IV-18.4, India Meteorological Department.

Rapid Weakening of Very Severe Cyclonic Storm 'Lehar' – A Case Study

R.P. Sharma and M. Mohapatra

1 Introduction

A depression formed over south Andaman Sea on 23rd evening and it intensified into a cyclonic storm, Lehar in the early morning of 24 November 2013 near Latitude 10.0° N and longitude 95.0° E. Moving northwestward, it crossed Andaman and Nicobar Islands near Port Blair around 0000 UTC of 25 November as a severe cyclonic storm. On 25th morning, it emerged into southeast Bay of Bengal and moved west-northwestward, intensified into a very severe cyclonic storm in the early hours of 26 November. However, while moving west-northwestwards over westcentral Bay of Bengal, it rapidly weakened from 27th afternoon and crossed Andhra Pradesh coast close to south of Machilipatnam around 0830 UTC of 28 November 2013 as a depression.

The salient features of this system are given below:

1. It was the first severe cyclonic storm to cross Andaman and Nicobar Islands in the month of November after 1989.
2. It had second landfall near Machilipatnam as a depression.
3. It rapidly weakened over the sea from the stage of very severe cyclonic storm (75 knots) at 0600 UTC of 27th to depression (25 knots) at 0000 UTC of 28th November 2013 (in 18 h) (RSMC, New Delhi 2014).

The very severe cyclonic storm, 'Lehar' was monitored mainly with satellite supported by meteorological buoys, coastal and island observations and Doppler Weather Radar (DWR), Machilipatnam. The half hourly INSAT/KALPANA imageries, hourly coastal observations and every 10 min DWR imageries and products

R.P. Sharma (✉) • M. Mohapatra
Cyclone Warning Division, India Meteorological Department, Mausam Bhawan, Lodi Road, 110003 New Delhi, India
e-mail: rpsharma1980@gmail.com

© Capital Publishing Company 2017

M. Mohapatra et al. (eds.), *Tropical Cyclone Activity over the North Indian Ocean*,
DOI 10.1007/978-3-319-40576-6_9

were used for monitoring of cyclonic storm as per standard operation procedure (IMD 2013).

Various numerical weather prediction (NWP) models and dynamical–statistical models including IMD's global and meso-scale models were utilised to predict the track and intensity of the storm. The Tropical Cyclone Module in the digitised forecasting system of IMD was utilised for analysis and comparison of various NWP models and decision making process. However, there was large divergence in NWP model guidance with respect to genesis and intensification of the system. There was more unanimity in the NWP models with respect to track prediction. Moving northwestwards, it rapidly weakened from the stage of very severe cyclonic storm at 0600 UTC of 27th to depression at 0000 UTC of 28 November 2013 (75–25 kts within 18 h) over the westcentral Bay of Bengal off Andhra Pradesh coast. This rapid weakening of the system could not be anticipated well by the forecasters and most of the numerical weather prediction models. Though there is dynamical statistical model for prediction of rapid intensification, there is no such model to predict rapid weakening (IMD 2013). Considering all these, there is a need for in-depth understanding of the process leading to rapid weakening. Rapid weakening is more common in northeast Pacific than in north Atlantic (33 % opposed to 11 %) in part due to lower areal extent of sea surface temperature (SST) at a temperature greater than 26 °C (Wood and Ritchie 2014). Rapid weakening frequency in northeast Pacific is positively correlated with SST anomalies associated with El-Nino Southern Oscillation (ENSO), likely as a result of stronger (weaker) TCs in El-Nino (La-Nino) seasons (Wood and Ritchie 2014). The strongest contributor to accelerated TC decay appeared to be from decreasing 400 to 700 hPa relative humidity (RH) in southern part of 33 % of all TCs from 1972 to 2012. Rapid weakening of TCs over Bay of Bengal is not very common. During the period 2003–2013 only 7 out of 40 (17.5 %) TCs underwent rapid weakening (decrease of intensity by 30 knots or more in 24 h) over the north Indian Ocean. Compared to track forecasting, the intensity forecast is still challenging as demonstrated by Mohapatra et al. (2013). Mc Gauley and Nolan (2011) have analysed the threshold values of environmental parameters favourable for tropical cyclogenesis to understand the dynamics of intensification. Therefore, a study is undertaken to analyse the environmental conditions leading to rapid weakening of Lehar.

2 Brief Life History of Lehar

A remnant of tropical depression over South China Sea moved across Malay peninsula and lay as a low pressure area over south Andaman Sea on 21 November 2013. It became well marked over the same area on 22nd and concentrated into a depression over south Andaman Sea near latitude 08.5° N and longitude 96.5° E, about 550 km south-southeast of Port Blair at 1200 UTC of 23 November 2013. The genesis was detected with the OceanSat-II winds and the observation from the coast of Thailand in addition to satellite imageries. The genesis was associated with upper troposphere ridge running along 13° N which provided adequate upper level divergence through Poleward outflow. The lower level convergence and relative vorticity increased over the area from 22 to 23 November. The sea surface temperature was 28–29 °C and Ocean thermal energy was 60–80 kJ/cm^2. The vertical wind shear of horizontal wind was moderate (10–20 knots). The Madden Julian oscillation (MJO) index lay in Phase 3 i.e. over east equatorial Indian Ocean. Past studies indicate that the Phase 3 is favourable for genesis of depression (Mohapatra and Adhikary 2011).

Due to the favourable atmospheric and Oceanic condition as mentioned above, the depression over south Andaman Sea moved northwestwards, intensified into a deep depression at 1800 UTC of 23rd and further into a cyclonic storm, 'Lehar' at 0300 UTC of 24th November and lay centred near latitude 10.0° N and longitude 95.0° E. It further intensified into a severe cyclonic storm, continued to move northwestwards and crossed Andaman and Nicobar Island near Port Blair around 0000 UTC of 25 November 2013. It then emerged into southeast Bay of Bengal, moved west-northwestwards and intensified into a very severe cyclonic storm at 2100 UTC of 26 November 2013 over southeast Bay of Bengal near latitude 12.5° N and longitude 91.0° E. It attained the maximum intensity of 75 knots at 1800 UTC of 26 November 2013 and the same intensity continued till 0300 UTC of 27 November 2013 when it lay over central Bay of Bengal.

As the westcentral Bay of Bengal was colder with Ocean thermal energy less than 50 kJ/cm^2 and there was entrainment of dry and cold air from central and northern parts of India into the cyclone field and vertical wind shear of horizontal wind increased and became high, the very severe cyclonic storm started to weaken rapidly from the afternoon of 27 November 2013. It weakened into a severe cyclonic storm at 1200 UTC of 27 November and lay centred near latitude 14.5° N and longitude 85.0° E. It further weakened into a cyclonic storm at 1800 UTC of 27 November 2013 with centre near latitude 15.0° N and longitude 84.0° E over westcentral Bay of Bengal. It weakened into a deep depression at 0000 UTC of 28 November 2013 with centre near latitude 15.5° N and longitude 82.0° E. At this time, the vertical wind shear was high (about 20 knots). It weakened into a depression and crossed Andhra Pradesh coast near latitude 15.9° N and longitude 81.1° E (close to south of Machilipatnam) around 0830 UTC of 28 November 2013. It weakened into a well marked low pressure area over coastal Andhra Pradesh and adjoining Telengana at 1800 UTC of 28 November 2013.

The system moved northwestwards/west-northwestwards as it lay to the south of the upper tropospheric steering ridge which moved northward from its position near latitude 13.0° N on the day of genesis to latitude 17° N on the day of landfall. The track of cyclone 'Lehar' is shown in Fig. 1. Typical satellite and radar pictures indicating the rapid weakening of the system are shown in Figs. 2 and 3a, b.

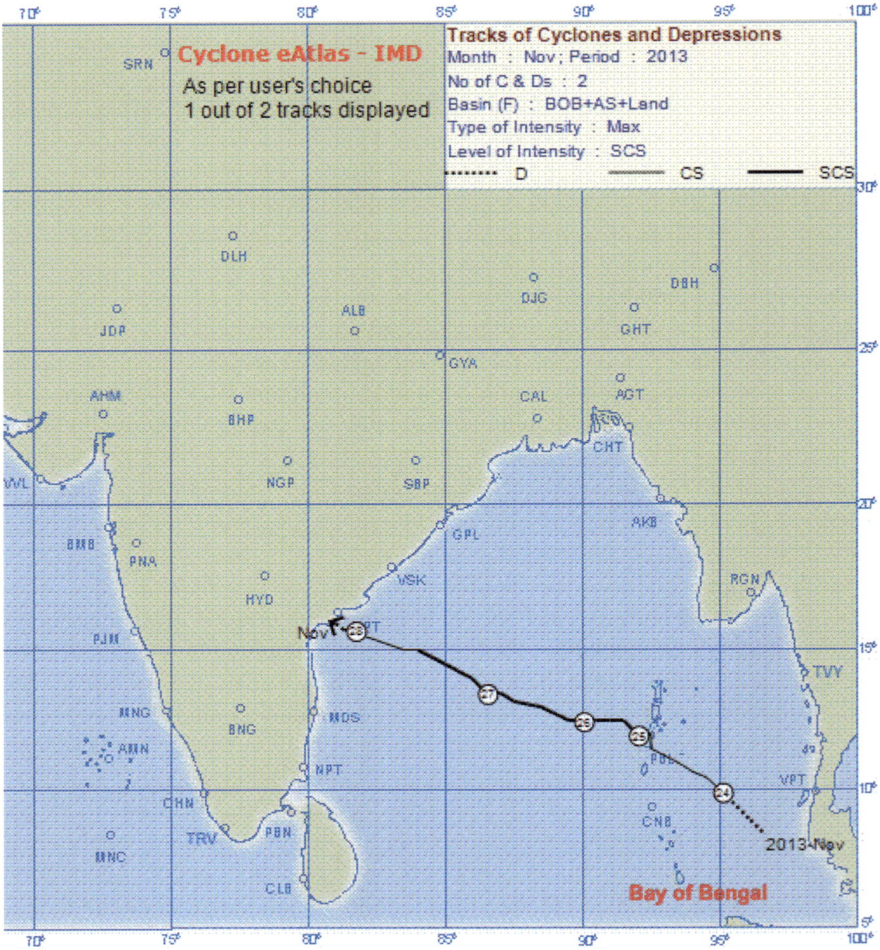

Fig. 1 Track of cyclone 'Lehar'

3 Results and Discussion

3.1 Performance of NWP Models for Rapid Weakening Prediction

IMD utilises dynamical statistical models viz statistical cyclone intensity prediction (SCIP) model for the prediction of TC intensity (Kotal et al. 2008), along with the Hurricane Weather Research and Forecasting model (HWRF) intensity predictions for intensity forecasting. The average 12-hrly intensity forecast errors (absolute and

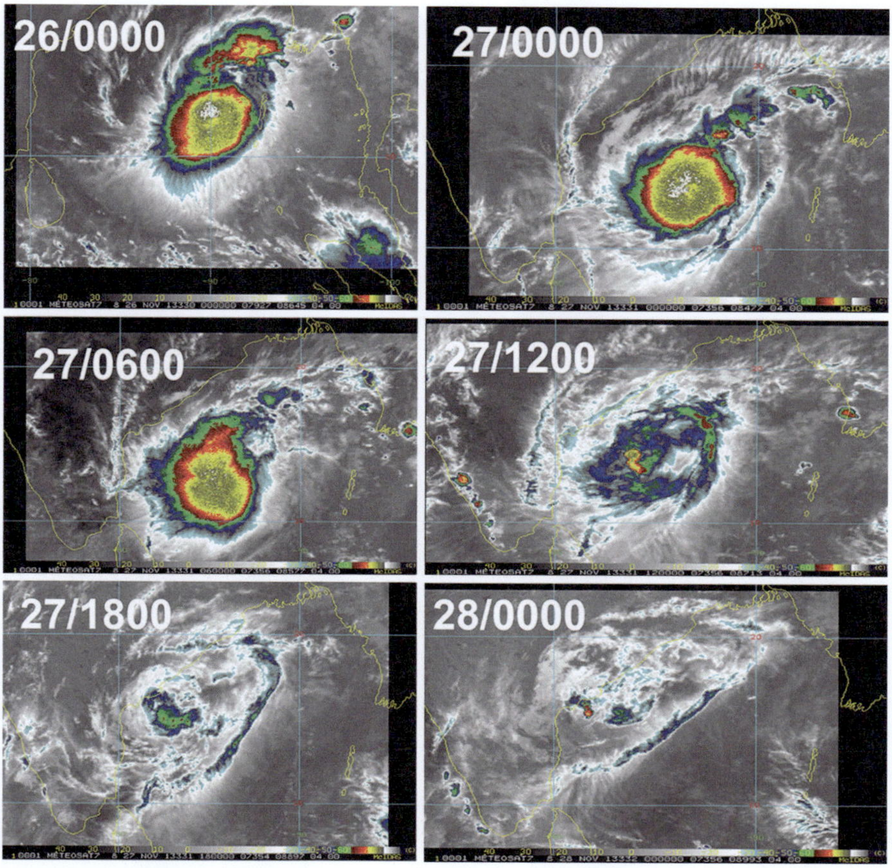

Fig. 2 Typical satellite imageries based on 00 UTC of 26 November, 00, 06, 12 and 18 UTC of 27 November and 00UTC of 28 November 2013

root mean square) of IMD-SCIP and IMD-HWRF intensity forecasts are presented in Tables 1 and 2. The forecast of SCIP model based on 00 UTC of 23 November 2013 for TC Lehar is shown in Fig. 4. It can be observed that the rapid weakening could not be predicted by this model (RSMC, New Delhi 2014). The performance of HWRF model intensity predictions based on 00 UTC of 23 November 2013 is also shown in Fig. 5 (RSMC, New Delhi 2014). This model also could not predict reasonably the rapid weakening of cyclone, Lehar.

All the above results suggest the need for detailed analysis of physical processes associated with rapid weakening. These are discussed in following sections.

3.2 *SST and Ocean Thermal Energy (OTE) in Relation to Rapid Weakening*

The SST over north Indian Ocean (NIO) and the Ocean Thermal Energy (OTE) are shown in Figs. 6 and 7, respectively. It is found that SST rapidly decreased and became less than 28 °C over the central Bay of Bengal on 27 November. On 28 November, it was even less than 26 °C over the part of central Bay of Bengal near Andhra Pradesh coast. Therefore SST was not favourable for the intensification over the Bay of Bengal from 27 November. Also TCs have the highest probability to intensify rapidly and little chance to weaken rapidly when SST is above 29 °C, low probability to intensify rapidly when SST is lower than 28 °C and

a

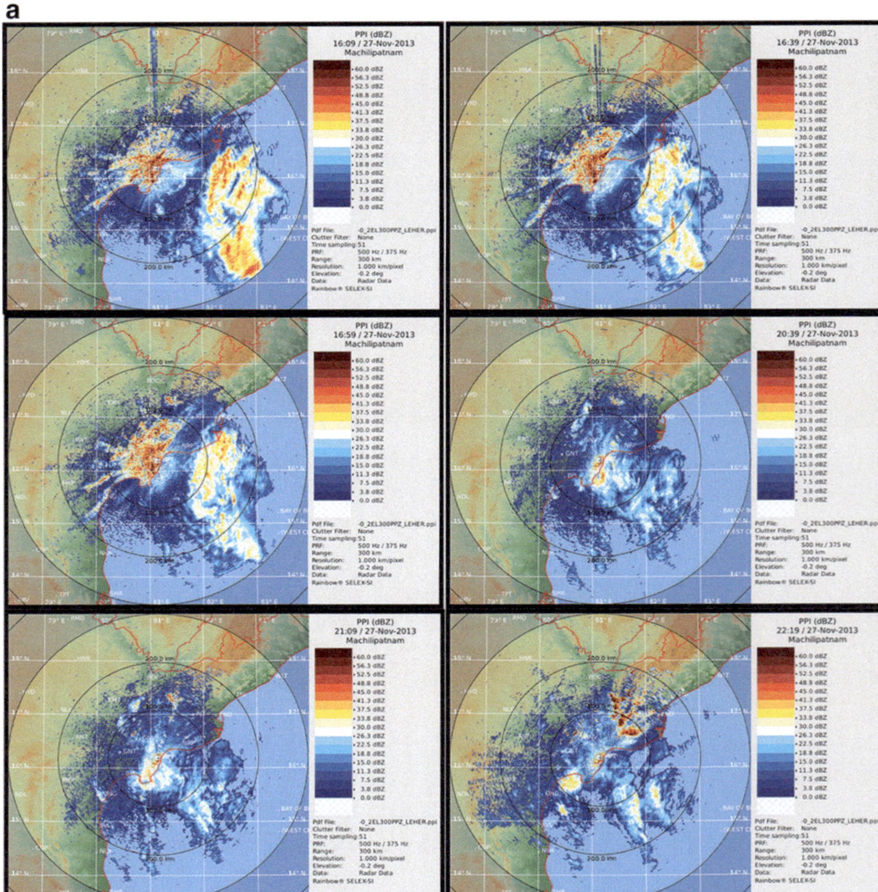

Fig. 3 (**a**) Machilipatnam RADAR imageries on 27 November 2013. (**b**) Machilipatnam RADAR imageries on 28 November 2013

b

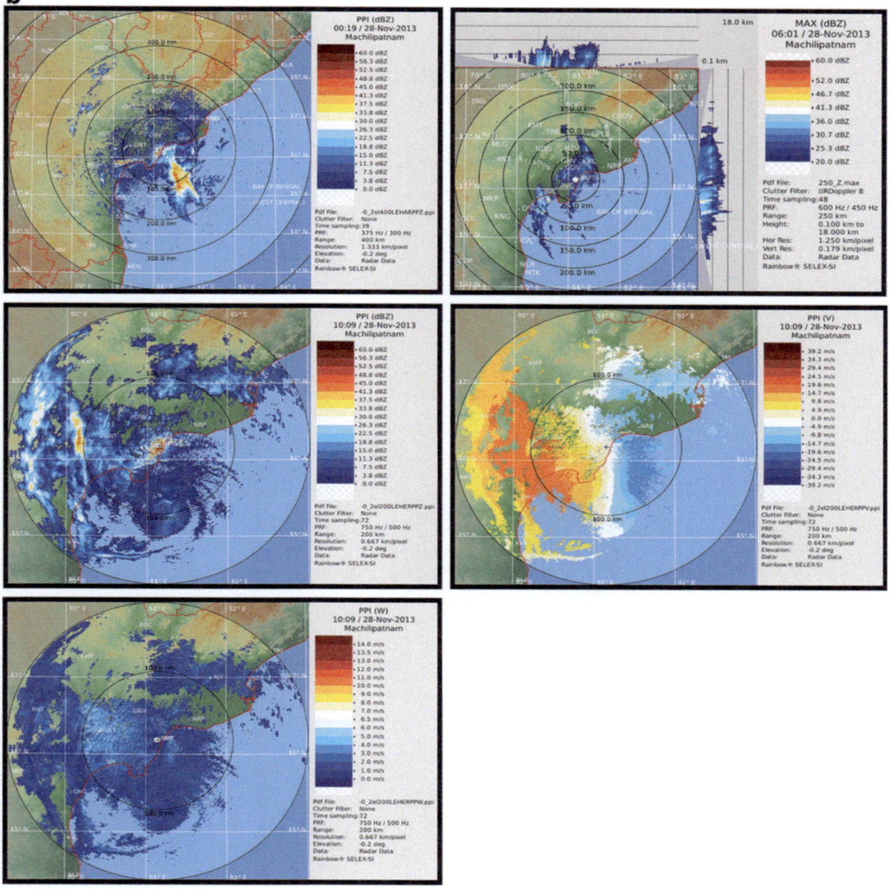

Fig. 3 (continued)

very low probability to intensify when SST is below 27 °C (Wang et al. 2014). Considering the OTE, it was less than 50 kJ/cm^2 over northwest, westcentral and adjoining southwest Bay of Bengal during the entire period 25–28 November. According to past studies, the OTE more than 50 kJ/cm^2 is favourable for intensification (Goni et al. 2009).

3.3 Relative Vorticity in Relation to Rapid Weakening

The relative vorticity in lower level (850 hPa level) at 0600 UTC of 23–28 November over the Bay of Bengal are shown in Fig. 8. It can be found that there was gradual increase in relative vorticity upto 25 November alongwith the increase

Table 1 Average absolute errors (knots) in intensity prediction based on IMD-SCIP and IMD-HWRF models

Lead time →	12 h	24 h	36 h	48 h	60 h	72 h	84 h	96 h	108 h	120 h
IMD-SCIP	5.6(10)	13.0(9)	16.9(8)	19.6(7)	20.3(6)	19.6(5)	–	–	–	–
IMD-HWRF	23.4(9)	12.9(8)	12.4(7)	12.7(7)	7.3(6)	13.6(5)	21.3(4)	22.7(3)	30.5(2)	57(1)

Number of forecasts verified is given in the parentheses.
(intensity forecasts prior to landfall are considered)

Table 2 Root mean square errors (knots) in intensity prediction based on IMD-SCIP and IMD-HWRF models

Lead time →	12 h	24 h	36 h	48 h	60 h	72 h	84 h	96 h	108 h	120 h
IMD-SCIP	6.6(10)	16.6(9)	19.7(8)	22.1(7)	24.0(6)	22.9(5)	–	–	–	–
IMD-HWRF	25.7(9)	17.1(8)	15.5(7)	13.6(7)	9.7(6)	19.2(5)	28.3(4)	29.5(3)	31.2(2)	57(1)

Number of forecasts verified is given in the parentheses.
(intensity forecasts prior to landfall are considered)

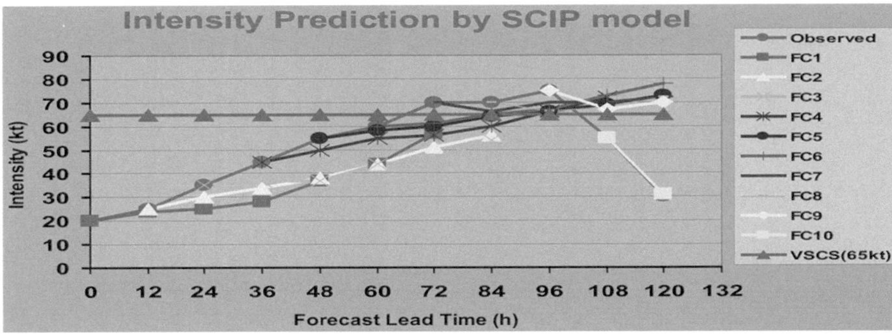

Fig. 4 Intensity forecasts by SCIP model (from 00 UTC of 23 November to 12 UTC of 27 November) in every 12 h interval

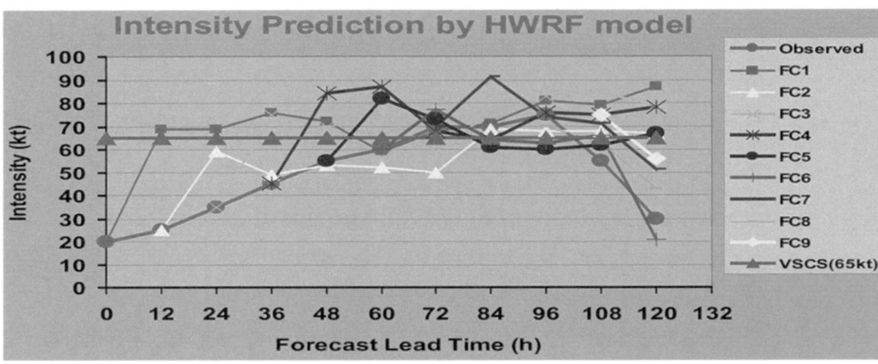

Fig. 5 Intensity forecasts by HWRF(IMD) model (from 00 UTC of 23 November to 00 UTC of 27 November) in every 12 h interval

in size of vorticity maxima in association with TC. On 26 November, the shape of vorticity maxima became elongated in north–south direction and the area coverage also decrease alongwith decrease in magnitude on 27 November. On 28 November the magnitude and area coverage significantly decreased leading to weakening of TC.

3.4 Vertical Wind Shear in Relation to Rapid Weakening

The vertical wind shear of horizontal wind is shown in Fig. 9. From the figures it is found that the vertical wind shear of horizontal wind was low to moderate during 23–26 November around the system centre. It became moderate on 27 November and high (more than 20 knot) on 28 November. The high vertical wind shear is not

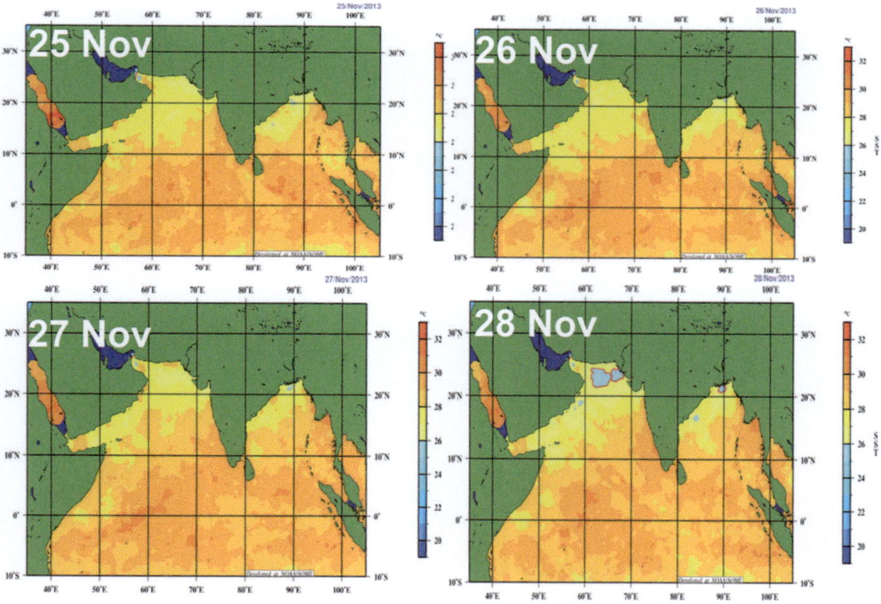

Fig. 6 SST over north Indian Ocean during 25–28 November 2013

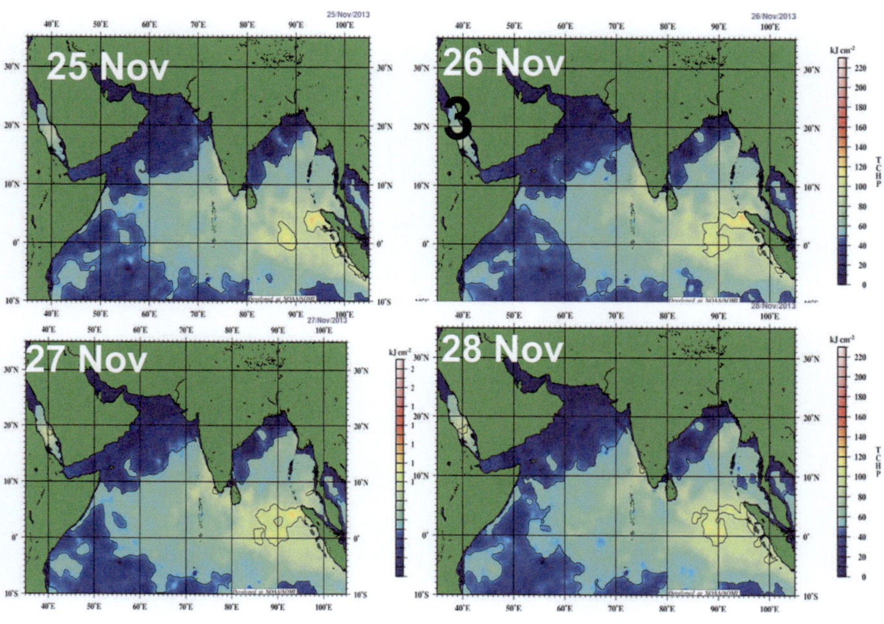

Fig. 7 Ocean Thermal Energy (OTE) over north Indian Ocean during 25–28 November 2013

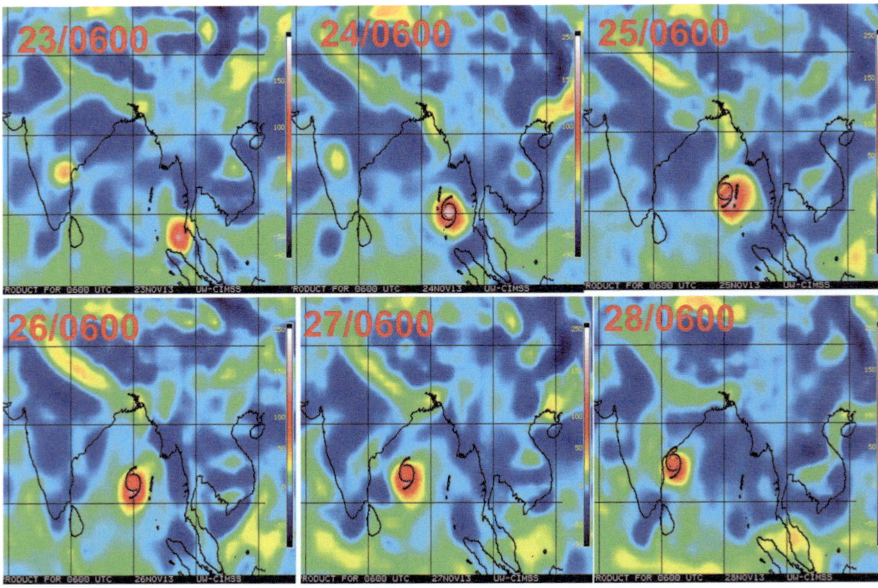

Fig. 8 Relative vorticity based on 0600 UTC of 23–28 November 2013

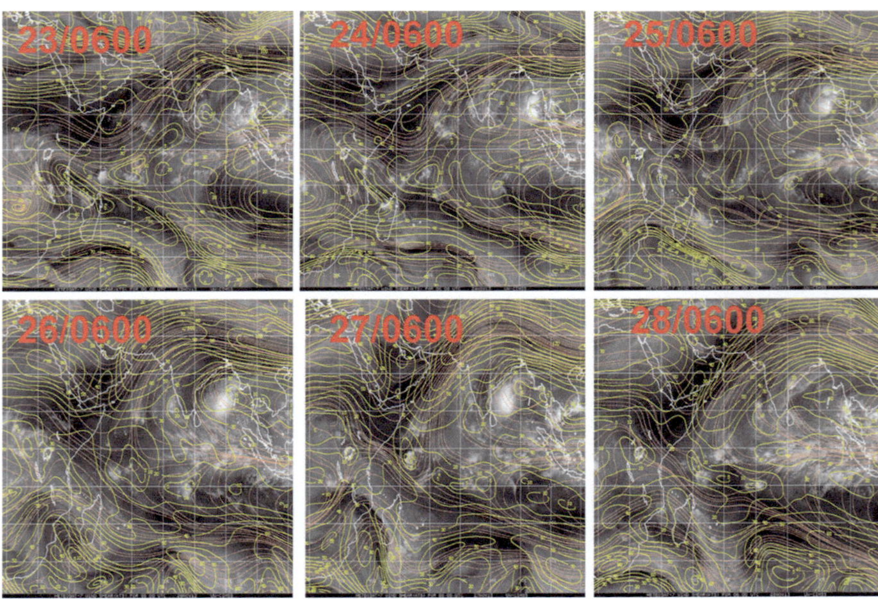

Fig. 9 Vertical wind shear based on 0600 UTC of 23–28 November 2013

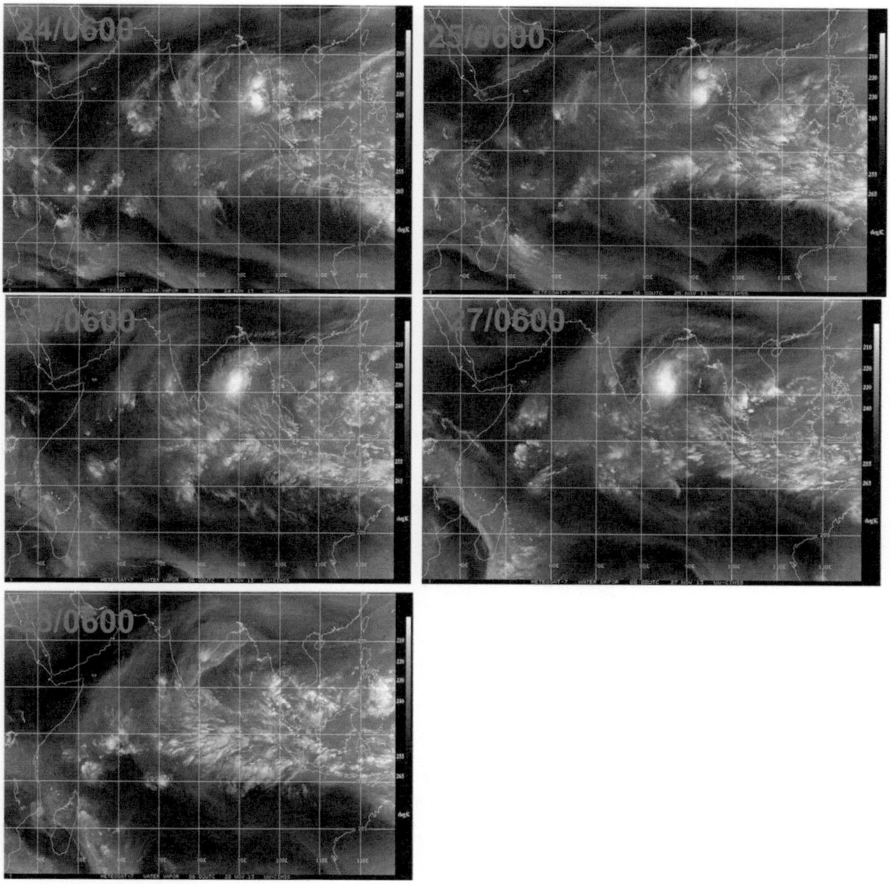

Fig. 10 Kalpana-I water vapour imageries based on 0600 UTC of 23–28 November 2013

favourable to maintain the intensity of TC. Over northwest Pacific, TCs drifting between 3 and 8 m/s have a relatively higher probability than those with translational speed less than 3 m/s to intensify rapidly under low shear (e.g. $VWS_{300-850} < 11$ m/s and $VWS_{700-1000} < 5$ m/s) primarily due to weaker negative Ocean feedback (Wang et al. 2014).

3.5 Humidity in Relation to Rapid Weakening

The water vapour imageries at 0600 UTC of 24–28 November are shown in Fig. 10. As the water vapour imagery represents the magnitude of water vapour in middle troposphere level, it can be observed that there was dry air over entire south India during the period 26–27 November. Further the dry air was engulfing the TC through semicircular band from dry northeast region of India towards the southwest sector of TC. As per Gray (1968), the relative humidity at middle tropospheric level is an important factor for cyclogenesis, intensification/weakening of TC. The relative humidity should be atleast 50 % at middle troposphere level for sustaining the intensity of TC. The strongest contributor to the accelerated TC decay appear to be from decreasing 400–700 hPa relative humidity in southern part of all the rapidly weakening TCs over eastnorth Pacific during 1979–2005 (Wood and Ritchie 2014). Tao and Zhang (2014a, b) based on sensitivity experiment have shown that reduction in environmental moisture content eventually leads to weakened convection and hence weakening of TC.

3.6 Liquid Water Content in Relation to Rapid Weakening

The liquid water content based on the microwave imageries during 23–28 November are shown in Fig. 11. It indicates the continuous feeding of warm and moist air from southern latitude towards the core of TC during its genesis and intensification phase 23–26 November. On 27 November the supply of warm and moist air started reducing and there was increase of dry and cold air. While the warm and moist air from southern latitude was continuously decreasing, the dry continental air gradually penetrated towards the core of TC. By 00 UTC of 28 November, the entire core was separated from spiral feeding of warm and moist air. It resulted in the rapid weakening of Lehar on 27–28 November.

The dry and cold air incursion from north India towards the core of TC can be visualised from the IMD, GFS model analysis as shown in Fig. 12. Further it can be seen that the dry and cold air was prominent in lower tropospheric levels (850 hPa).

4 Conclusion

The very severe cyclonic storm ‘Lehar’ which formed over the Bay of Bengal on 23 November 2013 moved west-northwestwards and rapidly weakened near the coast before landfall as a depression over Andhra Pradesh coast. The analysis of the associated physical features leading to rapid weakening indicates that rapid weakening was mainly due to (i) high vertical wind shear of horizontal wind (>20 knot) (ii) system entering into colder sea where the ocean thermal energy was less than 50 kJ/cm^2 and (iii) incursion of dry and cold air from central and northern parts of India into the core of the system.

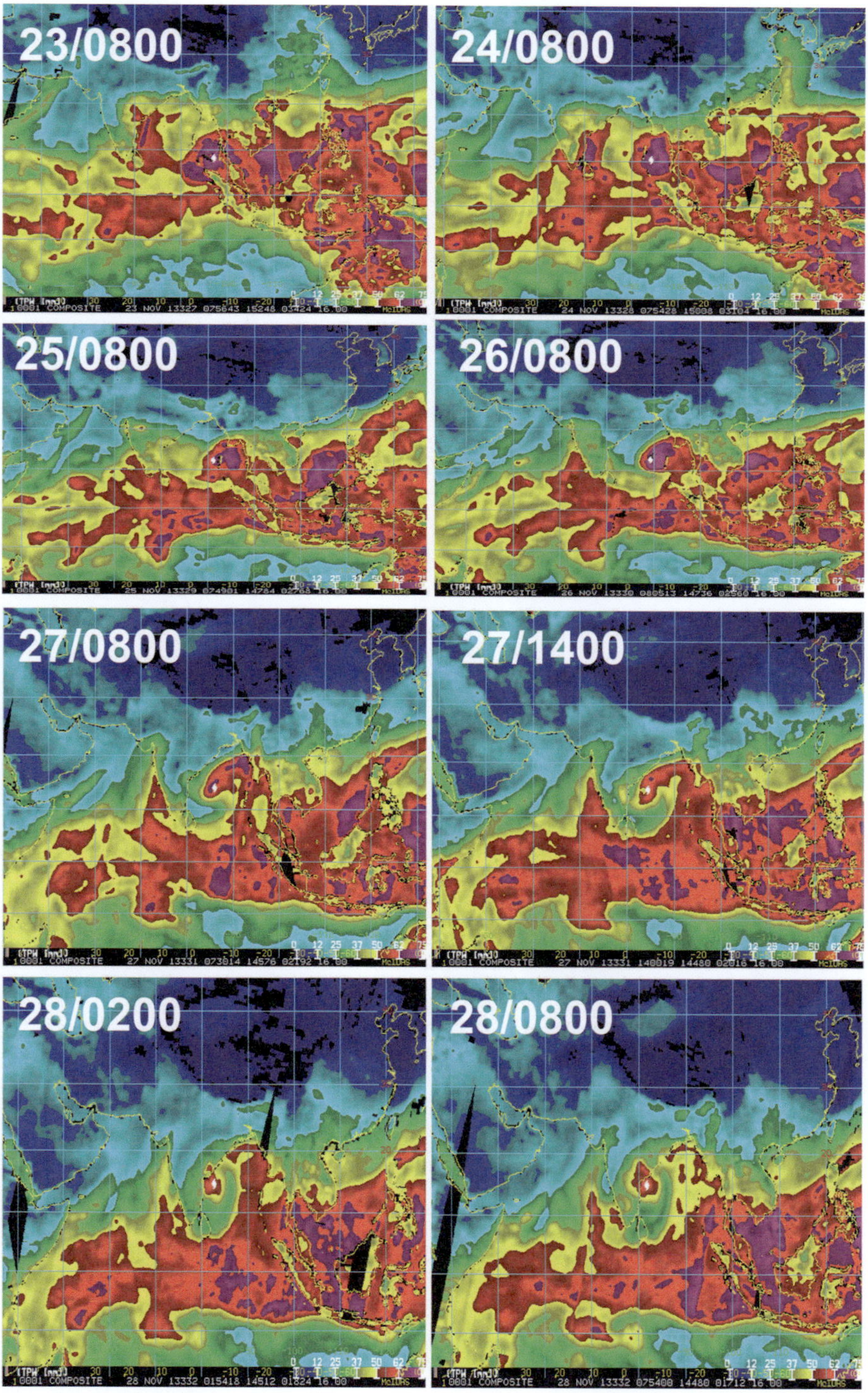

Fig. 11 Liquid water content based on microwave imageries of 23–28 November 2013

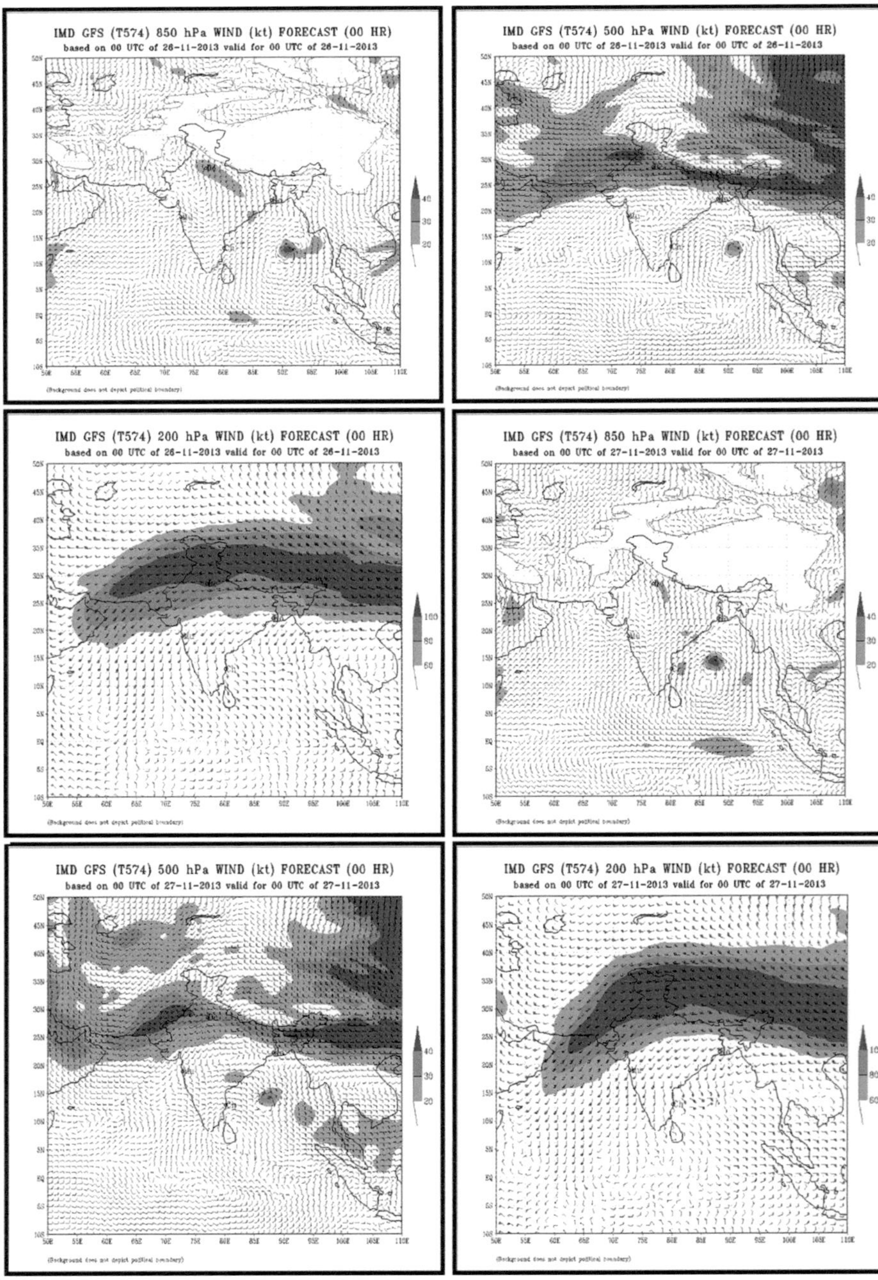

Fig. 12 IMD GFS wind analysis at 850, 500 and 200 hPa levels based on 00 UTC of 26–27 November 2013

Acknowledgement We are thankful to Director General of Meteorology, IMD for his encouragement and support for this work.

References

Goni, G.J, DeMaria, M, Knaff, J, Sampson, C, Ginis, I, Bringas, F, Mavume, A, Lauer, C, Lin, I-I, Ali, M.M, Sandery, P, Ramos-Buarque, S, Kang, K, Mehra, A, Chassignet, E. and Halliwell, G. (2009). Applications of satellite-derived ocean measurements to tropical cyclone intensity forecasting. *Oceanography*, **22**, 176–183.

Gray, M.W. (1968). Global view of the origin of tropical disturbances and storms. *Monthly Weather Review*, **96**, 669–700.

IMD (2013). Cyclone Warning Services: Standard Operation Procedure. Cyclone Warning Division, IMD, New Delhi.

Kotal, S.D, Roy Bhowmik, S.K, Kundu, P.K. and Das, A.K. (2008). A statistical cyclone intensity prediction model for Bay of Bengal. *Journal of Earth System Sciences,* **117**, 157–168.

McGauley, M.G. and Nolan, D.S. (2011). Measuring environmental favorability for tropical cyclogenesis by statistical analysis of threshold parameters. *Journal of Climate*, **24**, 5968–5997.

Mohapatra, M, Bandyopadhyay, B.K. and Nayak, D.P. (2013). Evaluation of operational tropical cyclone intensity forecasts over north Indian Ocean issued by India Meteorological Department. *Nature Hazards*, **68**, 433–451, doi: 10.1007/s11069-013-0624-z

Mohapatra, M. and Adhikary, S. (2011). Modulation of cyclonic disturbances over the north Indian Ocean by Madden–Julian oscillation, *Mausam*, **62**, 375–390.

RSMC, New Delhi (2014). Report Cyclonic Disturbances over north Indian Ocean during 2013. RSMC, New Delhi, pp. 1–235.

Tao, D. and Zhang, F. (2014a). Effect of environmental shear, sea-surface temperature, and ambient moisture on the formation and predictability of tropical cyclones: An ensemble-mean perspective. *Journal of Advances in Modeling Earth Systems*, **6**, 384–404.

Tao, D. and Zhang, F. (2014b). Effect of environmental shear, sea-surface temperature and ambient moisture on the formation and predictability of tropical cyclones: ensemble sensitivity and intrinsic error growth dynamics. *Quarterly Journal of the Royal Meteorological Society*, submitted.

Wang, Y, Rao, Y. and Tan, Z.-M. (2014). A statistical analysis of vertical wind shear effect on tropical cyclone intensity change over the western North Pacific. *Monthly Weather Review*, submitted.

Wood, K.M. and Ritchie, E.A. (2014). Large-scale moisture and thermodynamic properties associated with rapid weakening in the eastern North Pacific. Preprints, 31st Conf. on Hurricanes and Tropical Meteorology, San Diego, CA, American Meteorological Society.

Rapid Movement of Cyclone Viyaru Just Before Landfall-A Case Study

D.P. Nayak and M. Mohapatra

1 Introduction

A depression formed over southeast Bay of Bengal at 1430 h IST of 10 May 2013 near latitude 5.0° N and longitude 92.0° E. It moved northwestwards and intensified into a deep depression in the evening of the same day. Continuing its northwestward movement, it further intensified into a cyclonic storm, Viyaru in the morning of 11 May 2013. Under the influence of the anti-cyclonic circulation lying to the east, the cyclonic storm changed its direction of movement initially from northwestward to northward and then to north-northeastward on 13 and 14 May, respectively. On 15 May, it further came under the influence of the mid-latitude westerly trough running roughly along 77° E, which further helped in enhancing the north-northeastward movement of the cyclonic storm. As this trough came closer on 16th, the north-northeastward speed of the cyclonic storm significantly increased, becoming about 40–50 kmph. The cyclonic storm crossed Bangladesh coast near latitude 22.8° N and longitude 91.4° E, about 30 km south of Feni around 1330 h IST of 16 May 2013 with a sustained maximum surface wind speed of about 85–95 kmph. Maximum surface wind of 92 kmph has been reported over Patuakhali, Bangladesh during the time of landfall. Widespread rainfall with isolated heavy to very heavy falls occurred over Bangladesh. Fairly, widespread rainfall with isolated heavy rainfall also occurred over Mizoram, Manipur and Tripura. A storm surge of height of about 1 m has been reported in section of media. After the landfall, it continued to move north-northeastwards and weakened gradually due to interaction with land surface. It weakened into a deep depression over Mizoram in the evening and into a depression over Manipur around mid-night of 16th. It further weakened

D.P. Nayak (✉) • M. Mohapatra
Cyclone Warning Division (NWFC), India Meteorological Department, Mausam Bhawan, Lodi Road, 110003 New Delhi, India
e-mail: dpnayak1956@gmail.com

© Capital Publishing Company 2017
M. Mohapatra et al. (eds.), *Tropical Cyclone Activity over the North Indian Ocean*,
DOI 10.1007/978-3-319-40576-6_10

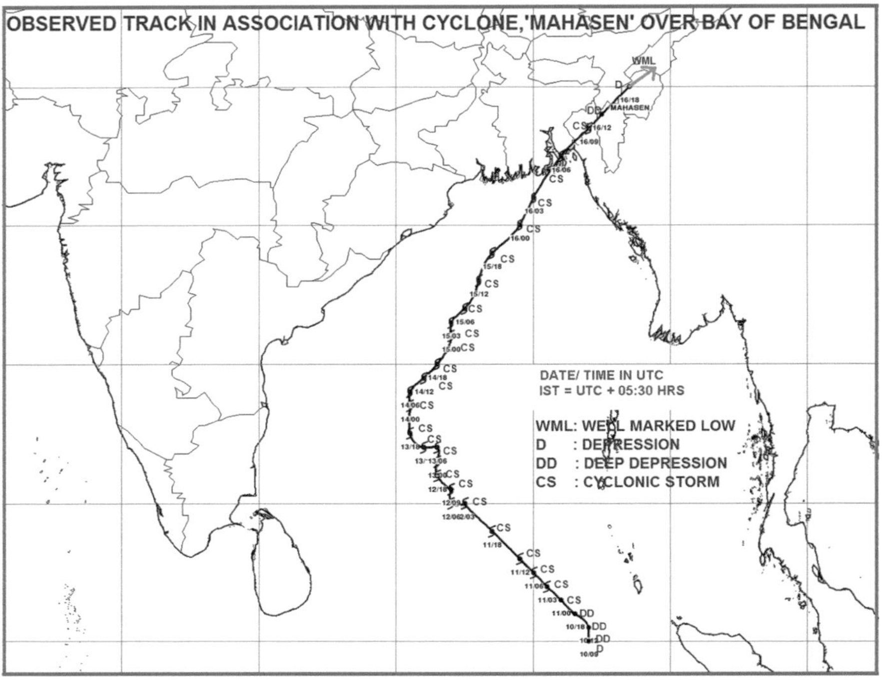

Fig. 1 Track of TC, Viyaru (10–16 May 2013)

into a well-marked low pressure area over Nagaland in the early morning and moved away towards Myanmar as a low pressure area in the morning of 17th.

The track of the system is shown in Fig. 1. The typical satellite and radar imageries are shown in Figs. 2 and 3.

The salient features of this storm are as follows:

1. The genesis of the disturbance took place in lower latitude, near 5° N.
2. It was one of the longest track over north Indian Ocean in recent period after the very severe cyclonic storm, Phet over the Arabian Sea (31 May–07 June 2010).
3. The cyclonic storm moved very fast (about 40–50 kmph) on the day of landfall, i.e. on 16 May 2013. Such type of fast movement of the cyclonic storm is very rare as normal translational speed is about 15 kmph.
4. Due to the faster movement, the adverse weather due to the cyclonic storm was relatively less.
5. It led to relatively higher track and landfall forecast error, especially with respect to time as actual landfall occurred well before the predicted time of landfall. As a result the lead time for the disaster managers were less.

Therefore a study has been undertaken to analyse the characteristic features associated with this rapid movement so that forecast can be improved in such cases in future. For this purpose, the best track data of IMD has been utilised to compare the movement of cyclone Viyaru with the past cyclones during 1990–2012 (RSMC,

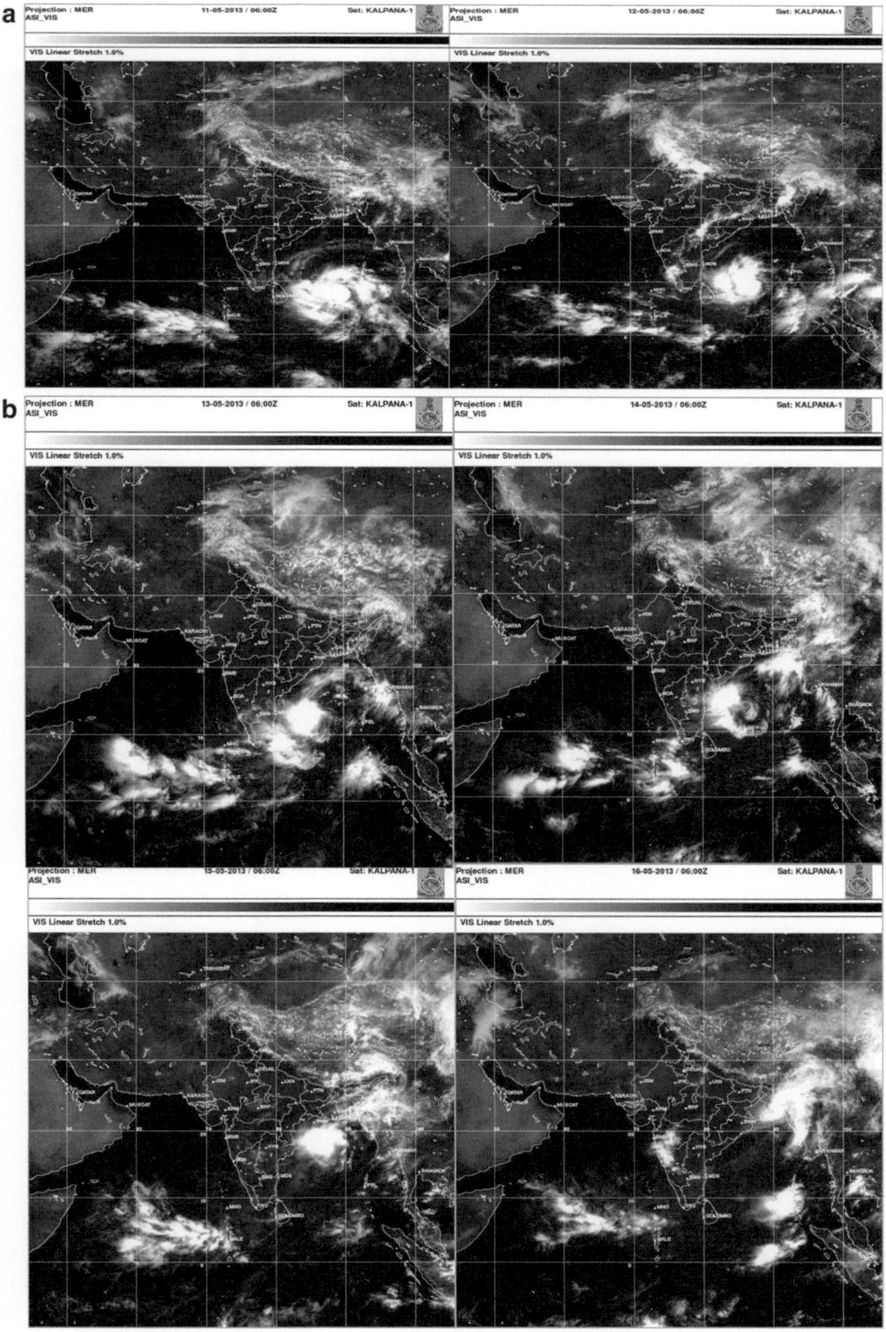

Fig. 2 (**a**) Typical Kalpana-1 Satellite imageries of cyclonic storm VIYARU at 0600 UTC of 11–12 May 2013. (**b**) Typical Kalpana-1 Satellite imageries of cyclonic storm VIYARU at 0600 UTC of 13–16 May 2013

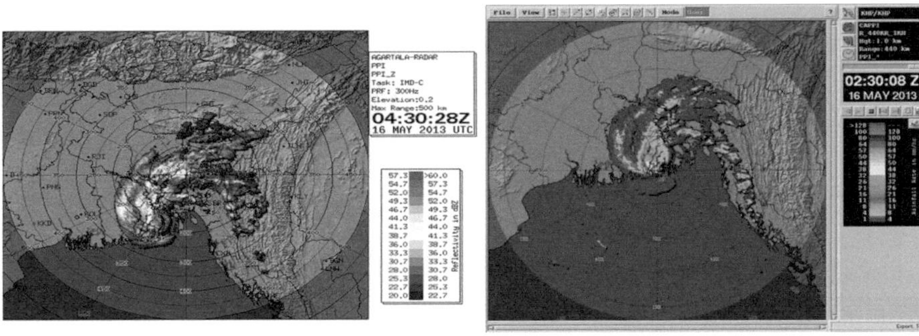

Fig. 3 Typical DWR imageries of Agartala and Khepupara (Bangladesh) in forenoon of 16 May 2013

New Delhi 2014). Details about the best track parameters are given by Mohapatra et al. (2012a). Further the synoptic and environmental features governing the movement of cyclone have been analysed. It is observed that the system got rapid steering for the northeastward movement due to upper air trough in middle latitude westerlies which ran along longitude 77° E on 15th and came further closer to the cyclone on 16 May. An anti-cyclonic circulation in upper troposphere which lay near latitude 14° N and longitude 95° E on 12 May moved gradually westwards leading to more northward component of movement of the cyclone. Due to the rapid movement, the cyclone did not intensify though it had one of the longest travel over the warmer sea. Further details are presented in this paper.

2 Data and Methodology

The data on cyclonic storm Viyaru has been collected from the 'Report on Cyclonic Disturbances' published by RSMC, New Delhi (2014). To find out movement characteristics, the translational speed of the cyclonic storm Viyaru has been calculated and analysed. The forecast errors for track and landfall forecast issued by IMD and various global and regional numerical weather prediction models has been compared and analysed to find out their performance and limitations in predicting the rapid movement of Viyaru. The satellite and radar imageries alongwith the various parameters influencing the track of the cyclone was also analysed for this purpose. The upper tropospheric wind at 200 hPa and middle tropospheric wind at 500 hPa level, the mean steering wind based on 700–200 hPa levels, vertical wind shear between 200 and 850 hPa levels based on CIMSS (http://tropic.ssec.wisc.edu/) during the life period of cyclone Viyaru (10–16 May 2013) were analysed. We also analysed the winds at 850, 500 and 200 hPa levels based on 0000 UTC of 11–16 May 2013 of IMD GFS model analysis.

The results are presented in Sect. 3 and the broad conclusions in Sect. 4.

3 Result and Discussions

3.1 *Translational Speed of Cyclone Viyaru*

The 6-h and 12-h average translational speed during the life period of cyclone Viyaru presented in Fig. 4. It is seen that the translational speed was near normal during the initial stages. It gradually increased from 15 May and reached the pick on 16th before landfall. The maximum 6-h translational speed was 52 kmph against the normal translational speed of 15 kmph. This exceptionally high translational speed lead to early landfall at 0800 UTC than the predicted time of landfall by NWP and operational forecast by IMD.

3.2 *Track Forecast Error of Cyclonic Storm Viyaru*

3.2.1 Operational Track and Landfall Forecast Error

In the first bulletin issued in the afternoon of 10 May (6 days in advance of landfall), when the system was a depression over southeast Bay of Bengal, located at 1900 km south of Chittagong, it was predicted that the system would intensify into a cyclonic storm and move towards Bangladesh–Myanmar (RSMC, New Delhi 2014). The average track forecast error is shown in Table 1. It was 152, 205 and 268 km,

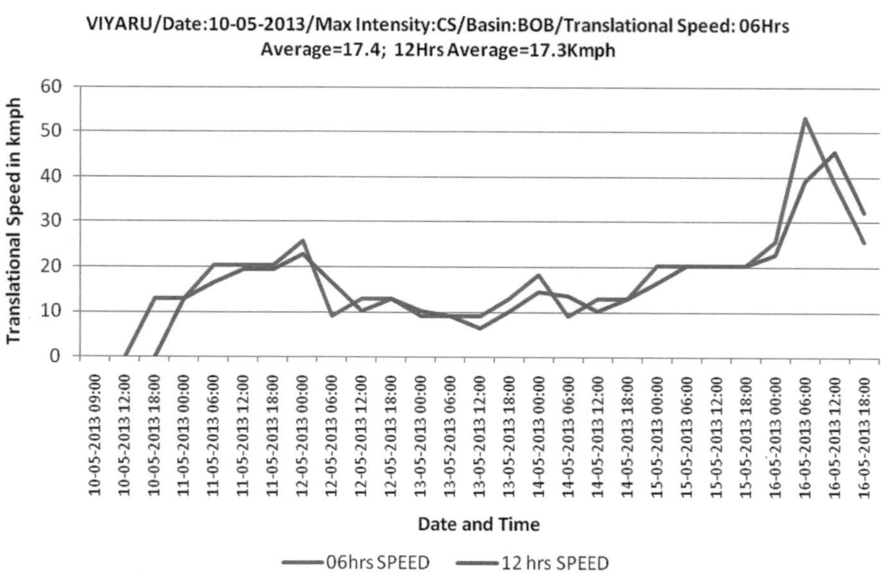

Fig. 4 Six and twelve hourly average translational speed of cyclonic storm, Viyaru (10–16 May 2013)

Table 1 Average track forecast error and skill of cyclonic storm, Viyaru

Lead period (h)	Track forecast error (km)	Track forecast skill (%)	Number of 6-h forecasts verified
12	90	28.6	24
24	152	31.8	22
36	189	46.5	20
48	205	60.0	18
60	208	68.0	16
72	268	69.5	14
84	251	75.2	10
96	308	76.9	8
108	291	80.8	6
120	222	86.3	4

respectively for 24, 48 and 72 h. forecast period against the long period average of 146, 254 and 376 km based on the period of 2003–2012. Further, for the first time the track forecast was issued for 120 h lead period. The 96 and 120 h average track forecast errors were about 308 and 222 km, respectively (Table 1).

The performance of the operational forecast has been compared with forecast by climatology and persistence (CLIPER) model as per international practice (Mohapatra 2015) CLIPER model is taken as a reference model to find out the relative performance of NWP models and operational forecasts of cyclones in different Ocean basins.

The gain in skill on operational forecast in relation to CLIPER, is quantified in percentage terms by;

$$\text{Gain in skill} = \frac{(\text{CLIPER} - \text{DPE})}{\text{CLIPER DPE}} \times 100\%$$

The DPE is the direct position error or simply track forecast error. The results are shown in Table 1. It is found that the operational forecasts were highly skillful compared to CLIPER model forecast.

Usually, the track forecast errors are higher world-wide for the recurving cyclones like Viyaru. Hence, the track forecast errors of Viyaru has been compared with the average errors of recurving cyclones during 2003–2012. It is found that the errors in case of Viyaru are significantly less than the average errors in case of recurving cyclones, as the average errors of recurving cyclones are about 91, 167, 249, 325, 398 and 474 km for 12, 24, 36, 48, 60 and 72 h lead period respectively (Mohapatra et al. 2012a, b, 2013).

Considering the landfall forecast error, the landfall on Bangladesh coast, near Chittagang was predicted in the first bulletin itself. The landfall point forecast errors 5 days before the landfall was about 100 km. It was 72, 125, and 165 km 24, 48 and 72 h, respectively before landfall (details of landfall forecast verification and accuracy are presented by Mohapatra et al. 2015). However, the landfall time

Table 2 Landfall point and landfall time forecast error (forecast-actual) of cyclonic storm, Viyaru

Lead period (h) of forecast from the time of landfall	Landfall point forecast error (km)	Landfall time forecast error (h)
12	57	+7
24	72	+8
36	94	+13
48	125	+10
60	89	+10
72	165	+5
96	184	+10
120	100	−8

error was relatively higher in different forecast times (Table 2). It was mainly due to the fact that the cyclonic storm, Viyaru moved very fast on the day of landfall which could not be predicted by most of the numerical weather prediction (NWP) models.

3.2.2 Track and Landfall Forecast Error of NWP Models

The track and landfall forecast error of different NWP models are presented in Tables 3 and 4, respectively.

The consensus forecast Multi Model Ensemble (MME) outperformed all the forecasts upto 72 h, and it ranged from 79 km at 12 h to 169 km at 72 h. ECMWF model forecast was superior to other model forecasts for 84 h to 108 h forecast (104–255 km) and again MME forecast error (176 km) is lowest at 120 h. The details of the MME technique is given by RSMC, New Delhi. The MME forecast tracks based on various initial conditions alongwith the observed track of TC Viyaru are presented in Fig. 5.

From Table 4, the landfall could not be predicted by IMD-GFS, IMD-WRF and IMD-QLM models 56 h in advance. For other models the landfall time forecast errors varied from 4 to 10 h. Based on 0000 UTC of 15 May 2013 (32 h lead period), the landfall time forecast errors varied from 4 h for IMD-HWRF and NCEP-GFS to 13 h for MME and 15 h for IMDQLM. Similarly, based on 1200 UTC of 15 May 2013 (20 h lead period), the landfall time forecast error ranged from 2 h for IMD-HWRF and IMD-WRF to 5 h for IMD-MME and NCEP-GFS. It clearly indicates that the deterministic models as well as the MME failed to predict the landfall time accurately though they could predict the track and landfall point reasonably (Fig. 5).

Table 3 Average track forecast errors (km) of NWP models

Lead time →	12 h	24 h	36 h	48 h	60 h	72 h	84 h	96 h	108 h	120 h
IMD-GFS	69 (7)	114 (7)	217 (6)	257 (4)	425 (4)	388 (3)	573 (3)	553 (2)	433 (1)	453 (1)
IMD-WRF	101 (8)	196 (7)	318 (6)	373 (5)	536 (5)	402 (4)	–	–	–	–
IMD-QLM	163 (7)	230 (6)	292 (5)	252 (4)	289 (4)	238 (3)	–	–	–	–
JMA	110 (8)	150 (7)	191 (6)	158 (5)	207 (5)	196 (4)	339 (4)	–	–	–
NCEP-GFS	102 (8)	81(7)	143 (6)	154 (5)	210 (5)	190 (4)	232 (4)	229 (3)	300 (2)	307 (2)
ECMWF	105 (8)	149 (7)	202 (6)	153 (5)	199 (5)	173 (4)	224 (4)	104 (3)	255 (3)	233 (2)
IMD-MME	79 (8)	120 (7)	168 (6)	122 (5)	189 (5)	169 (4)	253 (4)	245 (3)	295 (3)	176 (2)
IMD-HWRF	77 (10)	186 (9)	299 (8)	392 (7)	516 (6)	635 (5)	753 (4)	836 (3)	892 (2)	1008 (1)
OFFICIAL	90 (24)	152 (22)	189 (20)	205 (18)	208 (16)	268 (14)	251 (10)	308 (8)	291 (6)	222 (4)

Figures in paranthesis indicate number of forecasts verified

3.3 Rapid Track Changes Over North Indian Ocean

Based on data of 2003–2013 the track forecast errors has been calculated for all the cyclones and separately for those cyclones showing rapid changes in track. The results are shown in Table 5. It is found that the track forecast errors in case of cyclones with rapid changes of tracks are significantly higher than the normal errors for the all the forecast time scales from 12 to 72 h. The error is higher by about 11 %, 14 % and 19 % for 24, 48 and 72 h forecast, respectively.

3.4 Dynamical Features Associated with Rapid Movement of Cyclone VIYARU

The mean steering wind based on 700–200 level at 0000 UTC of 10–16 May 2013 is presented in Fig. 6. It can be seen that initial steering suggested northwestward movement of cyclone followed by north-northwestward upto 12th. It then suggested northward movement on 13th and northeastward movement on 14th onward leading to landfall over Bangladesh. Considering the upper tropospheric wind at 0000 UTC of 10–16 May 2013 (Fig. 7), it also suggested northwestward movement till 12th followed by northward movement on 13th and north-northeast ward movement from 14th onward. The wind analysis of 500 hPa based on

Table 4 Landfall forecast time errors (hr) of NWP models in case of TC, Viyaru

	FC based on 00 UTC/14.05.2013	FC based on 00 UTC/15.05.2013	FC based on 12 UTC/15.05.2013	FC based on 00 UTC/16.05.2013
Model	Lead time: 56 h	Lead time: 32 h	Lead time: 20 h	Lead time: 8 h
IMD-GFS	NO LF	NO LF	−4	Dissipated
IMD-WRF	NO LF	+5	+2	+4
IMD-QLM	NO LF	+15	+4	+4
JMA	+10	+12	+4	0
NCEP-GFS	−4	+4	−5	−4
ECMWF	+4	+12	+4	+1
IMD-MME	+10	+13	+5	+1
IMD-HWRF	+10	+4	−2	–

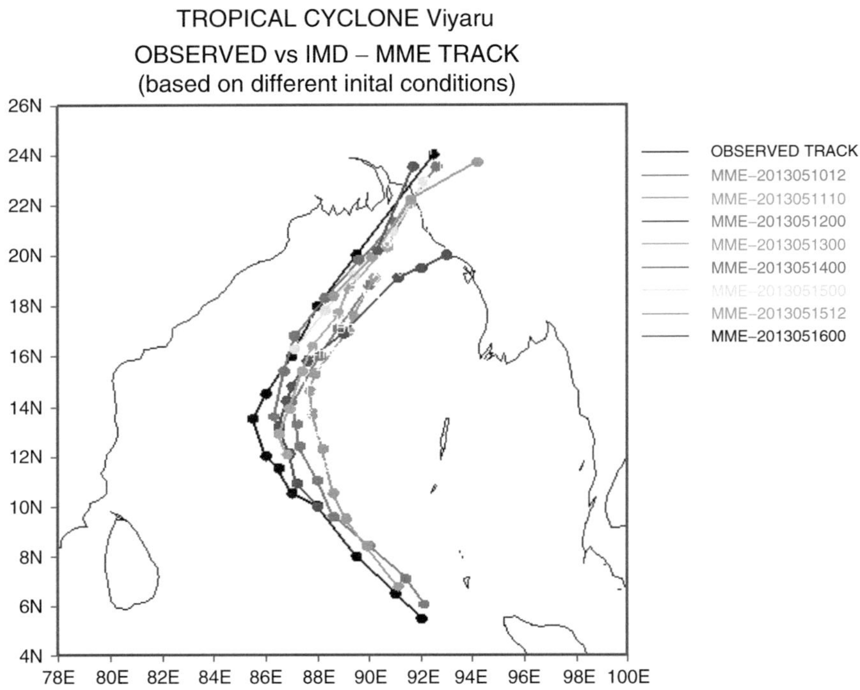

Fig. 5 The multi-model ensemble forecasts of CS Viyaru

Table 5 Track forecast errors in respect of Rapid track changes over north Indian Ocean

Lead period (h)	Track forecast error (km) for all cyclones (A)	Track forecast error (km) for cyclones with rapid changes of tracks (B)	Increase in error (in %) in case of cyclones with rapid changes of tracks ((B – A)/A × 100)
12	79.9	83.6	4.6
24	134.6	148.7	10.5
36	163.7	187.0	14.2
48	202.1	230.3	14.0
60	233.7	272.0	16.4
72	268.2	319.7	19.2

IMD-GFS model (Fig. 8) suggested also north-northeastward movement from 14th onwards in association with trough in westerly lying to the west. The 200 hPa analysis of IMD-GFS (Fig. 9) suggested initial northwestward movement upto 12th northwards movement on 13th and north-northeastward movement on 14th onwards. The translational speed on 16 increased significantly under the influence of middle to upper tropospheric westerly trough lying to the west of cyclone centre. Hence the cyclone Viyaru was mainly steered by the middle to upper tropospheric winds over the region (IMD 2003, 2013).

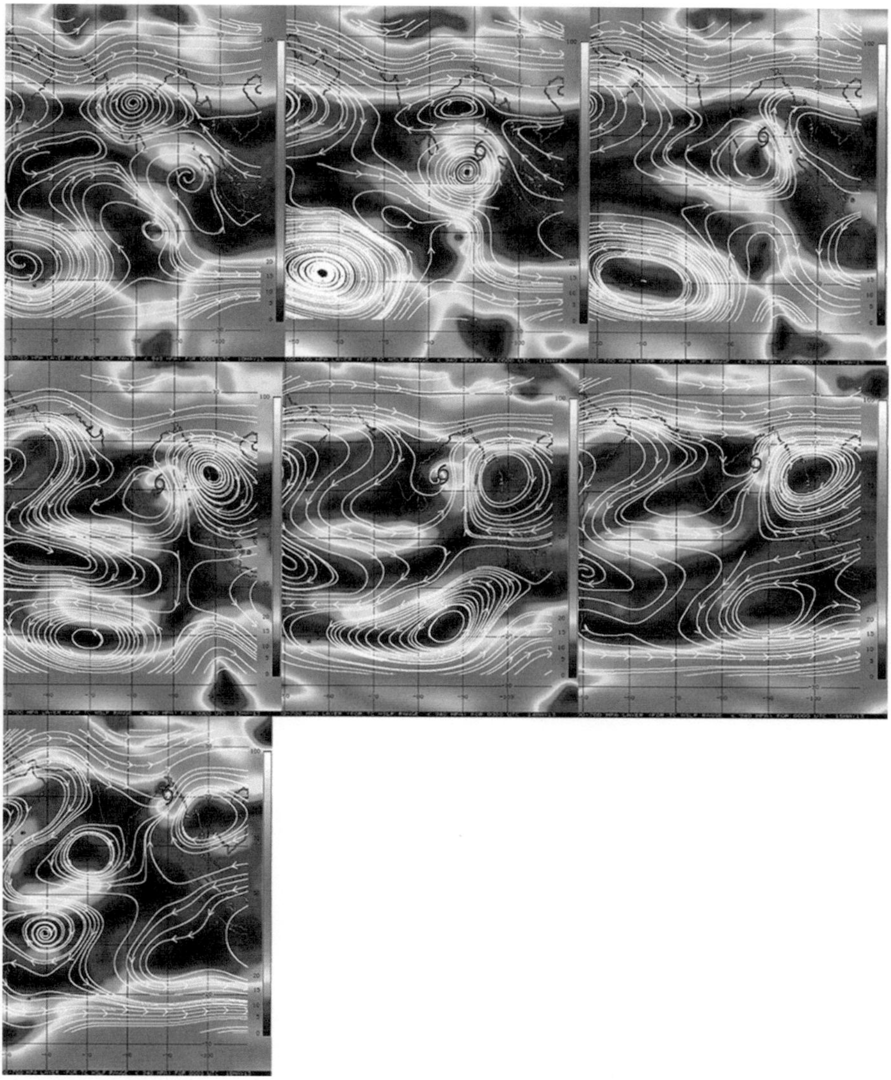

Fig. 6 Mean steering wind based on 700–200 hPa levels based on 0000 UTC of 10–16 May 2013

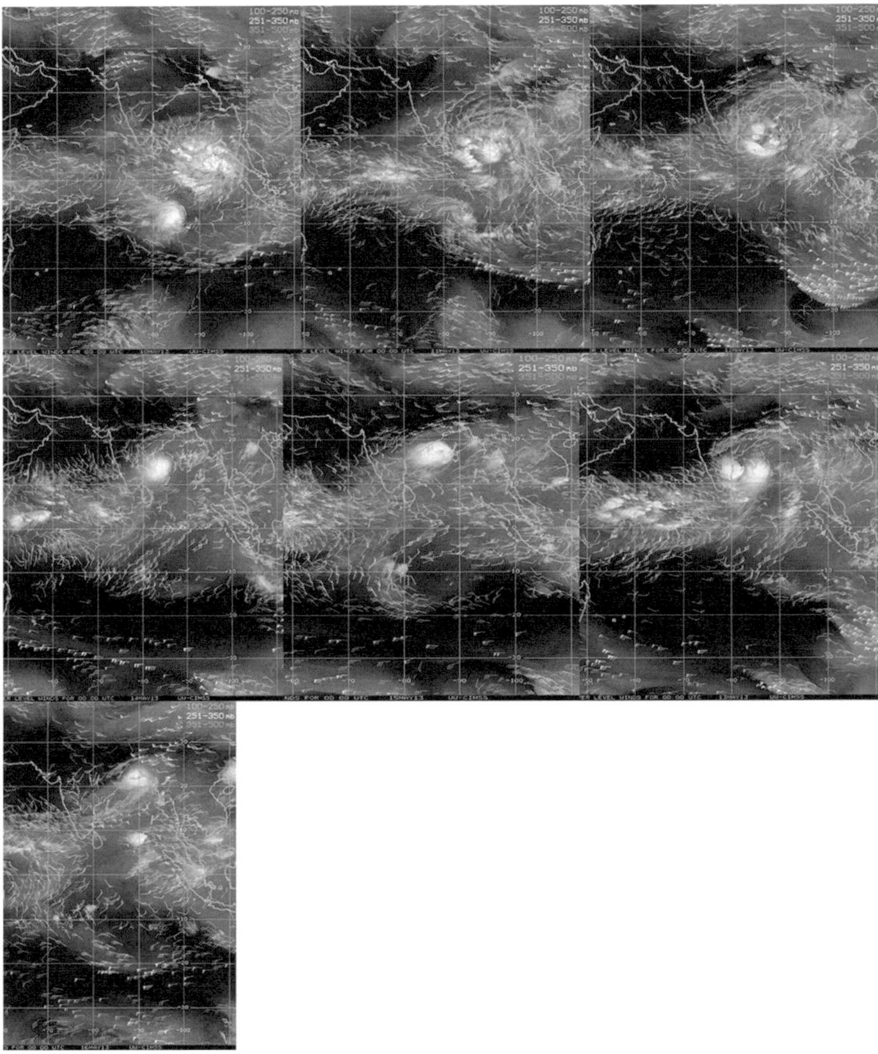

Fig. 7 Upper tropospheric winds at 0000 UTC of 10–16 May 2013

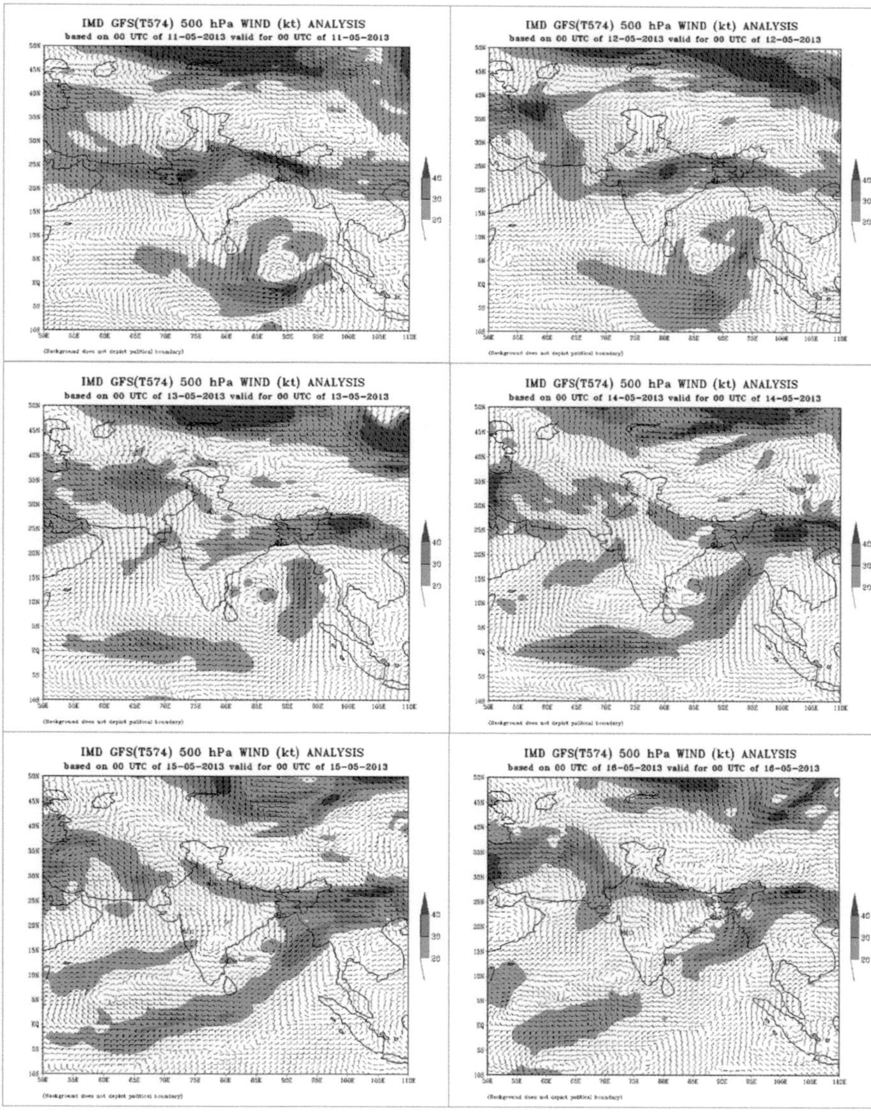

Fig. 8 IMD-GFS wind analyses at 500 hPa level based on 0000 UTC of 11–16 May 2013

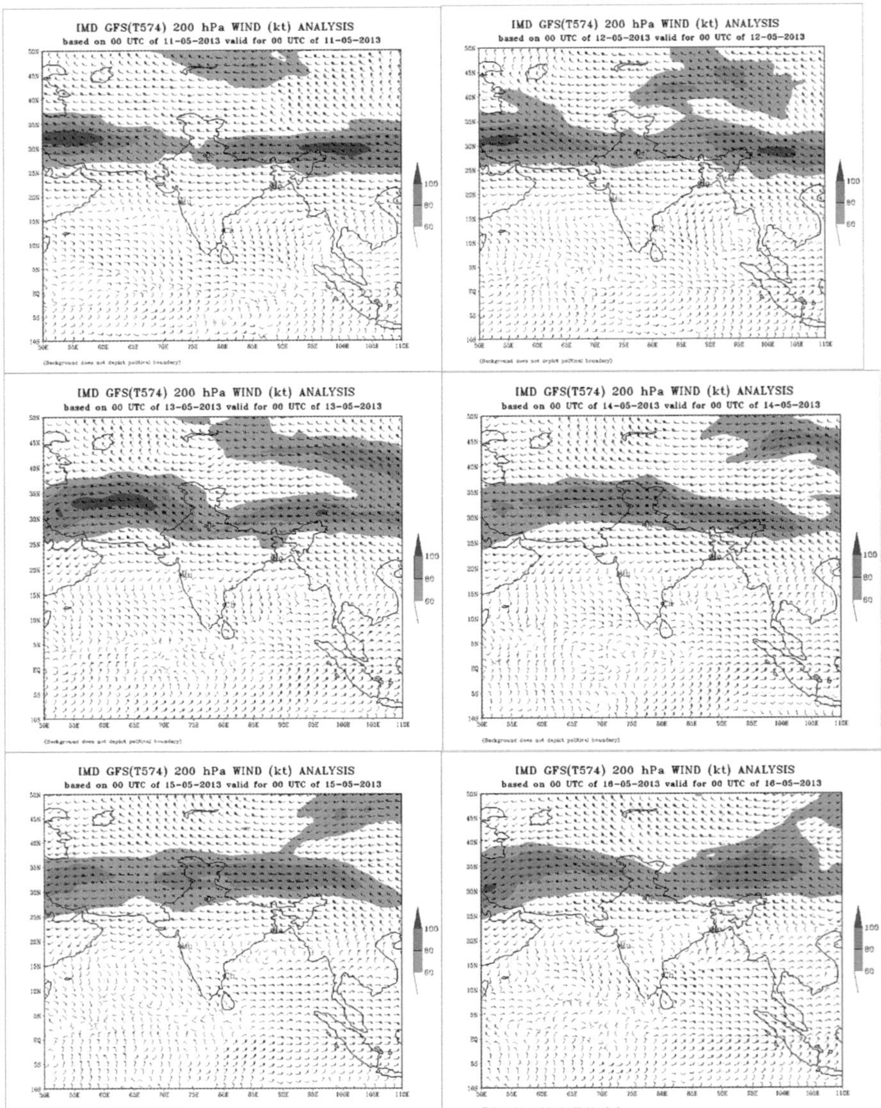

Fig. 9 IMD-GFS wind analyses at 200 hPa level based on 0000 UTC of 11–16 May 2013

4 Conclusions

From the above results and discussions, following broad conclusions are drawn

The TC Viyaru moved with a speed of 40–50 kmph on the day of landfall. The rapid movement of Viyaru on the day of landfall occurred in association with middle to upper tropospheric westerly trough lying to the west of cyclone centre. Most of the numerical models could not predict this rapid movement resulting in higher track forecast error and landfall time forecast error. The operational track and landfall forecast errors of RSMC, New Delhi were 150, 205 & 268 km and 72, 125 & 165 km for 24, 48 and 72 h lead period for the TC, Viyaru. Considering the data of 2009–2013, the increase in error in case of rapid track changes compared to the average track errors of all the TCs over the NIO, it is about 11, 14 and 19 % for 24, 48 and 72 h lead periods respectively.

Acknowledgement Authors are thankful to Director General of Meteorology, India Meteorological Department for his encouragement and support for this study.

References

IMD (2003). Cyclone Manual. IMD, India.

IMD (2013). Cyclone Warning Services: Standard Operation Procedure. Cyclone Warning Division, IMD, India.

Mohapatra, M, Bandyopadhyay, B.K. and Tyagi, A. (2012a). Best track parameters of tropical cyclones over the North Indian Ocean: A review. *Nature Hazards*, **63**, 1285–1317.

Mohapatra, M, Nayak, D.P. and Bandyopadhyay, B.K. (2012b). Evaluation of cone of uncertainty in tropical cyclone track forecast over north Indian Ocean issued by India Meteorological Department. *Tropical Cyclone Research and Review*, 2, 331–339.

Mohapatra, M, Nayak, D.P, Sharma, R.P. and Bandyopadhyay, B.K. (2013). Evaluation of official tropical cyclone track forecast over north Indian Ocean issued by India Meteorological Department. *Journal of Earth System Sciences*, **122**, 589–601.

Mohapatra, M. (2015). Tropical cyclone forecast verification by India Meteorological Department for north Indian Ocean: A review. *Tropical Cyclone Research and Review*, **03**, 1–14.

Mohapatra, M, Nayak, D.P, Sharma, M, Sharma, R.P. and Bandyopadhyay, B.K. (2015). Evaluation of landfall forecast over North Indian Ocean issued by India Meteorological Department, *Journal of Earth System Sciences*, **124**, 861–874.

RSMC, New Delhi (2014). Report on Cyclonic Disturbances over the north Indian Ocean during 2013. Cyclone Warning Division, India Meteorological Department, New Delhi, India, pp. 1–235.

Some Characteristics of Translational Speed of Cyclonic Disturbances Over North Indian Ocean in Recent Years

P.S. Chinchole and M. Mohapatra

1 Introduction

Tropical cyclone (TC) is one of the severe disastrous weather events causing loss of property and life, especially for north Indian Ocean (NIO) basin. According to Mohapatra et al. (2012b), entire east and west coast of India is prone for TC activity. However, the degree of proneness varies from district to district depending upon the number of TCs and severe TCs crossing/affecting the district as well as, associated adverse weather like heavy rain, gale wind and storm surge over the district due to the landfalling TCs (Mohapatra et al. 2012b). About 11 cyclonic disturbances (CDs) with maximum sustained wind speed (MSW of 17 knots or more) including depressions (MSW of 17–33 knots) and TCs (MSW of 34 knots or more) develop over the NIO during a year including 9 and 2 over the Bay of Bengal (BOB) and Arabian Sea (AS), respectively based on data of 1961–2010 (Mohapatra et al. 2014). Out of these, about five intensify into TCs including about four over BOB and one over the AS. About three severe TCs (MSW of 48 knots or more) are formed over the NIO during a year. It includes two over the BOB and one over the AS. Considering the frequency of very severe TCs (MSW of 64 knots or more), there have been about two very severe TCs per year over the NIO. The frequency is maximum during post-monsoon season (October–December) followed by pre-monsoon (March–May) and monsoon (June–September) season.

Considering the above, it is essential to provide accurate track and intensity forecast of the CDs with considerable lead time so that effective measures can be taken by the disaster managers. Though there have been significant improvement in track and intensity forecast in recent years due to modernisation programmes of

P.S. Chinchole (✉) • M. Mohapatra
India Meteorological Department, Mausam Bhawan, Lodi Road, New Delhi 110003, India
e-mail: pschinchole.imd@gmail.com

© Capital Publishing Company 2017

M. Mohapatra et al. (eds.), *Tropical Cyclone Activity over the North Indian Ocean*,
DOI 10.1007/978-3-319-40576-6_11

IMD and other measures, there is still scope for further improvements (Mohapatra et al. 2013a, b).

The review of past studies indicates that the translational speed apart from the other factors of the TC controls the environment and hence affects track and intensity changes of TCs. For forecasting TC intensity, the roles of two dominant factors that determine the extent of intensification, viz., vertical wind shear and TC translational speed need to be understood clearly. Several studies relating shear and speed of movement to intensification of TCs have been reported for various oceanic basins worldwide. It has been shown that TC translational speed influences asymmetry in TC structure and rainfall distribution (Chen et al. 2006). Regarding the role of translational speed on TC intensification, generally, it is understood that if TC movement is too slow, oceanic cooling due to turbulent mixing generated by whirling winds would disrupt the intensification process while on the other hand, if the TC movement is fast, the resulting asymmetric structure will also inhibit intensification.

The mean life of CDs over the BOB and the AS are 4.2 and 4.4 days respectively and the normal speeds of movement of CDs over BOB and AS are 12.9 and 12.8 kmphr (Raj 2012). Based on decadal variation of translational speed during 1961–2010, Geetha and Balachandran (2014) have shown that the translational speed of CDs over the BOB and AS has increased in recent years. However, these studies are not based on 6-h observations which are available in recent years since 1990. Considering all these, a study has been undertaken, to analyse various characteristics of translational speed and their effect on track and intensity of CDs over the NIO. For this purpose, the 6-h best track data from 1990 to 2013 (24 years) of India Meteorological Department (IMD) have been considered. Mean translational speed for 06-, 12-, 24-h period ending at 0000, 0600, 1200 and 1800 UTC during the CD period has been calculated and analysed.

2 Data and Methodology

For analysing movement characteristics, total 215 CDs during 1990–2013 over the NIO have been considered. After quality test, 6-h best track data from 1990 to 2013 of IMD (2008) have been considered. Mean translational speed for 06-, 12-, 24-h period ending at 0000, 0600, 1200 and 1800 UTC during the CD period has been calculated and analysed. The translational speed for a given period, say 6-h is defined as the speed of movement of CDs calculated from the great circle distance between the initial location of the CD in the beginning of 6-h period and the final location of the CD at the end of the 6-h period. The 12- and 24-h average translational speed is calculated in the similar method.

Mean annual translational speed, standard deviation in translational speed of TCs, over BOB, AS and NIO as whole and translational speed during different seasons for different CDs in BOB, AS and NIO with respect to the maximum intensity of the system viz. Depression (D), Deep Depression (DD), Cyclonic Storm

(CS), Severe Cyclonic Storm (SCS), Very Severe Cyclonic Storm (VSCS) and Super Cyclonic Storm (SuCS) has been calculated. The student's *t*-test has been applied to find out the significant difference, if any, between the mean values of the above mean translational speeds over BOB and AS for a given season and also the difference if any between any two seasons for a given Ocean basin.

Based on quality and availability, the whole period of best track information has been broadly classified by (Mohapatra et al. 2012a) into four phases, viz., (i) pre-1877, (ii) 1877–1890, (iii) 1891–1960 and (iv) 1961–2010. The period of 1961–2010 has been further classified into (a) 1961–1973, (b) 1974–1989 and (c) 1990 onwards. Though optimum observational network including satellite based observations leading to better estimation of location and intensity without missing of CDs was available since 1961, the climatology of CDs and TCs based on best track parameters like location and intensity (MSW) are published by IMD for every 6 h (00, 06, 12 and 18 UTC) throughout the life period of CD only since 1990 (Mohapatra et al. 2012a). Hence, translational speeds are calculated and analysed based on the data of 1990–2013. The tracks of these CDs are shown in Fig. 1. The CD which includes depression and TC over the NIO is defined as per the criteria adopted by the IMD (2013). The same classification is used in the study and is shown in the Table 1.

Currently, the intensity estimation takes into consideration (a) satellite INSAT/ METSAT, NOAA, TRMM, SSMIS, scatterometer wind etc., (b) Radar and (c) synoptic analysis (IMD 2013). Like the estimation of location of the system, when the system is far away from the coast and not within the radar range, satellite estimated intensity based on Dvorak's technique (Dvorak 1984) gets maximum weight. When the system comes closer to the coast, radar estimated intensity is considered along with satellite estimated intensity. When the system is very close to coast or over the land surface, the coastal observations get the highest preference followed by radar and satellite observations for estimating the intensity. The average error in MSW estimation has reduced over the years. Based on recent data, Goyal et al. (2013) have shown that there is a difference of T0.5 in the estimation of intensity by satellite method and best track estimates. Thus the quality of best track intensity of CDs and TCs is better during 1990–2013 for the study of translational speed.

There are various agencies including IMD and JTWC which provide the best track information over the NIO. However, there is difference in estimation of location and intensity of CDs by these two centers due to the subjectivity involved in the estimation (Mohapatra et al. 2012a). However, the difference in estimation decreases with increase in intensity of CDs like other Ocean basins. Here in this study, data of IMD' best track parameters have been utilised. Calculation of translational speed and direction may thus vary depending upon the source of best track data.

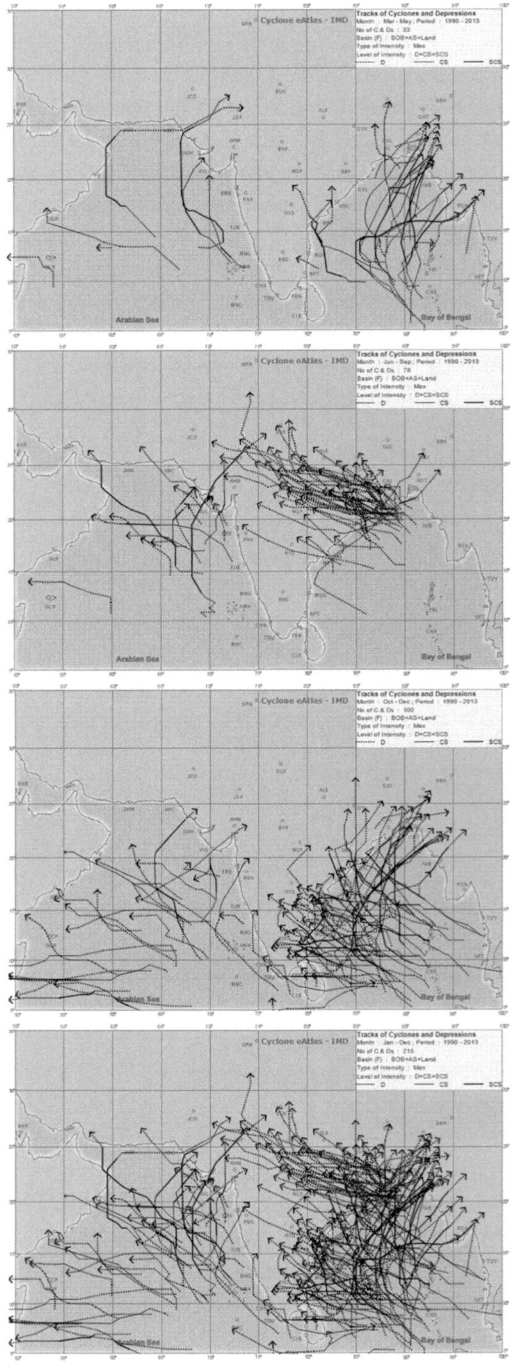

Fig. 1 Tracks of cyclonic disturbances over the basins of the NIO during (**a**) pre-monsoon (**b**) monsoon (**c**) post-monsoon seasons and (**d**) year as a whole over the period of 1990–2013

Table 1 Classification of cyclonic disturbances over the NIO. IMD (2013). Cyclone warning in India; Standard operational PROCEDURE. IMD, New Delhi

Type of Cyclonic Disturbance	Wind speed in knots		
Depression (D)	17–27	31–49	09–14
Deep depression (DD)	28–33	50–61	15–17
Cyclonic storm (CS)	34–47	62–88	18–24
Severe cyclonic Storm (SCS)	48–63	89–118	25–32
Very severe cyclonic storm (VSCS)	64–119	119–221	33–61
Super cyclone (SuCS)	120 and above	222 or more	62 or more

3 Result and Discussions

3.1 Mean Translational Speed of CDs During a Year

The annual translational speed for 06-, 12- and 24-h over the BOB, AS and NIO during 1990–2013 are given in Annexure I (Table 10–13). The number of data set considered for 6-, 12- and 24-h average translational speed is shown in Table 2. The average translational speed of CDs based on the above data for 06-, 12- and 24-h average period over the BOB, AS and NIO are shown in Table 3. The 06-, 12- and 24-h average translational speeds of CD over the NIO are about 13.9, 13.6 and 13.0 kmph, respectively. The average speed is higher over the BOB than over the AS, as 06-, 12- and 24-h average speed is about 14.3, 13.9 and 13.4 kmph over the BOB against 13.1, 12.8 and 12.5 kmph over the AS, respectively.

Considering the translational speed ending at 00, 06, 12 and 18 UTC, 6-h average translational speed is maximum as derived at 0600 UTC, followed by 1200 UTC and then at 0000 UTC over the BOB, AS and NIO as a whole. Considering the 12-h average translational speed, it is maximum at 1200 UTC. The 24-h average translational speed does not show significant variation with respect to time of estimation over BOB, AS and NIO as a whole as it varies from 13.2 to 13.5, 12.3 to 12.6 and 12.8 to 13.2 kmph, respectively. All these results indicate that the translational speed shows diurnal variation.

The standard deviation of translation speed for 06, 12 and 24 h over the different ocean basin are also analysed not shown. It indicates that the standard deviation is slightly higher over the BOB than over the AS.

3.2 Mean Translational Speed of CDs of Different Intensities During a Year

The mean translational speeds of CDs with different intensities are shown in Table 4. It is found that the speed varies from 13.5 to 14.3 kmph for 6-h averaging period for D to SCS stage and becomes 15.2 and 19.3 kmph for VSCS and SuCS.

Table 2 Number of data set considered for 6-, 12- and 24-h average translational speed over the basin AS, BOB and NIO

Season	Intensity	AS			BOB			NIO		
		06 h	12 h	24 h	06 h	12 h	24 hs	06 h	12 h	24 h
Winter	D	2	3	0	4	3	3	6	6	3
	DD	1	0	0	3	2	3	4	2	3
	CS	0	0	0	9	8	4	9	8	4
	SCS	0	0	0	0	0	0	0	0	0
	VSCS	0	0	0	0	0	0	0	0	0
	SUCS	0	0	0	0	0	0	0	0	0
Pre-monsoon	D	15	12	9	56	43	24	71	55	33
	DD	18	16	12	54	45	25	72	61	37
	CS	35	34	30	100	99	90	135	133	120
	SCS	11	11	11	37	37	37	48	48	48
	VSCS	27	27	27	64	64	64	91	91	91
	SUCS	0	0	0	7	7	7	7	7	7
Monsoon	D	32	26	22	234	209	159	281	248	203
	DD	41	34	23	180	167	132	221	201	155
	CS	37	36	30	31	30	27	68	66	57
	SCS	22	22	20	1	1	1	23	23	21
	VSCS	28	28	28	0	0	0	28	28	28
	SUCS	1	1	1	0	0	0	1	1	1
Post-monsoon	D	98	84	66	234	190	131	332	274	197
	DD	112	108	94	205	188	134	317	296	228
	CS	75	72	61	223	219	197	298	291	258
	SCS	25	25	24	65	65	63	90	90	87
	VSCS	3	3	3	107	107	107	110	110	110
	SUCS	0	0	0	3	3	3	3	3	3
Annual total		583	542	461	1617	1487	1211	2215	2042	1694

Table 3 06-, 12-, 24-Average translational speeds of CDs over BOB, AS and NIO during a year

Translational speed in kmph									
Based time	BOB			AS			NIO		
UTC	6	12	24	6	12	24	6	12	24
0000	13.1	12.1	13.4	12.6	11.0	12.5	12.9	11.8	13.1
0600	17.3	14.8	13.5	16.5	14.4	12.6	17.0	14.7	13.2
1200	15.0	15.5	13.3	13.8	14.4	12.3	14.6	15.2	12.8
1800	12.0	13.4	13.2	9.9	11.6	12.6	11.5	12.9	13.0
All timings	14.3	13.9	13.4	13.1	12.8	12.5	13.9	13.6	13.0

Considering the 12-h averaging period, it varies from 13.1 to 14.3 kmph for D to VSCS stage and becomes 17.4 kmph for SuCS. Considering the 24-h averaging period, it varies from 12.2 to 13.9 kmph for D to VSCS stage and becomes

Table 4 Translational speed (kmph) of CDs of different intensities over the NIO

	Translational speed (kmph) of CDs for different averaging periods		
Intensity	06 h	12 h	24 h
D	14.3	14.2	13.9
DD	14.0	13.1	12.2
CS	14.2	13.7	13.0
SCS	13.5	13.1	12.7
VSCS	15.2	14.3	13.5
SuCS	19.3	17.4	16.9

16.9 kmph for SuCS. In general, the translational speed is higher for D, VSCS and SuCS.

Considering the averages of 06, 12 and 24 h data as obtained from the Table 5, it is observed that as the intensity increases from CS to SuCS, translational speed also increases over the BOB. It increases only in VSCS and SuCS stage over the AS.

The average translational speed with respect to different intensity of CDs over the BOB and AS are shown in Fig. 2. It is found that there is significant difference in average translational speed of CDs with increase in intensity over Bay of Bengal. It is increasing in the stage of VSCS and SUCS stage in both the basins.

3.3 Inter-seasonal Variation of Translational Speed of CDs Over the NIO

The average translational speeds during different seasons are shown in Tables 6, 7 and 8 respectively for CDs of different intensities over the AS, BOB and NIO. Considering the CDs over the AS in different seasons (Table 6), the translational speed during pre-monsoon season increases as the D intensifies into DD. It then decreases, as the DD intensifies gradually into SCS. Thereafter, the speed increases with intensification into VSCS/SuCS. During monsoon season, the speed is minimum in DD stage and higher in D and CS stage. The speed decreases in SCS stage and then increases in VSCS and SuCS stages. During post-monsoon season, the translational speed gradually increases as the intensity of the disturbance increases. Comparing the translational speeds in different seasons over the AS, the speed is higher in monsoon followed by pre-monsoon season in case of D, higher in post-monsoon followed by monsoon season in case of CS or higher intensity of the disturbance. In case of DD, it is higher in post-monsoon followed by pre-monsoon season.

Considering BOB (Table 7), the translational speed is maximum in monsoon followed by pre-monsoon season in case of D, DD, CS and followed by post-monsoon in case of SCS. The speed is minimum in winter season in case of all disturbances. During pre-monsoon season, the speed increases with increase in intensity from D to CS. It then decreases with intensification into SCS and then

Table 5 Translational speed at different intensities for the basin: BOB and AS

Basin	BOB			AS		
Intensity stage	06 h	12 h	24 h	06 h	12 h	24 h
D and above	14.6	14.0	13.4	13.4	13.0	12.6
DD and above	14.4	13.7	12.9	13.6	13.0	12.5
CS and above	14.5	13.9	13.1	14.0	13.6	13.1
SCS and above	15.0	14.3	13.5	13.7	13.0	12.6
VSCS and above	15.5	14.7	13.7	15.0	13.8	13.4
SUCS	19.9	17.5	16.9	13.0	16.5	16.2

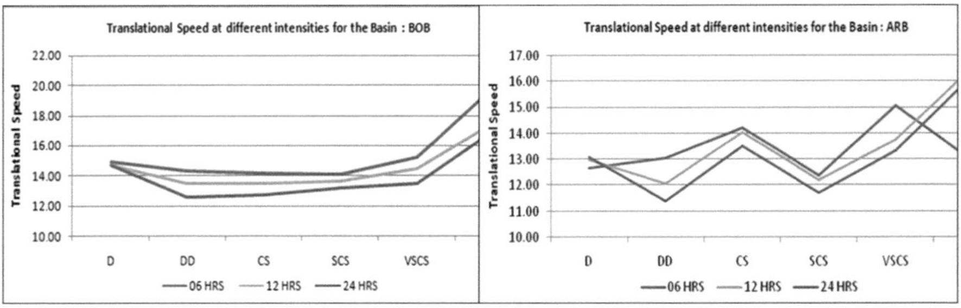

Fig. 2 Translational speed of CDs at different intensities over the BOB and AS

Table 6 Season-wise translational speed at different intensities of CDs over the AS

Maximum intensity of CD	Averaging period (h)	Translational speed (kmph) in different seasons			
		Winter	Pre-monsoon	Monsoon	Post-monsoon
D	6	17.5	9.3	15.8	12
	12	19.6	9.5	17.5	11.8
	24	–	9.6	15.9	12.6
DD	6	–	12.1	10	14.1
	12	–	11.9	8.6	13.2
	24	–	11.9	8.6	12
CS	6	–	10.9	14.7	15.5
	12	–	10.6	14.3	15.5
	24	–	9.4	13.7	15.5
SCS	6	–	6.3	10.5	16.7
	12	–	6.7	10.5	16.2
	24	–	6.7	10.5	15
VSCS	6	–	13.9	15.6	20.9
	12	–	11.8	15.1	19.3
	24	–	11.8	14.5	17.1
SuCS	6		–	13	–
	12		–	16.5	–
	24		–	16.2	–

Table 7 Season-wise translational speed at different intensities of CDs over the BOB

Maximum intensity of CD	Averaging period (h)	Translational speed (kmph) in different seasons			
		Winter	Pre-monsoon	Monsoon	Post-monsoon
D	6	2.3	11.9	17.9	12.9
	12	5.2	13.0	17.4	12.3
	24	7.7	13.0	17.2	12.1
DD	6	10.4	15.2	14.8	13.6
	12	5.5	14.1	14.4	12.6
	24	2.6	11.9	13.8	11.7
CS	6	6.4	15.8	16.3	13.5
	12	6.6	15.0	15.1	12.9
	24	4.6	14.1	14.6	12.1
SCS	6	–	11.3	29.0	15.5
	12	–	11.4	26.7	14.7
	24	–	11.4	21.6	14.1
VSCS	6	–	15	–	15.4
	12	–	13.7	–	15.0
	24	–	12.2	–	14.3
SuCS	6	–	20.7	–	17.9
	12	–	18.4	–	15.4
	24	–	16.8	–	17.4

increases with intensification into VSCS/SuCS. There is no difference in translational speed in case of D, DD and CS during post-monsoon season and is higher in case of SCS/VSCS/SuCS. During monsoon season, the speed is higher in case of D and SCS stage and less in case of DD and CS stage. The translational speed is significantly less during winter season like that over AS.

Considering the NIO as a whole, the translational speed is also minimum during winter season. It is maximum in monsoon season in case of D and CS, in post-monsoon season in case of SCS and SuCS and higher and almost equal during monsoon and post-monsoon seasons in case of VSCS (Table 8).

Considering the monthly translational speed of CDs (Table 9) the speed is higher in December and January over the AS. It is higher in July and August (about 17 kmph) followed by June (15 kmph) for the CDs over BOB. This again clearly indicates that the depression/deep depression over the BOB during the main monsoon month moves faster than the CDs in other months. The speed is minimum in the month of February. It may be due to the fact that the CDs during these months develop over lower latitude and hence experience less impact of planetary vorticity leading to less northward component of movement. Further they do not experience any impact of the mid-latitude westerlies. The distribution of translational speed over the NIO is similar to that over the BOB. Comparing the pre-monsoon and post-monsoon months, there is no significant difference in translational speed during May and October over the AS and the speed is higher during November. Over the

Table 8 Season-wise translational speed at different intensities of CDs over the NIO

Maximum intensity of CD	Averaging period (h)	Translational speed (kmph) in different seasons			
		Winter	Pre-monsoon	Monsoon	Post-monsoon
D	6	7.4	11.4	17.1	12.6
	12	12.4	12.2	16.8	12.2
	24	7.7	12.1	15.8	12.3
DD	6	18.1	14.4	13.9	13.8
	12	5.5	13.5	13.4	12.8
	24	2.6	11.9	13	11.8
CS	6	6.4	14.5	15.4	14
	12	6.6	13.9	14.7	13.5
	24	4.6	12.9	14.1	12.9
SCS	6	–	10.1	11.3	15.8
	12	–	10.3	11.2	15.1
	24	–	10.3	11.1	14.3
VSCS	6	–	14.6	15.6	15.6
	12	–	13.1	15.1	15.1
	24	–	12.1	14.5	14.4
SUCS	6	–	20.7	13	17.9
	12	–	18.4	16.5	15.4
	24	–	16.8	16.2	17.4
All	6	8.7	13.4	15.1	13.5
	12	8.2	13.0	14.8	13.2
	24	4.9	12.2	14.2	12.8

BOB, the speed in the months of October, November and May are similar and higher than that during April.

4 Conclusions

Following broad conclusions are drawn from the above results and discussion

The 06-, 12- and 24-h average translational speeds of CDs over the NIO are about 13.9, 13.6 and 13.0 kmph, respectively. The average speed is higher over the BOB than over the AS, as 06, 12 and 24-h average speed is about 14.3, 13.9 and 13.4 kmph over the BOB against 13.1, 12.8 and 12.5 kmph over the AS respectively.

There is significant difference in average translational speeds of CDs with increase in intensity over BOB and AS. The translational speed is higher in the stage of VSCS and SUCS stage in both the basins and in different seasons. Comparing the translational speeds in different seasons, it is minimum during winter seasons over all types of disturbances in both the Ocean basins.

Table 9 Monthly average translational speed (km/hr) of CDs over the BOB, AS and NIO for 6-, 12- and 24-h averaging period

	BOB			AS			NIO		
Basin/Month	06 h	12 h	24 h	06 h	12 h	24 h	06 h	12 h	24 h
Jan	3.1	3.1	2.6	25.4	19.6	–	8.6	7.6	2.6
Feb	8.8	9.2	10.5	–	–	–	8.8	9.2	10.5
Mar	12.6	12.5	10.4	–	–	–	12.6	12.5	10.4
Apr	13.4	13.1	12.4	–	–	–	13.4	13.1	12.4
May	14.5	14.1	13.2	11.2	10.5	10.1	13.5	13.0	12.2
Jun	14.7	14.4	13.5	14.3	14.3	14.1	14.5	14.3	13.7
Jul	17.2	16.5	16.3	–	–	–	17.3	16.6	15.4
Aug	17.4	17.1	17.0	–	–	–	17.4	17.1	17.0
Sep	14.0	14.2	14.0	6.8	6.4	6.2	11.2	11.2	10.4
Oct	14.1	13.6	13.4	11.6	11.0	10.7	13.2	12.7	12.4
Nov	14.5	14.0	13.7	14.6	14.3	14.0	14.5	14.1	13.8
Dec	11.3	11.0	10.4	17.4	17.9	18.3	12.2	12.0	11.6

Comparing the translational speeds in different seasons over the AS, the speed is higher in monsoon followed by pre-monsoon season in case of D, higher in post-monsoon followed by monsoon season in case of CS or higher intensity of the disturbance. In case of DD, it is higher in post-monsoon followed by pre-monsoon season.

Considering BOB, the translational speed is maximum in monsoon followed by pre-monsoon season in case of D, DD, CS and followed by post-monsoon in case of SCS. There is no significant difference in translational speeds in case of VSCS and SuCS during pre-monsoon and post-monsoon seasons.

Acknowledgement Authors are thankful to Director General of Meteorology, IMD for his encouragement and support for this work.

Annexure I

Table 10 06-, 12- and 24-h averages annual translational speed (km/hr) over BOB, AS and NIO during 1990–2013

Period	Annual								
Basin	BOB			AS			NIO		
Averaging period/year	06 h	12 h	24 h	06 h	12 h	24 h	06 h	12 h	24 h
1990	15.9	14.9	14.0	7.6	7.5	6.8	15.4	14.4	13.6
1991	14.3	14.6	14.3	–	–	–	14.3	14.6	14.3
1992	15.5	15.3	14.9	17.3	17.0	17.1	15.9	15.8	15.4
1993	16.6	16.3	16.5	15.4	15.2	15.8	16.0	15.7	16.1
1994	14.8	14.4	13.7	16.7	17.3	17.6	15.7	15.9	15.7
1995	18.1	18.5	18.5	6.8	6.7	6.3	15.4	15.5	15.2
1996	13.4	13.1	11.8	8.8	9.3	9.5	12.1	12.0	11.2
1997	14.1	13.6	13.2	17.8	16.1	12.7	14.4	13.8	13.2
1998	16.6	16.1	15.9	15.7	14.8	13.8	16.1	15.4	14.6
1999	14.8	14.3	14.3	15.3	14.6	14.2	14.9	14.4	14.3
2000	14.0	12.8	11.8	–	–	–	14.0	12.8	11.8
2001	12.7	11.9	11.6	9.8	8.7	8.0	10.4	9.2	8.6
2002	13.1	12.2	11.5	16.3	16.1	15.7	13.9	13.1	12.6
2003	12.3	11.9	12.2	16.2	15.1	14.4	12.9	12.5	12.4
2004	15.0	14.7	14.0	10.3	9.9	10.1	11.9	11.5	10.7
2005	9.5	9.4	9.4	–	–	3.7	9.5	9.4	8.9
2006	16.8	16.4	15.7	6.9	6.3	3.2	13.9	13.4	12.5
2007	15.6	15.6	15.4	12.0	12.3	12.6	14.3	14.4	14.4
2008	12.2	12.1	11.1	16.7	17.4	15.8	12.8	12.7	12.0
2009	14.7	14.5	13.5	19.8	19.9	19.5	15.6	15.4	14.3
2010	14.8	14.1	13.3	12.6	12.6	12.6	13.9	13.5	13.0
2011	9.7	9.6	9.5	12.1	11.7	12.2	11.2	10.9	11.1
2012	11.5	11.1	10.8	22.8	23.3	25.3	13.6	13.6	13.5
2013	13.8	13.5	13.3	10.4	10.1	9.0	13.6	13.2	13.0

Table 11 06-, 12- and 24-h averages annual translational speed (km/hr) over BOB, AS and NIO during pre-monsoon season of 1990–2013

Period	Pre-monsoon season								
Basin	BOB			AS			NIO		
Averaging period/year	06 h	12 h	24 h	06 h	12 h	24 h	06 h	12 h	24 h
1990	13.18	10.64	8.41	–	–	–	13.18	10.64	8.41
1991	14.14	14.17	13.32	–	–	–	14.14	14.17	13.32
1992	17.17	17.22	16.75	–	–	–	17.17	17.22	16.75
1993	–	–	–	–	–	–			
1994	14.68	14.06	13.11	–	–	–	14.68	14.06	13.11
1995	12.72	13.07	12.79	–	–	–	12.72	13.07	12.79
1996	27.33	25.10		–	–	–	27.33	25.10	

(continued)

Table 11 (continued)

Period	Pre-monsoon season								
Basin	BOB			AS			NIO		
Averaging period/ year	06 h	12 h	24 h	06 h	12 h	24 h	06 h	12 h	24 h
1997	16.91	17.01	16.92	–	–	–	16.91	17.01	16.92
1998	15.16	15.42	16.19	–	–	–	15.16	15.42	16.19
1999	–	–	–	15.28	14.65	14.16	15.28	14.65	14.16
2000	13.66	14.43	12.82	–	–	–	13.66	14.43	12.82
2001	–	–	–	9.58	7.93	7.17	9.58	7.93	7.17
2002	18.33	18.33	–	16.29	16.07	15.70	16.61	16.34	15.70
2003	10.74	10.14	9.31	–	–	–	10.74	10.14	9.31
2004	12.19	12.03	9.30	8.39	8.51	8.20	9.66	9.64	8.52
2005	–	–	–	–	–	–	–	–	–
2006	13.59	13.21	12.35	–	–	–	13.59	13.21	12.35
2007	13.00	13.36	13.30	–	–	–	13.00	13.36	13.30
2008	11.61	11.45	10.97	–	–	–	11.61	11.45	10.97
2009	17.93	17.59	17.39	–	–	–	17.93	17.59	17.39
2010	13.23	13.05	12.89	6.34	6.05	6.03	9.89	9.78	9.87
2011	–	–	–	–	–	–	–	–	–
2012	–	–	–	–	–	–	–	–	–
2013	15.97	16.02	15.74	–	–	–	15.97	16.02	15.74

Table 12 06-, 12- and 24-h averages annual translational speed (km/hr) over BOB, AS and NIO during monsoon season of 1990–2013

Period	Monsoon season								
Basin	BOB			AS			NIO		
Averaging period/ year	06 h	12 h	24 h	06 h	12 h	24 h	06 h	12 h	24 h
1990	17.38	17.59	17.66	–	–	–	17.38	17.59	17.66
1991	16.45	16.82	16.28	–	–	–	16.45	16.82	16.28
1992	18.78	18.20	18.01	11.39	10.26	9.77	16.32	15.56	15.26
1993	–	–	–	–	–	–		–	–
1994	–	–	–	18.26	18.93	19.33	18.26	18.93	19.33
1995	11.84	12.53	11.63	–	–	–	11.84	12.53	11.63
1996	13.36	13.16	12.44	10.96	11.79	11.39	12.33	12.58	11.96
1997	15.08	14.49	14.17	–	–	–	15.08	14.49	14.17
1998	22.80	22.77	21.85	17.25	17.12	16.38	18.31	18.11	16.95
1999	19.05	18.31	21.30	–	–	–	19.05	18.31	21.30
2000	25.47	21.85		–	–	–	25.47	21.85	
2001	–	–	11.46	9.78	9.80	9.83	9.78	9.80	9.98
2002	–	–		–	–	–	–	–	–
2003	–	–	20.85	–	–	–	–	–	20.85

(continued)

Table 12 (continued)

Period	Monsoon season								
Basin	BOB			AS			NIO		
Averaging period/ year	06 h	12 h	24 h	06 h	12 h	24 h	06 h	12 h	24 h
2004	15.78	15.08	16.40	6.10	3.34	2.67	10.68	9.21	7.93
2005	15.40	13.59	13.14	–	–	3.69	15.40	13.59	9.59
2006	18.60	18.12	17.44	2.70	2.92	3.20	13.97	13.57	12.82
2007	16.01	16.01	15.86	14.18	14.30	14.88	15.45	15.48	15.58
2008	11.07	11.55	12.38	16.67	17.38	17.21	14.80	15.26	15.03
2009	16.73	17.98	17.25	18.83	16.81		17.22	17.74	17.25
2010				16.25	15.92	15.40	16.25	15.92	15.40
2011	7.91	8.15	8.00	6.48	8.64	6.48	8.99	9.23	8.34
2012	–	–	–	–	–	–	–	–	
2013	18.92	17.57	19.52	–	–	–	18.92	17.57	19.52

Table 13 06-, 12- and 24-h averages annual translational speed (km/hr) over BOB, AS and NIO during post-monsoon season of 1990–2013

Period	Post-monsoon season								
Basin	BOB			AS			NIO		
Averaging period/ year	06 h	12 h	24 h	06 h	12 h	24 h	06 h	12 h	24 h
1990	16.11	14.75	14.44	7.62	7.45	6.78	14.80	13.66	13.41
1991	11.41	11.72	11.86	–	–	–	11.41	11.72	11.86
1992	13.84	13.88	13.35	20.49	20.87	21.83	15.62	15.69	15.37
1993	16.55	16.32	16.47	15.41	15.17	15.76	15.95	15.72	16.09
1994	15.17	15.39	15.18	15.47	15.98	16.39	15.36	15.78	16.02
1995	25.77	25.48	24.96	6.79	6.67	6.30	17.80	17.48	16.77
1996	12.73	12.59	11.57	7.38	7.72	8.27	11.41	11.37	10.84
1997	7.57	7.06	6.46	17.77	16.11	12.69	10.86	9.87	8.20
1998	16.18	15.55	15.42	15.00	13.89	12.74	15.51	14.59	13.78
1999	13.38	13.20	13.41	–	–	–	13.38	13.20	13.41
2000	13.50	12.32	11.72	–	–	–	13.50	12.32	11.72
2001	12.69	11.86	11.65	11.43	10.61	10.59	12.32	11.50	11.38
2002	12.80	11.94	11.53	–	–	–	12.80	11.94	11.53
2003	17.03	17.83	17.10	16.18	15.10	14.40	16.64	16.53	16.24
2004	18.04	18.13	18.48	14.01	13.81	13.59	15.02	14.80	14.34
2005	9.58	9.59	9.33	–	–	–	9.58	9.59	9.33
2006	6.54	6.38	6.23	–	–	–	6.54	6.38	6.23
2007	15.89	15.62	14.83	9.53	10.25	10.40	12.54	12.79	12.54
2008	13.05	12.90	11.07			13.40	13.05	12.90	11.57
2009	9.08	8.66	7.97	20.16	20.72	19.54	12.77	12.52	11.27
2010	15.35	14.46	13.42	–	–	–	15.35	14.46	13.42

(continued)

Table 13 (continued)

Period	Post-monsoon season								
Basin	BOB			AS			NIO		
Averaging period/ year	06 h	12 h	24 h	06 h	12 h	24 h	06 h	12 h	24 h
2011	12.82	11.74	11.70	12.46	11.88	12.31	12.56	11.84	12.15
2012	11.49	11.10	10.81	22.78	23.35	25.29	13.62	13.62	13.49
2013	12.36	12.05	11.84	10.41	10.11	9.03	12.15	11.84	11.55

References

Chen, S.S, Knaff, J.A. and Marks, F.D. (2006). Effects of vertical wind shear and storm motion on tropical cyclone rainfall asymmetries deduced from TRMM. *Monthly Weather Review*, **134**, 3190–3208.

Dvorak, V.F. (1984). Tropical cyclone intensity analysis using satellite data. Technical Report (NOAA TR NESDIS 11), National Oceanic and Atmospheric Administration, National Environmental Satellite, Data, and Information Service, 47 pp.

Geetha, B. and Balachandran, S. (2014). Decadal variations in translational speed of cyclonic disturbances over north Indian Ocean. *Mausam*, **65**, 115–118.

Goyal, S, Mohapatra, M. and Sharma, A.K. (2013). Comparison of best track parameters of RSMC, New Delhi with satellite estimates over north Indian Ocean, *Mausam*, **64**, 25–34.

IMD (2013). Cyclone Warning in India; Standard Operational Procedure. IMD, New Delhi, India.

IMD (2008). Tracks of Cyclones and Depressions (1891–2007), Electronic Version 1.0/2008. IMD, Chennai, India.

Mohapatra, M, Bandyopadhyay, B.K. and Tyagi, A. (2012a). Best track parameters of tropical cyclones over the North Indian Ocean: A review. *Natural Hazards*, **63**, 1285–1317.

Mohapatra, M, Mandal, G.S, Bandyopadhyay, B.K, Tyagi, A. and Mohanty, U.C. (2012b). Classification of cyclone hazard prone districts of India. *Natural hazards*, **63**, 1601–1620.

Mohapatra, M, Bandyopadhyay, B.K. and Nayak, D.P. (2013a). Evaluation of operational tropical cyclone intensity forecasts over north Indian Ocean issued by India Meteorological Department. *Nature Hazards*, doi: 10.1007/s11069-013-0624-z.

Mohapatra, M, Nayak, D.P, Sharma, R.P. and Bandyopadhyay, B.K. (2013b). Evaluation of official tropical cyclone track forecast over north Indian Ocean issued by India Meteorological Department. *Journal of Earth System Sciences*, **122**, 589–601.

Mohapatra, M. Bandyopadhyay, B.K. and Tyagi, A. (2014). Construction and quality of best tracks parameters for study of climate change impact on Tropical Cyclones over the North Indian Ocean during satellite era. *In*: Mohanty, U.C, Mohapatra, M, Singh, O.P, Bandyopadhyay, B.K. and Rathore, L.S. (eds.) Monitoring and Prediction of Tropical Cyclones over the Indian Ocean and Climate Change.Springers and Capital publishers, New Delhi, India, pp. 3–17.

Raj, Y.E.A. (2012). Climatology and forecasting based on climatology of tropical cyclones of Indian Seas. India Met. Dept., Met. Monograph No. Cyclone Warning – 9/2011.

Life Period of Cyclonic Disturbances Over the North Indian Ocean During Recent Years

S.V.J. Kumar, S.S. Ashthikar, and M. Mohapatra

1 Introduction

Tropical cyclones (TCs) are one of the most devastating disastrous weather events worldwide and especially over the North Indian Ocean (NIO). About 75 % of all TCs which killed 5000 or more human population had developed over this basin during the past 300 years (Dube et al. 2013). Considering this, there is always a need to strengthen monitoring, prediction and warning services for the TCs apart from preparedness and planning. The preparedness and planning for TC mitigation needs climatological information about the TCs to assess the hazard potential. Mohapatra et al. (2012a) and Mohapatra (2015) have analysed TC hazard proneness of various coastal districts of India in terms of frequency, intensity and associated adverse weather like heavy rain, gale winds and storm surge etc., However, the damage potential and the lead time available for management of the TC disaster depend crucially on the life period of the TCs, especially its period of stay over oceanic region before the landfall as the track and landfall forecast can be provided accurately to the disaster managers after the genesis of the TC only. The TCs over the NIO show large-scale spatio-temporal variations in terms of genesis, track, intensity and landfall (Mohapatra et al. 2014; Tyagi et al. 2010; IMD 2008, 2013). Therefore, there is a need to find out the average life period of the cyclonic disturbances (CDs) with different intensities in different seasons and year as a whole over different ocean basins, namely, Bay of Bengal (BOB), Arabian Sea (AS) and the North Indian Ocean (NIO) as a whole. There are many studies over the Northwest Pacific and North Atlantic Ocean on the above climatological aspects. However, studies are limited over the NIO.

S.V.J. Kumar (✉) • S.S. Ashthikar • M. Mohapatra
India Meteorological Department, Lodhi Road, New Delhi 110003, India
e-mail: kumarsvj@yahoo.com

© Capital Publishing Company 2017

M. Mohapatra et al. (eds.), *Tropical Cyclone Activity over the North Indian Ocean*,
DOI 10.1007/978-3-319-40576-6_12

According to Emanuel (2005a, b), the life period of TC depends on the sea surface temperature (SST) which may or may not be caused by global warming. It is an established fact that SST > 26 °C is a requirement for genesis of TC (Gray 1968; Lighthill et al. 1994). TCs form only over warm oceans from which they gain energy (Anthes 1982; Holland 1997). Several studies have pointed out that global and tropical atmospheric temperatures are increasing (Parker et al. 2004; Smith and Reynolds 2005). Tropical ocean SSTs have increased by 0.5 °C in recent years since 1970 (Elsner and Kocher 2000). On one hand, there has also been a significant rise in SSTs over the NIO including BOB and AS also (Rajeevan and Mc Phaden 2004). On the other hand, the frequency of TCs over this region has decreased in the recent years. Based on the data during satellite period (1961–2010), Mohapatra et al. (2014) have shown that there has been a significant decrease in the frequency of TCs, severe TCs, very severe TCs over the NIO and the BOB. However, there has been no significant decrease in frequency of TCs during the recent years over the AS. The CD which includes depression and TC over the NIO are defined as per the criteria adopted by the IMD (2013). The same classification is used in the study and is shown in the Table 1.

Considering all the above, a study is undertaken to analyse the average life period of the CDs of different intensities based on the data of 1990–2013 for different seasons, year as a whole and also over different ocean basins like BOB, AS and NIO. The results of the study can be utilised by the disaster managers in their planning and preparedness for TC management. It can also be used to assess the hazard potential of the TCs over this region. It will help in better understanding of the physical processes associated with genesis, track, intensity and life cycle of TCs over the NIO.

2 Data and Methodology

As a whole, 215 CDs which developed during the period 1990–2013 including 64, 60, 38, 20, 29 and 4 numbers of D, DD, CS, SCS, VSCS and SuCS, respectively have been considered. Details of CDs considered in the study are shown in Table 2 and their tracks are shown in the Fig. 1. The past studies by Mohapatra et al. (2014) based on the data of 1961–2010 (satellite era) indicate that about 11 CDs have developed per year over the NIO (9 over the BOB and 2 over the AS). Thus, it indicates that the frequency of CDs (D and above) has decreased by about 2 in 1990–2013 as compared to 1961–1990. However, considering CS and above, about 3.8 CDs per year have developed over the NIO including 2.7 over the BOB and 1.2 over the AS during 1990–2013 compared to 5, 4, 1, respectively over the NIO, the BOB and the AS during 1961–1990. Considering SCS and above, the average frequency is 2.2 over the NIO including 1.5 over the BOB and 0.7 over the AS during 1990–2013 against 3, 2 and 1 over the NIO, the BOB and the AS respectively during 1961–1990. It indicates that during 1990–2013, the frequency of all categories of CDs have decreased significantly as compared to 1961–1990. There

Table 1 Classification of cyclonic disturbances over the NIO

Type of cyclonic disturbance	Wind speed		
	knots	km/h	mps
Depression (D)	17–27	31–49	09–14
Deep depression (DD)	28–33	50–61	15–17
Cyclonic storm (CS)	34–47	62–88	18–24
Severe cyclonic storm (SCS)	48–63	89–118	25–32
Very severe cyclonic storm (VSCS)	64–119	119–221	33–61
Super cyclone (SuCS)	120 and above	222 or more	62 or more

Source: IMD (2013)

are many studies about the decreasing trend in recent years which suggest that the decreasing trend may be attributed to (i) increase in vertical wind shear, (ii) decrease in mid-tropospheric Relative Humidity (RH) and (iii) decrease in the Ocean thermal energy though there is a significant increase in the SST.

The data set considered in the study is based on the best track data developed by the IMD Regional Specialized Meteorological Centre ((RSMC), New Delhi, 1991–2014). Details of the best track procedure have been described by Mohapatra et al. (2012b). Over the years, there has been a significant improvement in monitoring the location and intensity of TCs due to upgradation in observational tools and techniques and monitoring procedures, over the NIO region, like other ocean basins. According to Mohapatra et al. (2012b), depending upon the quality and availability of information, the best tack data from 1960 onwards (Satellite era) has been classified into 3 phases: (i), 1961–1973, (ii) 1974–1989 and (iii) 1990 onwards. It is interesting to mention that though optimal observational network including satellite leading to better estimation of location, track and intensity is available from 1960, the best track data with more detailed information on location and intensity are published by the IMD for every 6 h based on 00, 06, 12, 18 UTC throughout the life period of the CDs only since 1990. Therefore, the life period of the CDs is more specifically defined from 1990 onwards. Hence, the IMD best-track information from 1990 to 2013 which provides the minute details about the life cycle of CDs over the NIO is used for the study.

There are two agencies, namely, the India Meteorological department (IMD) and Joint Typhoon warning Centre (JTWC) which determine the best track over the NIO. However, IMD is the nodal agency and acts also as an RSMC for the NIO. Further, the local data and information which are available to the RSMC, New Delhi may not be available to the JTWC. Therefore, due to the difference in availability of information and also the subjectivity involved in estimation of location and intensity, there can be differences in estimation by the IMD and the JTWC. Therefore, the life period of the CDs calculated by these two agencies may differ to a certain extent.

Average life period of CDs over the region has been calculated by stratifying the CDs based on the maximum intensity of the CDs during their life period. Further, the duration of each stage of the CDs such as D, DD, CS, SCS, VSCS, SuCS etc.,

Table 2 Season-wise distribution of CDs considered for the study (1990–2013)

Basin->	BOB					AS					Land					NIO + Land				
Type of CD	W	PrM	Mon	PsM	Annual	W	PrM	Mon	PsM	Annual	W	PrM	Mon	PsM	Annual	W	PrM	Mon	PsM	Annual
D	1	3	23	15	42	0	1	7	2	10	0	1	10	1	12	1	5	40	18	64
DD	0	5	25	17	47	1	0	1	11	13	0	0	0	0	0	1	5	26	28	60
CS	2	6	2	16	26	0	2	3	7	12	0	0	0	0	0	2	8	5	23	38
SCS	0	3	1	7	11	0	1	4	4	9	0	0	0	0	0	0	4	5	11	20
VSCS	0	6	0	17	23	0	3	1	2	6	0	0	0	0	0	0	9	1	19	29
SUCS	0	2	0	1	3	0	0	1	0	1	0	0	0	0	0	0	2	1	1	4
Total	3	25	51	73	152	1	7	17	26	51	0	1	10	1	12	4	33	78	100	215

W Winter, *PrM* Pre-monsoon, *Mon* Monsoon, *PsM* Post-monsoon seasons, *BOB* Bay of Bengal, *AS*, Arabian Sea, *NIO* North Indian Ocean

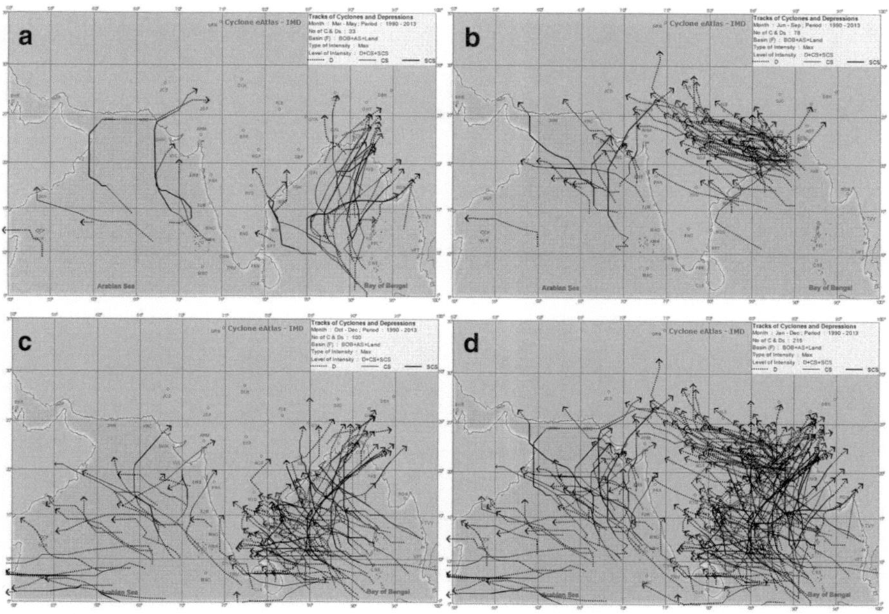

Fig. 1 Tracks of cyclonic disturbances over the basins of the NIO during (**a**) pre-monsoon, (**b**) monsoon, (**c**) post-monsoon seasons and (**d**) year as a whole over the period of 1990–2013

has been calculated. Based on the above, the average life period of CDs in different cumulative stages has also been calculated and analysed.

Mean life periods of CDs of different maximum intensities, duration of different intensity stages and cumulative duration of stages developed during the year as well as pre-monsoon (March–May), monsoon (June–September) and post-monsoon (October–December) seasons are subjected to the student's *t*-test to find significant difference, if any, in the mean life period of CDs developed over the two ocean basins viz., the BOB and the AS and also among different seasons. The results are presented and discussed in Sect. 3 and broad conclusions are presented in Sect. 4.

3 Results and Discussion

3.1 Average Life Period of the CDs

The average life period and its Standard deviation based on maximum intensity of a CD during pre-monsoon, monsoon and post-monsoon seasons as well as year as a whole over the BOB, AS and NIO as a whole are presented in Fig. 2. The life period over the AS, BOB and NIO as a whole during pre-monsoon season increases with increase in maximum intensity. Considering the NIO as a whole, the average life period is about 2 days in case of D and DD, 3.5 days in case of CS, 4 days in case of

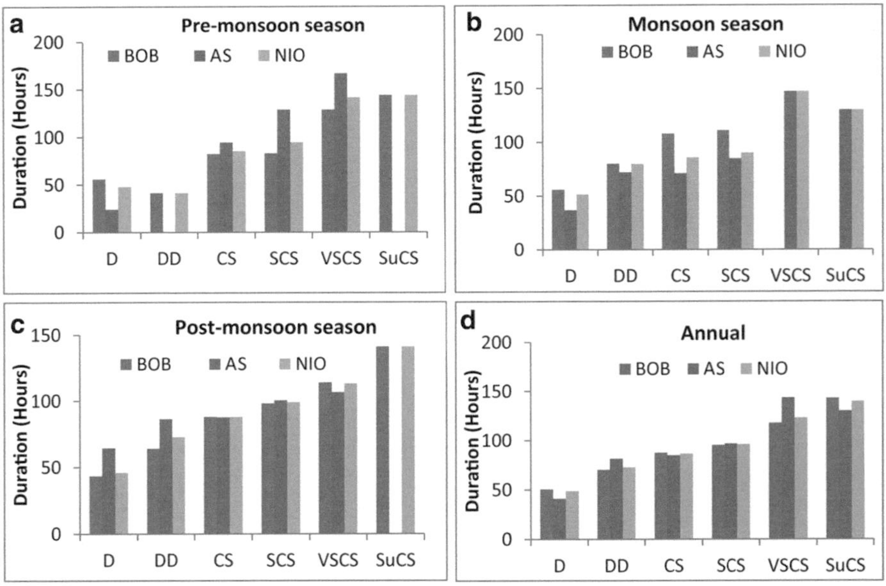

Fig. 2 Mean duration in the life cycle of CDs over the BOB, the AS and the NIO during (**a**) pre-monsoon season, (**b**) monsoon season, (**c**) post-monsoon season and (**d**) year as a whole based on the data of 1990–2013

SCS and 6 days in case of VSCS and SuCS. Comparing the mean life periods over BOB and AS, there is no significant difference (Table 3) during pre-monsoon season.

Considering the NIO as a whole, the life period during monsoon season is about 2 days in case of D, 3.25 days in case of DD, 3.5 days in case of CS, 3.75 days in case of SCS, 6 days in case of VSCS and 5.5 days in case of SuCS. During the monsoon season, the mean life period of Ds is significantly higher over the BOB than over the AS at 90 % confidence level (Table 3). It is mainly due to the fact that Ds over the BOB during monsoon season mostly develop over the northern part of the BOB and move west-northwestwards along the monsoon-trough (Rao 1976). Due to the prevalent moisture during the season over the landmass, these depressions could travel a long path without dissipation. However, the Ds over the AS develop mostly as onset vortex during the month of June (Fig. 1) over the east central AS and move northward with the progress of monsoon and have less duration of life.

The life period over the NIO is about 2 days in case of D, 3 days in case of DD, 3.7 days in case of CS, 4 days in case of SCS, 4.7 days in case of VSCS and 6 days in case of SuCS during post-monsoon season. The mean life period is higher over the AS in case of DD at 95 % level of confidence and almost the same over both the basins from CS onwards (Table 3).

Considering year as a whole, the life period of CDs over the NIO is about 2 days, 3 days, 3.5 days, 4 days, 5 days and 5.75 days respectively for D, DD, CS, SCS,

Table 3 Level of confidence (%) in difference in mean duration of life cycle of CDs over the BOB and the AS

	Level of confidence (%) in difference in mean duration of life cycles of CDs over the BOB and the AS during			
Maximum intensity	Pre-monsoon	Monsoon	Post-monsoon	Annual
D	–	90	–	–
DD	–	–	95	–
CS	–	–	–	–
SCS	–	–	–	–
VSCS	–	–	–	**90**

Bold, higher mean duration over the AS; Unbold, higher duration over the BOB

VSCS and SuCS. VSCSs have higher mean life period over the AS than over the BOB at 90 % confidence level (Table 3). The higher mean life period of VSCS over the AS as compared to that over the BOB may be due to the fact that CDs develop over southeast AS and southeast BOB (IMD 2008). Most of the CDs are recurving type inching with northward and northeastward movement as shown in Fig. 1 after reaching 20° N. As the area of coverage of AS is larger, the track length becomes longer over the AS. Further, as eastern AS is warmer, these CDs do not dissipate over the sea resulting in higher life period. However, due to northeastward recurvature, the sea area available to the VSCS over the BOB is less. After hitting the coast they usually dissipate.

3.2 Duration of Different Intensity Stages During the Life Period of CDs

The duration of each stage of CD during different seasons and year as a whole for the period of 1990–2013 are given in Annexure I. The mean duration of different stages of intensity of CDs is presented in Fig. 3 for different seasons and the year as a whole. During the pre-monsoon season and the year as a whole, the mean duration is higher over the AS than over the BOB in case of VSCS stage only (Table 4) at 90 % level of confidence. During the monsoon season, the duration D, DD and CS stages are significantly higher over BOB than over the AS at 99 %, 95 % and 99 % confidence levels, respectively (Table 4). During the post-monsoon season, while the mean duration of CDs in D, DD stages are significantly less over BOB than over the AS at 95 % and 90 % level of confidence respectively, the duration of VSCS is significantly higher over BOB than over the AS at 90 % confidence level. The significantly higher life period over AS during VSCS stage in pre-monsoon compared to that over the BOB may be attributed to the fact that the rate of intensification of CDs in VSCS stage is higher over the AS than over the BOB due to associated large scale southwest monsoon circulation in its onset phase. Also, most of these CDs with VSCS intensity are recurving type leading to longer duration in

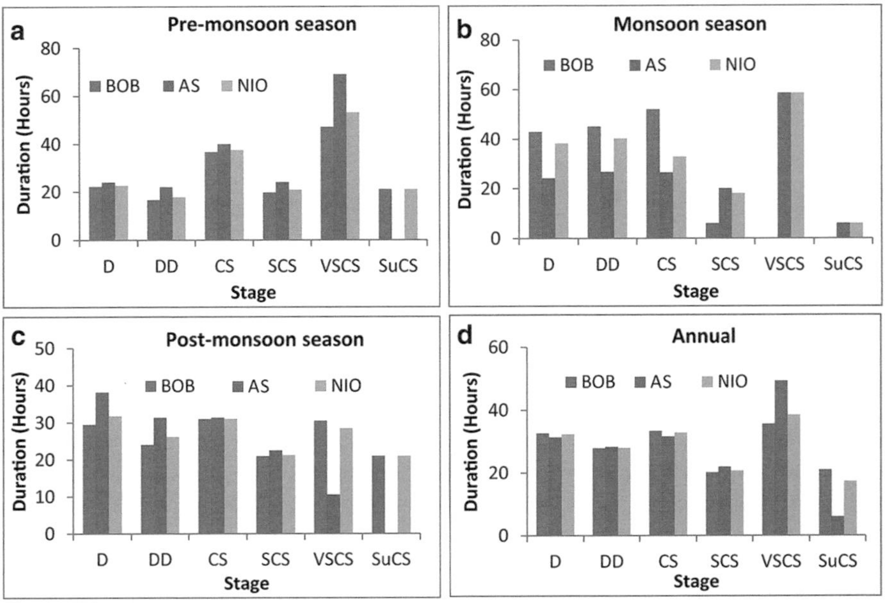

Fig. 3 Mean duration of different stages of intensity of CDs over the BOB, the AS and the NIO during (**a**) pre-monsoon season, (**b**) monsoon season, (**c**) post-monsoon season and (**d**) year as a whole based on the data of 1990–2013

Table 4 Level of confidence (%) in difference in mean duration of different stages of CDs over the BOB and the AS

	Level of confidence (%) in difference in mean duration of different stages of CDs over the BOB and the AS during			
Stage of CD	Pre-monsoon	Monsoon	Post-monsoon	Annual
D	–	99	**95**	–
DD	–	95	**90**	–
CS	–	99	–	–
SCS	–	–	–	–
VSCS	**90**	–	90	**90**

Bold, higher mean duration over the AS; Unbold, higher duration over the BOB

VSCS stage (Fig. 1). Further, the area of AS is larger than that of the BOB. The longer duration in D, DD and CS over the BOB during monsoon season may be attributed to the fact that monsoon depressions mostly develop over the head Bay of Bengal and move west-northwestwards along the monsoon trough. On many occasions, they can reach up to the west of 75° E as it can be observed in Fig. 1. However, the depressions over the AS during monsoon season usually develop as a vortex during the onset phase and move northwards and weaken gradually over the land. The duration of VSCS stage over the BOB is significantly higher than that over the AS during the post-monsoon season. It may be due to the fact that during post-monsoon season, most of the VSCS develop over the BOB and develop from the remnants of the CDs from the South China Sea which evolve over the South Andaman Sea. As these CDs mostly move northwestwards also in the warm central BOB and low vertical wind shear, they intensify into VSCS stage and higher intensity is maintained till the landfall in most cases (Fig. 1).

3.3 Cumulative Duration of Stages of CDs Over the NIO

Considering the cumulative duration of stages, the D and above to VSCS and above show higher durations over the BOB than over the AS in all the seasons and the year as a whole except CS and above, SCS and above, VSCS and above in monsoon season when AS has higher cumulative durations than the BOB (Fig. 4a,b). Considering the significant mean differences, it is found that pre-monsoon D and above, DD and above, CS and above, SCS and above have higher duration over the BOB than over the AS respectively at 99 %, 95 %, 95 % and 90 % level of confidence (Table 5). During monsoon season, D and above have higher duration over the BOB than over the AS at 99 % level confidence where as CS and above, SCS and above, VSCS and above have higher durations over the AS than over the BOB at 90 %, 95 % and 90 % levels of confidence respectively. During post-monsoon and year as a whole, D and above, DD and above, CS and above, SCS and above, VSCS and above have higher durations over the BOB than over the AS at 99 % level of confidence.

To summarise, cumulative duration of D and above to SCS and above during pre-monsoon, D and above during monsoon season D and above to VSCS and above during post-monsoon as well as the year as a whole are higher over the BOB than over the AS while the cumulative duration of stages of CS and above, SCS and above, VSCS and above are higher over the AS than over the BOB during monsoon season.

4 Conclusions

The study shows that the average life period of CDs over the NIO is about 2, 3, 3.5, 4, 5 and 5.75 days, respectively for D, DD, CS, SCS, VSCS and SuCS. VSCS have higher mean life period over both the AS and the BOB in pre-monsoon, post-monsoon and year as a whole. Although the VSCS stage has significantly higher duration over the AS than over the BOB in pre-monsoon and the year as a whole, it is significantly higher over the BOB than over the AS during post-monsoon season. During the monsoon season, the duration D, DD and CS stages are significantly higher over BOB than they are over the AS.

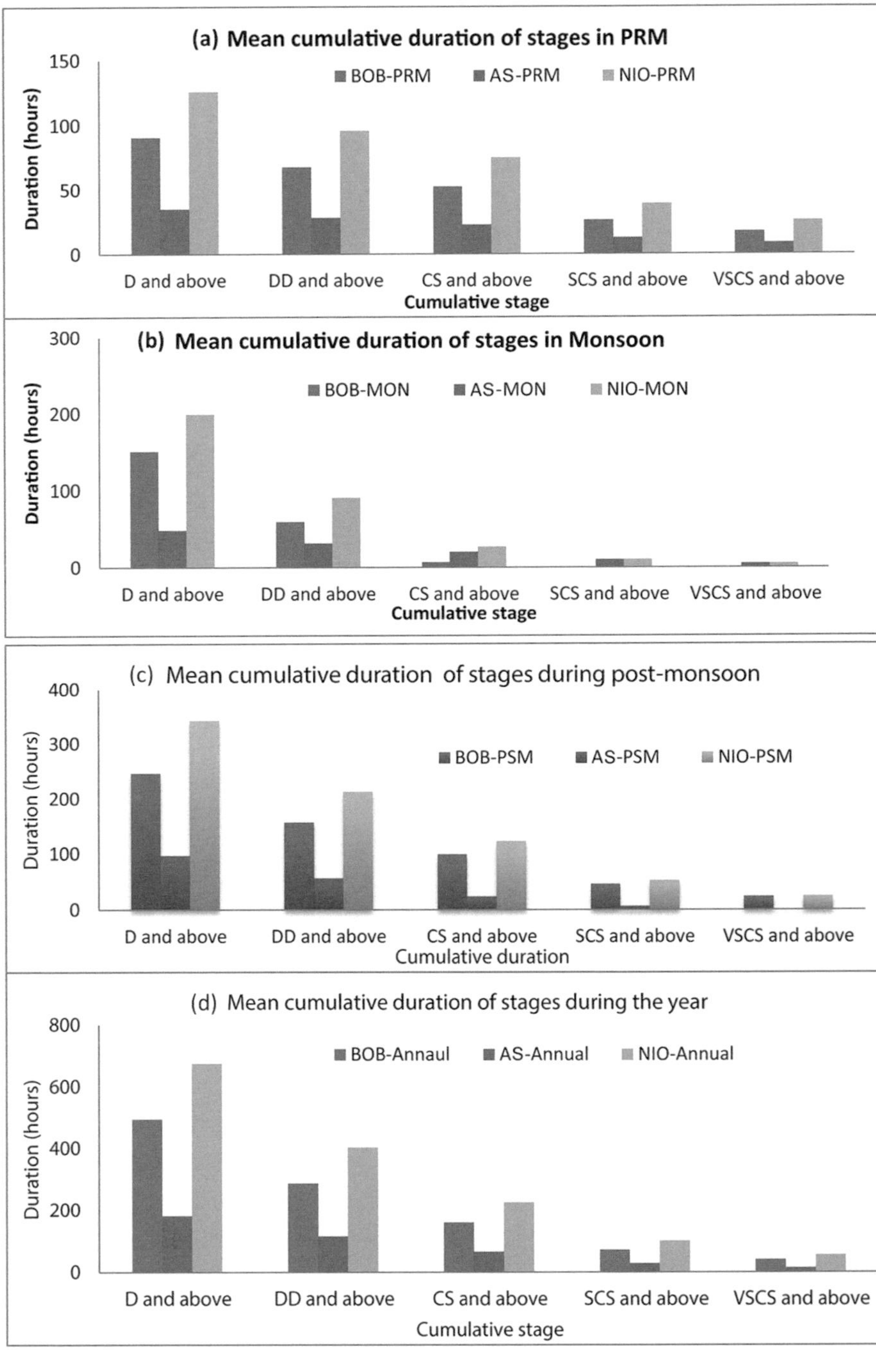

Fig. 4 Mean cumulative duration of stages (D & above, CS & above, SCS & above and VSCS & above) over the BOB, the AS and the NIO as a whole in (**a**) pre-monsoon (PRM), (**b**) monsoon, (**c**) post-monsoon (PSM) season and (**d**) year as awhole (Annual) based on the data of 1990–2013

Table 5 Level of confidence (%) in difference in mean cumulative duration of different stages of CDs over the BOB and the AS

	Level of confidence (%) in difference in mean duration of different cumulative stages of CDs over the BOB and the AS			
Cumulative stages of CD	Pre-monsoon	Monsoon	Post-monsoon	Annual
D and above	99 %	99 %	99 %	99 %
DD and above	95 %	–	99 %	99 %
CS and above	95 %	**90 %**	99 %	99 %
SCS and above	90 %	**95 %**	99 %	99 %
VSCS and above	–	**90 %**	99 %	99 %

Bold, higher mean duration over the AS; Unbold, higher duration over the BOB

In respect of cumulative duration of stages, D and above to SCS and above during pre-monsoon, D and above during monsoon season D and above to VSCS and above during post-monsoon as well as the year as a whole are higher over the BOB than over the AS while the cumulative duration of stages of CS and above, SCS and above, VSCS and above are higher over the AS than over the BOB during monsoon season.

Acknowledgement Authors are thankful to Director General of Meteorology, IMD for his encouragement and support for this work.

Annexure I

Duration (h) of cyclonic disturbances in different stages over the NIO from 1990 to 2013 in winter

	Bay of Bengal							Arabian Sea							NIO						
Year	D	DD	CS	SCS	VSCS	SuCS	Total	D	DD	CS	SCS	VSCS	SuCS	Total	D	DD	CS	SCS	VSCS	SuCS	Total
1990	0	0	0	0	0	0	0	0	0	0	0	0	0	0	0	0	0	0	0	0	0
1991	0	0	0	0	0	0	0	0	0	0	0	0	0	0	0	0	0	0	0	0	0
1992	0	0	0	0	0	0	0	0	0	0	0	0	0	0	0	0	0	0	0	0	0
1993	0	0	0	0	0	0	0	0	0	0	0	0	0	0	0	0	0	0	0	0	0
1994	0	0	0	0	0	0	0	0	0	0	0	0	0	0	0	0	0	0	0	0	0
1995	0	0	0	0	0	0	0	0	0	0	0	0	0	0	0	0	0	0	0	0	0
1996	0	0	0	0	0	0	0	0	0	0	0	0	0	0	0	0	0	0	0	0	0
1997	0	0	0	0	0	0	0	0	0	0	0	0	0	0	0	0	0	0	0	0	0
1998	0	0	0	0	0	0	0	0	0	0	0	0	0	0	0	0	0	0	0	0	0
1999	33	15	9	0	0	0	57	0	0	0	0	0	0	0	33	15	9	0	0	0	57
2000	0	0	0	0	0	0	0	0	0	0	0	0	0	0	0	0	0	0	0	0	0
2001	0	0	0	0	0	0	0	0	0	0	0	0	0	0	0	0	0	0	0	0	0
2002	0	0	0	0	0	0	0	0	0	0	0	0	0	0	0	0	0	0	0	0	0
2003	0	0	0	0	0	0	0	0	0	0	0	0	0	0	0	0	0	0	0	0	0
2004	0	0	0	0	0	0	0	0	0	0	0	0	0	0	0	0	0	0	0	0	0
2005	18	33	45	0	0	0	96	0	0	0	0	0	0	0	18	33	45	0	0	0	96
2006	0	0	0	0	0	0	0	27	9	0	0	0	0	36	27	9	0	0	0	0	36
2007	0	0	0	0	0	0	0	0	0	0	0	0	0	0	0	0	0	0	0	0	0
2008	0	0	0	0	0	0	0	0	0	0	0	0	0	0	0	0	0	0	0	0	0
2009	0	0	0	0	0	0	0	0	0	0	0	0	0	0	0	0	0	0	0	0	0
2010	0	0	0	0	0	0	0	0	0	0	0	0	0	0	0	0	0	0	0	0	0

2011	15	0	0	0	0	0	15	0	0	0	0	0	0	0	15	0	0	0	0	0	15
2012	0	0	0	0	0	0	0	0	0	0	0	0	0	0	0	0	0	0	0	0	0
2013	0	0	0	0	0	0	0	0	0	0	0	0	0	0	0	0	0	0	0	0	0
SUM	66	48	54	0	0	0	168	27	9	0	0	0	0	36	93	57	54	0	0	0	204

Duration (h) of cyclonic disturbances in different stages over the NIO from 1990 to 2013 in pre-monsoon

	Bay of Bengal							Arabian Sea							NIO						
Year	D	DD	CS	SCS	VSCS	SuCS	Total	D	DD	CS	SCS	VSCS	SuCS	Total	D	DD	CS	SCS	VSCS	SuCS	Total
1990	18	12	18	12	72	18	150	0	0	0	0	0	0	0	18	12	18	12	72	18	150
1991	30	24	48	24	42	24	192	0	0	0	0	0	0	0	30	24	48	24	42	24	192
1992	30	24	30	0	0	0	84	0	0	0	0	0	0	0	30	24	30	0	0	0	84
1993	0	0	0	0	0	0	0	0	0	0	0	0	0	0	0	0	0	0	0	0	0
1994	75	9	24	12	41	0	161	0	0	0	0	0	0	0	75	9	24	12	41	0	161
1995	90	69	0	0	0	0	159	0	0	0	0	0	0	0	90	69	0	0	0	0	159
1996	15	9	0	0	0	0	24	0	0	0	0	0	0	0	15	9	0	0	0	0	24
1997	12	15	30	18	39	0	114	0	0	0	0	0	0	0	12	15	30	18	39	0	114
1998	18	12	21	21	0	0	72	24	0	0	0	0	0	24	42	12	21	21	0	0	96
1999	0	0	0	0	0	0	0	6	18	30	6	84	0	144	6	18	30	6	84	0	144
2000	39	12	21	0	0	0	72	0	0	0	0	0	0	0	39	12	21	0	0	0	72
2001	0	0	0	0	0	0	0	18	12	45	30	78	0	183	18	12	45	30	78	0	183
2002	9	15	0	0	0	0	24	48	27	27	0	0	0	102	57	42	27	0	0	0	126
2003	9	78	75	36	36	0	234	0	0	0	0	0	0	0	9	78	75	36	36	0	234
2004	9	9	27	9	12	0	66	6	9	81	33	0	0	129	15	18	108	42	12	0	195
2005	0	0	0	0	0	0	0	0	0	0	0	0	0	0	0	0	0	0	0	0	0
2006	6	3	45	18	45	0	117	0	0	0	0	0	0	0	6	3	45	18	45	0	117
2007	57	6	21	0	0	0	84	0	0	0	0	0	0	0	57	6	21	0	0	0	84

(continued)

	Bay of Bengal							Arabian Sea							NIO						
Year	D	DD	CS	SCS	VSCS	SuCS	Total	D	DD	CS	SCS	VSCS	SuCS	Total	D	DD	CS	SCS	VSCS	SuCS	Total
2008	9	12	15	27	90	0	153	0	0	0	0	0	0	0	9	12	15	27	90	0	153
2009	51	21	75	9	0	0	156	0	0	0	0	0	0	0	51	21	75	9	0	0	156
2010	12	15	45	30	0	0	102	66	66	57	27	45	0	261	78	81	102	57	45	0	363
2011	0	0	0	0	0	0	0	0	0	0	0	0	0	0	0	0	0	0	0	0	0
2012	0	0	0	0	0	0	0	0	0	0	0	0	0	0	0	0	0	0	0	0	0
2013	66	21	129	0	0	0	216	0	0	0	0	0	0	0	66	21	129	0	0	0	216
	555	366	624	216	377	42	2180	168	132	240	96	207	0	843	723	498	864	312	584	42	3023

Duration (h) of cyclonic disturbances in different stages over the NIO from 1990 to 2013 in monsoon

	Bay of Bengal							Arabian Sea							NIO						
Year	D	DD	CS	SCS	VSCS	SuCS	Total	D	DD	CS	SCS	VSCS	SuCS	Total	D	DD	CS	SCS	VSCS	SuCS	Total
1990	264	6	0	0	0	0	270	0	0	0	0	0	0	0	264	6	0	0	0	0	270
1991	216	0	0	0	0	0	216	0	0	0	0	0	0	0	216	0	0	0	0	0	216
1992	84	102	0	0	0	0	186	30	48	24	0	0	0	102	114	150	24	0	0	0	288
1993	33	24	0	0	0	0	57	0	0	0	0	0	0	0	33	24	0	0	0	0	57
1994	63	15	0	0	0	0	78	21	21	42	6	0	0	90	84	36	42	6	0	0	168
1995	90	0	0	0	0	0	90	0	0	0	0	0	0	0	90	0	0	0	0	0	90
1996	15	21	60	0	0	0	96	21	21	24	9	0	0	75	36	42	84	9	0	0	171
1997	174	387	63	6	0	0	630	0	0	0	0	0	0	0	174	387	63	6	0	0	630
1998	12	24	0	0	0	0	36	30	30	27	36	48	0	171	42	54	27	36	48	0	207
1999	108	36	0	0	0	0	144	0	0	0	0	0	0	0	108	36	0	0	0	0	144
2000	21	0	0	0	0	0	21	0	0	0	0	0	0	0	21	0	0	0	0	0	21
2001	33	0	0	0	0	0	33	12	27	48	0	0	0	87	45	27	48	0	0	0	120
2002	0	0	0	0	0	0	0	0	0	0	0	0	0	0	0	0	0	0	0	0	0
2003	57	39	0	0	0	0	96	0	0	0	0	0	0	0	57	39	0	0	0	0	96
2004	24	54	0	0	0	0	78	36	81	27	15	0	0	159	60	135	27	15	0	0	237

2005	207	72	33	0	0	0	312	120	0	0	0	0	0	120	327	72	33	0	0	0	432
2006	297	102	0	0	0	0	399	9	24	21	33	0	0	87	306	126	21	33	0	0	486
2007	234	159	0	0	0	0	393	18	15	25	21	69	6	154	252	174	25	21	69	6	547
2008	96	54	0	0	0	0	150	48	0	0	0	0	0	48	144	54	0	0	0	0	198
2009	18	72	0	0	0	0	90	42	0	0	0	0	0	42	60	72	0	0	0	0	132
2010	0	0	0	0	0	0	0	0	0	0	0	0	0	0	0	0	0	0	0	0	0
2011	99	96	0	0	0	0	195	24	0	0	0	0	0	24	123	96	0	0	0	0	219
2012	0	0	0	0	0	0	0	0	0	0	0	0	0	0	0	0	0	0	0	0	0
2013	48	0	0	0	0	0	48	0	0	0	0	0	0	0	48	0	0	0	0	0	48
SUM	2193	1263	156	6	0	0	3618	411	267	238	120	117	6	1159	2604	1530	394	126	117	6	4777

Duration (h) of cyclonic disturbances in different stages over the NIO from 1990 to 2013 in post-monsoon

	Bay of Bengal							Arabian Sea							NIO						
Year	D	DD	CS	SCS	VSCS	SuCS	Total	D	DD	CS	SCS	VSCS	SuCS	Total	D	DD	CS	SCS	VSCS	SuCS	Total
1990	138	72	36	18	0	0	264	36	12	0	0	0	0	48	174	84	36	18	0	0	312
1991	132	12	42	0	0	0	186	0	0	0	0	0	0	0	132	12	42	0	0	0	186
1992	102	180	90	90	6	0	468	96	60	54	0	0	0	210	198	240	144	90	6	0	678
1993	30	42	15	9	24	0	120	54	24	39	9	12	0	138	84	66	54	18	36	0	258
1994	81	30	15	18	0	0	144	24	9	69	9	9	0	120	105	39	84	27	9	0	264
1995	27	24	45	30	51	0	177	39	27	60	0	0	0	126	66	51	105	30	51	0	303
1996	117	84	55	65	45	0	366	63	24	9	33	0	0	129	180	108	64	98	45	0	495
1997	3	6	120	0	0	0	129	12	54	0	0	0	0	66	15	60	120	0	0	0	195
1998	66	108	45	30	24	0	273	126	93	45	42	0	0	306	192	201	90	72	24	0	579
1999	84.5	27	57	36	57	21	283	0	0	0	0	0	0	0	84.5	27	57	36	57	21	282.5
2000	129	123	84	30	63	0	429	0	0	0	0	0	0	0	129	123	84	30	63	0	429
2001	42	9	24	0	0	0	75	9	12	18	0	0	0	39	51	21	42	0	0	0	114
2002	147	60	126	3	0	0	336	0	0	0	0	0	0	0	147	60	126	3	0	0	336
2003	144	42	30	33	0	0	249	18	15	18	30	0	0	81	162	57	48	63	0	0	330
2004	51	0	0	0	0	0	51	108	48	33	12	0	0	201	159	48	33	12	0	0	252

(continued)

Year	Bay of Bengal							Arabian Sea							NIO						
	D	DD	CS	SCS	VSCS	SuCS	Total	D	DD	CS	SCS	VSCS	SuCS	Total	D	DD	CS	SCS	VSCS	SuCS	Total
2005	204	162	138	0	0	0	504	0	0	0	0	0	0	0	204	162	138	0	0	0	504
2006	12	6	18	0	0	0	36	0	0	0	0	0	0	0	12	6	18	0	0	0	36
2007	42	9	15	6	75	0	147	18	120	0	0	0	0	138	60	129	15	6	75	0	285
2008	93	114	63	0	0	0	270	66	24	0	0	0	0	90	159	138	63	0	0	0	360
2009	33	48	33	0	0	0	114	24	21	18	0	0	0	63	57	69	51	0	0	0	177
2010	168	42	45	39	18	0	312	0	0	0	0	0	0	0	168	42	45	39	18	0	312
2011	33	42	39	6	39	0	159	225	93	15	0	0	0	333	258	135	54	6	39	0	492
2012	72	78	39	0	0	0	189	51	63	30	0	0	0	144	123	141	69	0	0	0	333
2013	204	75	96	111	147	0	633	0	78	0	0	0	0	78	204	153	96	111	147	0	711
SUM	2155	1395	1270	524	549	21	5914	969	777	408	135	21	0	2310	3124	2172	1678	659	570	21	8224

Duration (h) of cyclonic disturbances in different stages over the NIO from 1990 to 2013 during year as a whole

Year	Bay of Bengal							Arabian Sea							NIO						
	D	DD	CS	SCS	VSCS	SuCS	Total	D	DD	CS	SCS	VSCS	SuCS	Total	D	DD	CS	SCS	VSCS	SuCS	Total
1990	420	90	54	30	72	18	684	36	12	0	0	0	0	48	456	102	54	30	72	18	732
1991	378	36	90	24	42	24	594	0	0	0	0	0	0	0	378	36	90	24	42	24	594
1992	216	306	120	90	6	0	738	126	108	78	0	0	0	312	342	414	198	90	6	0	1050
1993	63	66	15	9	24	0	177	54	24	39	9	12	0	138	117	90	54	18	36	0	315
1994	219	54	39	30	41	0	383	45	30	111	15	9	0	210	264	84	150	45	50	0	593
1995	207	93	45	30	51	0	426	39	27	60	0	0	0	126	246	120	105	30	51	0	552
1996	147	114	115	65	45	0	486	84	45	33	42	0	0	204	231	159	148	107	45	0	690
1997	189	408	213	24	39	0	873	12	54	0	0	0	0	66	201	462	213	24	39	0	939
1998	96	144	66	51	24	0	381	180	123	72	78	48	0	501	276	267	138	129	72	0	882
1999	226	78	66	36	57	21	483.5	6	18	30	6	84	0	144	232	96	96	42	141	21	627.5
2000	189	135	105	30	63	0	522	0	0	0	0	0	0	0	189	135	105	30	63	0	522
2001	75	9	24	0	0	0	108	39	51	111	30	78	0	309	114	60	135	30	78	0	417
2002	156	75	126	3	0	0	360	48	27	27	0	0	0	102	204	102	153	3	0	0	462

2003	210	159	105	69	36	0	579	18	15	18	30	0	0	81	228	174	123	99	36	0	660
2004	84	63	27	9	12	0	195	150	138	141	60	0	0	489	234	201	168	69	12	0	684
2005	429	267	216	0	0	0	912	120	0	0	0	0	0	120	549	267	216	0	0	0	1032
2006	315	111	63	18	45	0	552	36	33	21	33	0	0	123	351	144	84	51	45	0	675
2007	333	174	36	6	75	0	624	36	135	25	21	69	6	292	369	309	61	27	144	6	916
2008	198	180	78	27	90	0	573	114	24	0	0	0	0	138	312	204	78	27	90	0	711
2009	102	141	108	9	0	0	360	66	21	18	0	0	0	105	168	162	126	9	0	0	465
2010	180	57	90	69	18	0	414	66	66	57	27	45	0	261	246	123	147	96	63	0	675
2011	147	138	39	6	39	0	369	249	93	15	0	0	0	357	396	231	54	6	39	0	726
2012	72	78	39	0	0	0	189	51	63	30	0	0	0	144	123	141	69	0	0	0	333
2013	318	96	225	111	147	0	897	0	78	0	0	0	0	78	318	174	225	111	147	0	975
SUM	4969	3072	2104	746	926	63	11,880	1575	1185	886	351	345	6	4348	6544	4257	2990	1097	1271	69	16,228

References

Anthes, R.A. (1982). Tropical Cyclones: Their Evolution Structure and Effects. Meteorological Monograph No. 41, American Meteorological Society, 208 pp.

Dube, S.K, Poulose, J. and Rao, A.D. (2013). Numerical simulation of storm surge associated with severe cyclonic storms in the Bay of Bengal during 2008–11. *Mausam*, **64**, 193–202.

Elsner, J.B. and Kocher, B. (2000). Global tropical cyclone activity: A link to the North Atlantic Oscillation. *Geophysical Research Letters*, **27**, 129–132.

Emanuel, K. (2005a). Increasing destructiveness of tropical cyclones over the past 30 years. *Nature*, **436**, 686–688, doi:10.1038/nature03906

Emanuel, K.A. (2005b). Emanuel replies. *Nature*, **438**, E13.

Gray, W.M. (1968). Global view of the origin of tropical disturbances and storms. *Monthly Weather Review*, **96**, 669–700.

Holland, G.J. (1997). The maximum potential intensity of tropical cyclones. *Journal of Atmospheric Sciences*, **54**, 2519–2541.

IMD (2013). Cyclone Warning in India; Standard Operational Procedure. IMD, New Delhi.

IMD (2008). 'Tracks of Cyclones and Depressions (1891–2007), Electronic Version 1.0/2008, Published by IMD, Chennai.

Lighthill, J, Holland, G.J, Gray, W.M. and Guard, C.P. (1994). Global climate change and tropical cyclones. *Bulletin of the American Meteorological Society*, **75**, 2147–2157.

Mohapatra, M. (2015). Cyclone hazard proneness of districts of India. *Journal of Earth System Sciences*, **124**, 515–526.

Mohapatra, M, Mandal, G.S, Bandyopadhyay, B.K, Tyagi, A. and Mohanty, U.C. (2012a). Classification of cyclone hazard prone districts of India. *Natural Hazards*, **63**, 1601–1620.

Mohapatra, M, Bandyopadhyay, B.K. and Tyagi, A. (2012b). Best track parameters of tropical cyclones over the North Indian Ocean: A review. *Natural Hazards*, **63**, 1285–1317.

Mohapatra, M, Bandyopadhyay, B.K. and Tyagi, A. (2014). Construction and Quality of best tracks parameters for study of climate change impact on Tropical Cyclones over the North Indian Ocean during satellite era. *In*: Mohanty, U.C, Mohapatra, M, Singh, O.P, Bandyopadhyay, B.K. and Rathore, L.S. (eds.) Monitoring and Prediction of Tropical Cyclones over the Indian Ocean and Climate Change. Springers and Capital publishers, New Delhi, India, pp. 3–17.

Parker, D.E, Alexander, L.V. and Kennedy, J. (2004). Global and regional climate of 2003. *Weather*, **59**, 145–152.

Rajeevan, M. and Mc Phaden, M.J. (2004). Tropical Pacific upper ocean heat content variations and Indian summer monsoon rainfall. *Geophysical Research Letters*, **31**, L18203, doi 10.1029/2004GL020631

Rao, Y.P. (1976). Southwest Monsoon, IMD, Met. Monograph, Synoptic Meteorology, No. 1/1976.

Smith, T.M. and Reynolds, R.W. (2005). A Global Merged Land–Air–Sea Surface Temperature Reconstruction Based on Historical Observations (1880–1997). *Journal of Climate*, **18**, 2021–2036.

Tyagi, A, Mohapatra, M, Bandyopadhyay, B.K. and Naresh Kumar (2010). Inter-annual variation of frequency of cyclonic disturbances landfalling over WMO/ESCAP Panel Member Countries, WMO/TD-No. **1541** on Ist WMO International conference on Indian Ocean Tropical Cyclones and climate change, Muscat, Sultanate of Oman, 08–11 March 2009', WWRP-**2010/2**, pp. 1–7.

Part III
Cyclogenesis, Monitoring and Prediction

Seasonal Forecast of Tropical Cyclogenesis Over Bay of Bengal During Post-monsoon Season

D.R. Pattanaik, O.P. Sreejith, D.S. Pai, and Madhuri Musale

1 Introduction

The extensive coastal belt of India is very vulnerable to the deadly storms known as tropical cyclones (Mohapatra et al. 2012; Mohapatra 2015). These systems form initially as low-pressure areas (when the maximum sustained surface wind speed is < 17 knots) over the north Indian Ocean (NIO) and then intensify into depressions (maximum sustained surface wind speed is between 17 and 33 knots) and occasionally become tropical cyclones (when the surface wind exceeds 33 knots). For the tropical cyclones only, the peak is observed during November, followed by October in the post-monsoon season (October–December; OND), followed by the month of May in pre-monsoon season (March–May; MAM) (Pattanaik 2005). Over the Bay of Bengal (BOB), the months of October–November are known to produce cyclonic disturbances (CDs; system of intensity depression and higher) of severe intensity, which after crossing the coast cause damages to life and property over many countries of south Asia surrounding the BOB. The strong winds, heavy rains and large storm surges associated with CDs are the factors that eventually lead to loss of life and property. Rains (sometimes even more than 30 cm/24 h) associated with cyclones are another source of damage. The combination of a shallow coastal plain along with the world's highest population density coupled with low socio-economic conditions in the region surrounding the BOB has resulted in several land falling CDs becoming devastating natural disasters. Thus, the seasonal forecast of frequency of CDs over the BOB is of great use to many users. Interest from the media and the general public in seasonal forecasts of CDs has increased tremendously.

D.R. Pattanaik (✉) • O.P. Sreejith • D.S. Pai • M. Musale
India Meteorological Department, Pune 411005, India
e-mail: drpattanaik@gmail.com

© Capital Publishing Company 2017

M. Mohapatra et al. (eds.), *Tropical Cyclone Activity over the North Indian Ocean*,
DOI 10.1007/978-3-319-40576-6_13

Seasonal forecasts of tropical cyclone activity in various regions have been developed since the first attempts in the early 1980s by Nicholls (1979) for the Australian region and Gray (1984a, b) for the North Atlantic region. Over time, forecasts for different regions, using differing methodologies, have been developed. Considering El-Nino Southern Oscillation (ENSO) as one major forcing for explaining the variability of CDs over the different global ocean basins, the relationship of ENSO and tropical cyclones was subsequently developed into statistical forecasts model for predicting seasonal activity of CDs over different ocean basins. Chan et al. (1998, 2001) developed seasonal forecast of annual Cyclone frequency for the season from May to December (i.e. main active season) over the Northwest Pacific based on ENSO as one predictor. There have been some attempts (Krishnakumar et al. 2007; Balachandran and Geetha 2012; Mohapatra and Adhikary 2010) to make some aspects of seasonal forecast of CD frequency over the NIO based on sea surface temperature (SST) data over Indian Ocean.

In addition to the empirical model for the seasonal forecast of tropical cyclone there have been recent developments to make seasonal forecast for CDs frequency based on dynamic models over different Ocean basins (Vitart et al. 1997; Camargo et al. 2005). However, as indicated over the North and South Indian Ocean, dynamical models usually perform poorly (Camargo et al. 2005). It is not clear to what extent this is due to model errors or to a lack of predictability. In view of this there is a need to develop a suitable empirical model for the seasonal forecast (deterministic and probabilistic) of CDs frequency over the BOB during OND season, which can be used in providing outlooks for seasonal frequency of CDs on operational basis. In the present study, an attempt is made to develop an empirical model based on the method of Principal Component Regression (PCR) for the seasonal forecast of CDs frequency over the BOB during the post-monsoon season OND. As the main cyclone season over BOB during OND is just after the monsoon season from June to September, the variability of CDs frequency could be related to the variability of large-scale features during the monsoon season. First part of the objective of the present paper is to investigate this variability of CDs frequency over the Bay of Bengal to identify suitable predictors for the seasonal forecast of CDs. The second objective of the present study is to develop a PCR model for the seasonal forecast of CDs over the BOB using these predictors.

2 Data and Methodology

The frequency of cyclonic disturbances (CDs) during the recent 43 years (1971–2013) over the BOB during OND is used in the present study. These data sets are collected from the storm e-atlas (available at: www.rmcchennai.tn.nic.in) published by the India Meteorological Department (IMD). The large-scale variables used for finding suitable predictors are obtained from the National Center for Environmental Prediction (NCEP) reanalysis (Kalnay et al. 1996). A Principal Component Regression model is developed for the seasonal forecast of CDs frequency over the BOB from the 43 years of data. The first 40 years of data (1971–2010) is used for the developing for forecast model and the whole periods of 43 years (1971–2013) including the latest 3 years (2011–2013) are used for the verification purpose. It is advantageous to use PCR model compared to multiple regression model when there is significant inter-correlation among the predictors. The PCR model avoids the inter-correlation and helps to reduce the degrees of freedom by restricting the number of predictors. In case of PCR model the Principal Component Analysis (PCA) is first applied on the predictor set and then few PCs (principal components) having highest CC with the predictant are selected. The selected PCs are than related to the CD frequency series using a multiple regression equation. Singh and Pai (1996) and Rajeevan et al. (2000) have used PCR model for the monsoon forecast. For statistical prediction methods, cross-validation (Michaelsen 1987) is needed to help reduce artificial skill that can exist in the training data sample but vanishes when the method is applied to a real-time forecast for the future. In this case, the cross validation of 1 year is used. Thus, in cross-validation forecast the PCR models are derived from all years except for one that are withheld, and that year is then used as the target of the prediction. This is repeated for all the season separately.

Different verification measures can be used both for the deterministic and probability forecast. The conventional (i.e. Pearson) correlation coefficient (CC) is also informative when applied to deterministic forecasts. In addition, the root mean square error (RMSE), mean absolute error (MAE) and mean bias can be used for the verification of deterministic forecast. The CC denotes how well the model forecasts agree with observations. The 'MAE' and 'RMSE' does not indicate the direction of the deviations. The other skill scores, like 'mean bias', mainly explain how the average forecast magnitude compares to the average observed magnitude. Positive (negative) values of 'mean bias' indicate the average tendency of the model to over-predict (or under-predict). There are many verification measures, which can be used for the verification of probability forecast (WMO 2002). The verification scores used here are Hit score (HS), Hit Skill Score (HSS) and Relative Operating Characteristics (ROC). In ROC graph (Swets et al. 2000) 'Hit rate' is plotted in y-axis and the 'false alarm' in x-axis. In this graph, any point that lies in the lower right triangle performs worse than reference forecast or say the climatology forecast, where the hit rate is lower than the false alarm rate. The area underneath the ROC curve is the ROC score. ROC scores above 0.5 reflect positive discrimination skill, 1.0 representing the maximum possible score. Thus, the ROC graph shows the ability of the classifier to rank the hit rates relative to the false alarm rates.

3 Results and Discussion

3.1 Variability of CDs Frequency

The year-to-year variability in frequency of CDs during the post-monsoon season OND over the BOB during the 43 years period is shown in Fig. 1. The mean and standard deviation (SD) of CDs frequency is found to be 3.4 and 1.4, respectively with a co-efficient of variability 41.2 %. Based on Fig. 1 the lower tercile, median and upper tercile values of CDs frequency over the BOB during the post-monsoon season (OND) are found to be 3, 4 and 4, respectively. Thus, the above (below) normal CDs years are identified, when the frequency of CDs over the BOB is higher (lower) than 4 (3). Thus, the 10 above normal CDs years (1972, 1973, 1975, 1976, 1985, 1988, 1992, 2005, 2010 and 2013) and 10 below normal CDs years (1979, 1986, 1995, 1997, 2001, 2004, 2006, 2007, 2009 and 2011) are identified from Fig. 1 with remaining years from the 43 years period coming under the category of normal CDs year.

3.2 Identification of Predictors (Parameters) for PCR Model

The predictor for the forecast of CDs during post-monsoon season over the BOB is identified from the variables during the monsoon season (June–August) based on the NCEP reanalysis data (Kalnay et al. 1996). It is found that the sea surface temperature over equatorial Pacific and Indian Ocean is one of the driving forces of year-to-year variability of frequency of CDs during the primary CDs season of OND. A study by Girishkumar and Ravichandran (2012) has shown that during OND season the accumulated cyclone energy in the BOB is negatively correlated with Niño3.4 SST anomaly. Similarly, another convectively coupled phenomenon like the Madden Julian Oscillation (MJO), which is nothing but the eastward migration of convective bands over the equatorial Indian and Pacific Ocean and its consequent influences on the large scale circulation. Some studies (Kikuchi and Wang 2010; Mohapatra and Adhikary 2011) have also indicated association of MJO activity with cyclogenesis over the BOB. As the tropical SST and associated convection is intimately connected with large scale circulation at lower and upper troposphere, the large-scale variables having significant correlation with CDs frequency are only considered. In order to find out suitable predictors the frequency of CDs during OND over the BOB during the period from 1971 to 2010 is correlated with the large-scale variables such as Zonal wind at 850 hPa level (U850) Meridianal wind at 850 hPa level (V850), Geopotential height at 500 and 200 hPa level (Z500 and Z200), Zonal wind at 200 hPa level (U200), sea level pressure (SLP), and SST for the period from July to August. The correlation maps of the five parameters (V850, U200, U850, SLP and SST) are shown in Fig. 2a–e respectively with regions of significant CCs (significant at 95 % level) are shaded.

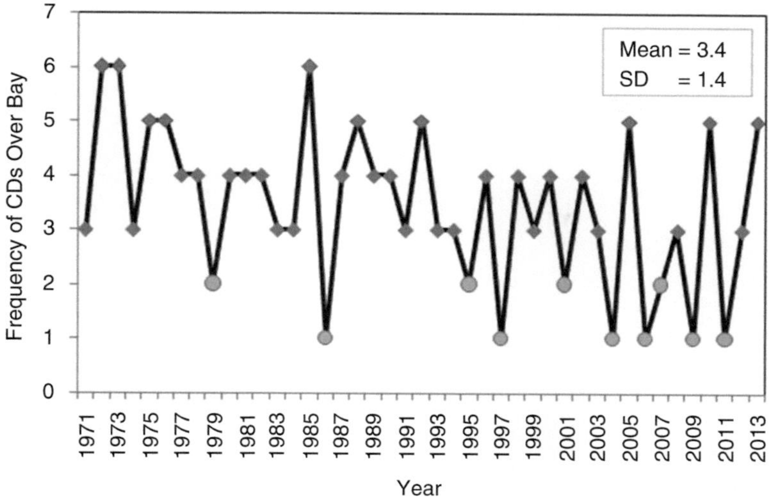

Fig. 1 Frequency of cyclonic disturbances (CDs) over Bay of Bengal (1971–2013)

In order to understand the significance of this correlation the composite anomalies of each of these five variables are analysed. The composite anomalies of meridional wind at 850 hPa level during June–August (Fig. not shown) prior to the excess and deficient seasons of CDs during OND shows the contrasting patterns over the region of significant CC shown in Fig. 2a. Similarly, the composites anomalies for other variables like U200, U850, SLP and SST during July–August prior to excess and deficient post-monsoon CDs seasons (Fig. not shown) also shows the contrasting patterns, which is consistent with the CC maps shown in Fig. 2b–e.

The composite anomalies of meridional wind shows northerly (southerly) component over the north-west parts of Australia and adjoining south-east equatorial Indian Ocean associated with stronger (weaker) Australian high during June to August prior to excess CD years over the BOB. Thus, the stronger Australian high during June–August is favourable for the above normal cyclogenesis during subsequent cyclone seasons over BOB i.e. during OND. The second variable i.e. the upper level easterly jet is one of the semi-permanent features of Indian monsoon originating from the anti-cyclonic flow over the Tibetan region and becoming easterly over southern part of India and extended westward upto Africa. As seen from Fig. 2b the significant negative CC during July–August is over the southern part of Africa and is associated with stronger (weaker) easterly anomalies over this region prior to above (below) normal CDs over the BOB during OND. The third variable is the low level monsoon westerly at 850 hPa during July–August which shows significant positive correlation with the frequency of CDs during post monsoon season (Fig. 2c). With respect to the correlation of CDs frequency during OND season with the Sea Level Pressure (SLP) during July–August, significant negative is CC is obtained for SLP over the southeast equatorial Indian Ocean (Fig. 2d). This indicates the lower SLP over the southeast equatorial Indian Ocean

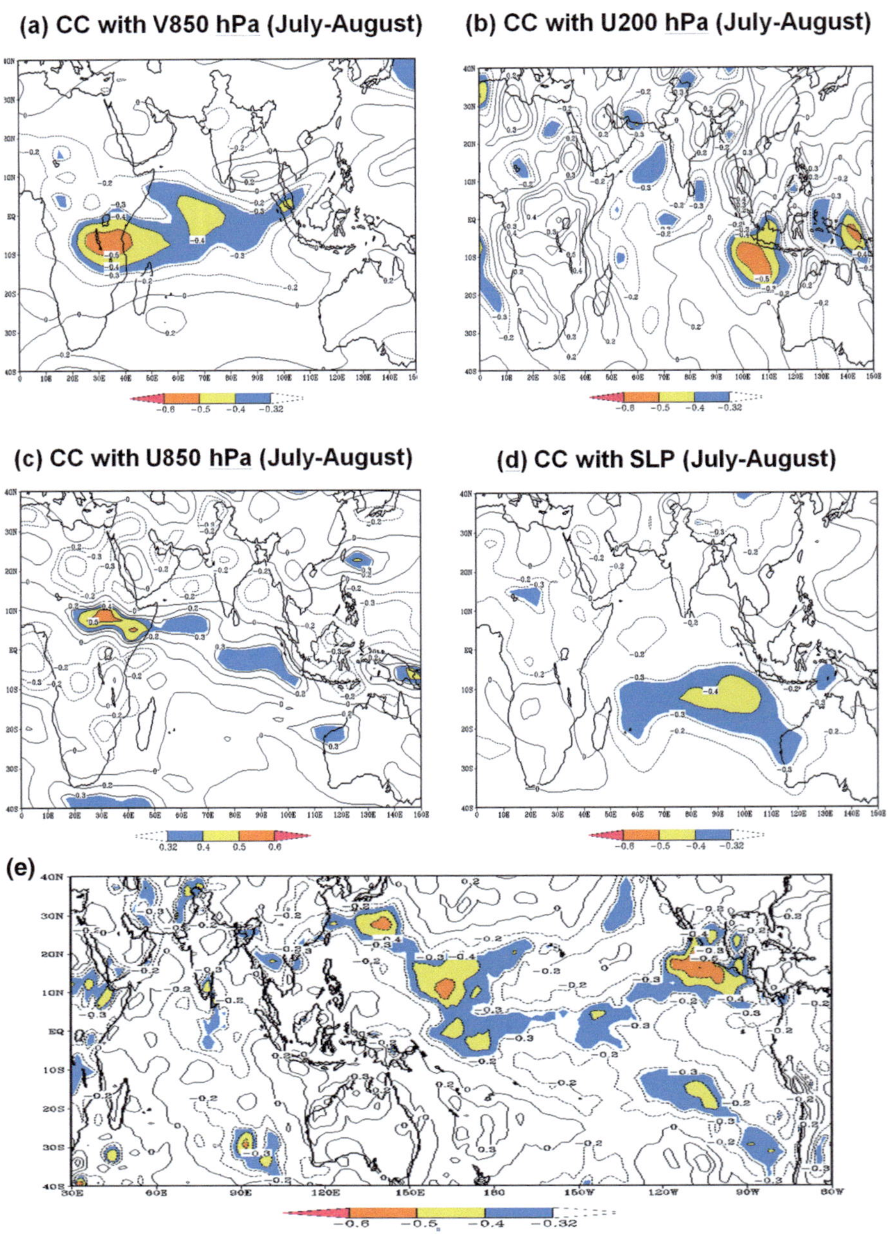

Fig. 2 Spatial Correlation coefficient (CC) maps between frequency of cyclonic disturbances during October-December and five variables: (**a**) V850, (**b**) U200, (**c**) U850, (**d**) SLP and (**e**) SST

(which is part of weak Australian high) during July–August is conducive for above normal CDs over the BOB during subsequent post-monsoon season. The final variable i.e. the SST composite anomaly during July–August prior to the year of excess (deficient) season of CDs shows contrasting patterns over the Pacific with cooler (warmer) than normal SST over the western, central and eastern equatorial Pacific, indicating the El-Nino during July–August is not favourable for the cyclogenesis over the BOB during subsequent post-monsoon season from October to December (Fig. not shown). The SST composites also indicate negative anomalies over the Northwest Pacific Ocean during July–August prior to the year of excess CDs during OND season. When the SST indices over Nino1 + 2, Nino3, Nino3.4 and Nino4 regions (http://www.cpc.ncep.noaa.gov/products/analysis_monitoring/ensostuff/nino_regions.shtml) during July–August are correlated with CDs frequency over BOB during OND season the significant CC is found to be with Nino4 SST (5 N–5S and 160E–150 W). The quantitative relationship between SST during July–August and CDs frequency during OND season can be seen further from the CC map shown in Fig. 2e, which show pockets of significant negative CC over the northeast Pacific, western Pacific and also northwest Pacific. Thus, in addition to SST anomaly over Nino4 region one more pocket of SST over the northwest Pacific Ocean is also considered as potential predictor. Based on the regions of significant CCs and contrasting large-scale variables noticed in the composite anomalies the six predictors are identified and is given in Table 1 along with its CC values for the period 1971–2010. As seen from Table 1 the CCs are highly significant (at least 95 %) for all the six predictors.

4 Development of PCR Model for the Forecast of CDs Frequency Over BOB

After identifying the suitable predictors the next step is to develop PCR forecast model for the training period data from 1971 to 2010 and verify the same for the whole period from 1971 to 2013. As shown in Table 1 the six predictors identified are highly correlated with the frequency of CDs during the post-monsoon season over the BOB. The multi-co-linearity reduces the multiple CC when all the predictors are used in a multiple regression (MR) model. To avoid this there is a need to develop a PCR model where the principal components (PCs) of the predictors which are mutually orthogonal in nature and hence independent from one another. The PCR model is developed for the deterministic forecast as well as probability forecast and the verification measures discussed above are used to determine the skill. When these six parameters are used in the PCR model the variance explained by six EOF modes are found to be 55.5 %, 16.6 %, 11.0 %, 7.9 %, 5.5 % and 3.4 %, respectively. Thus, the cumulative variance explained by leading EOF modes indicate the first 2 EOF modes explained more than 72 % of the total variance

Table 1 List of predictors (parameters) considered for the forecast of CDs frequency along with their correlation coefficient during training period (1971–2010)

Predictor No.	Parameters	Bounded area	CC (1971–2010)
1	South-east equatorial Indian Ocean 850 hPa meridional wind (June–August) V850	5S–15S, 100–110E	−0.62
2	Strength of upper level (200 hPa) easterly jet over Africa (July–August) U200	5S–10S, 27.5–40E	−0.54
3	Monsoon zonal wind at 850 hPa (July–August) U850	5 N–10 N, 27.5–37.5E	0.53
4	Strength of Australian high (July–August) SLP	10S–20S, 85–105E	−0.41
5	Sea surface temperature (SST) over the northwest Pacific (July–August) SST1	25 N–30 N, 136–144E	−0.54
6	SST over the Nino 4 region (Jul–August) SST2	5 N–5S, 160E–150 W	−0.41

and the first 3 EOF modes explained more than 83 % and first 4 EOF modes explained more than 90 % of the total variance.

4.1 *Verification of the Deterministic Forecast*

Based on the cross validated PCR model using the six predictors the frequency of CDs predicted during the post-monsoon season of October–December is compared with the observed CDs frequency during the whole period from 1971 to 2013 (Fig. 3). As seen from Fig. 3 the forecast frequency of CDs matches very well with the observed frequency of CDs during most of the years except 1972, 1995, 2002, 2004 and 2011, where the frequency of forecast CDs deviates by more than one from the observed CDs. However, the forecast and observed frequency of CDs matches in rest of the years.

To see the performance of the six parameters PCR model for the forecast of CDs frequency the verifications scores of the deterministic forecast are calculated and is given in Table 2. As seen in column 'a' in Table 2 the CC in case of six parameters PCR model is found to be highly significant both during the training period and also the whole period. Similarly, the other verification scores as shown in Table 2 also indicates negligible mean bias (almost close to zero). The very negligible mean bias indicates, on average the forecast climatology and observed climatology of CDs are very close to each other. The RMSE and MAE also found to be close to 1 or slightly less than 1. The deterministic forecast of CDs frequency for the test period from 2011 to 2013 is found to be 2.5, 3.5 and 4, respectively, which is very close to the observed CDs frequency of 1, 3 and 5 during 2011, 2012 and 2013, respectively.

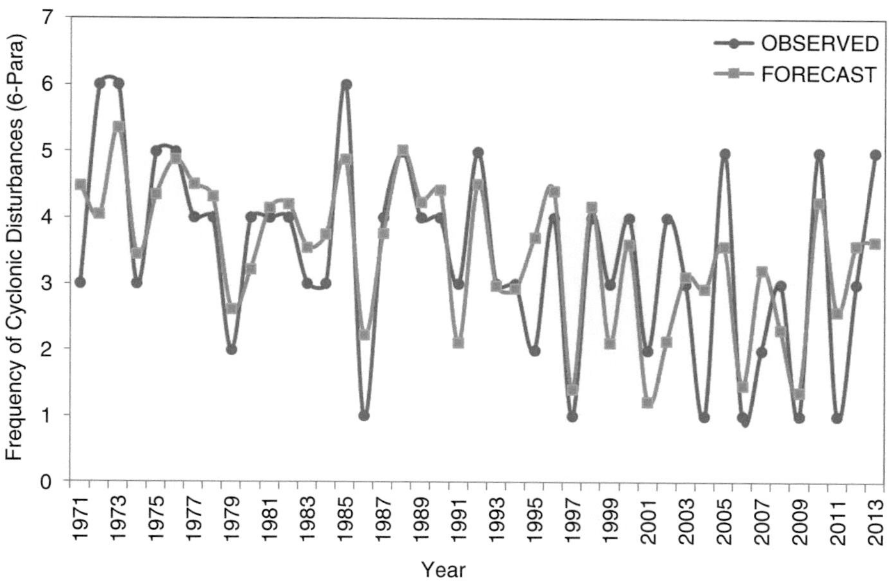

Fig. 3 Verification of deterministic forecast frequency of cyclonic disturbances (CDs) over the Bay of Bengal during post-monsoon season by six parameters PCR model along with corresponding observed frequency of CDs

Table 2 Different verification scores of deterministic and probability forecast indicating the performance of six parameters PCR model for forecast of seasonal frequency of cyclonic disturbances over the Bay of Bengal

Verification scores (training period 1971–2010)	Scores (a)	Verification scores (probability forecast)	Scores (b)
Pearson's correlation coefficient (CC)	0.69	Hit score	62.5 %
% Variance	48.1 %	Hit skill score	43.8 %
Root Mean Square Error (RMSE)	1.01	ROC area (above–normal)	0.87
Mean Absolute Error (MAE)	0.80	ROC area (below–normal)	0.86
Mean bias	−0.01		
CC between observed and forecast frequency during whole period (1971–2013)	0.77		

Thus, although the deterministic forecast captured the trend of the frequency of CDs during 3-year periods correctly there is difficulty in predicting the exact frequency of CDs and the probability forecast may provide some additional useful guidance to the forecasters.

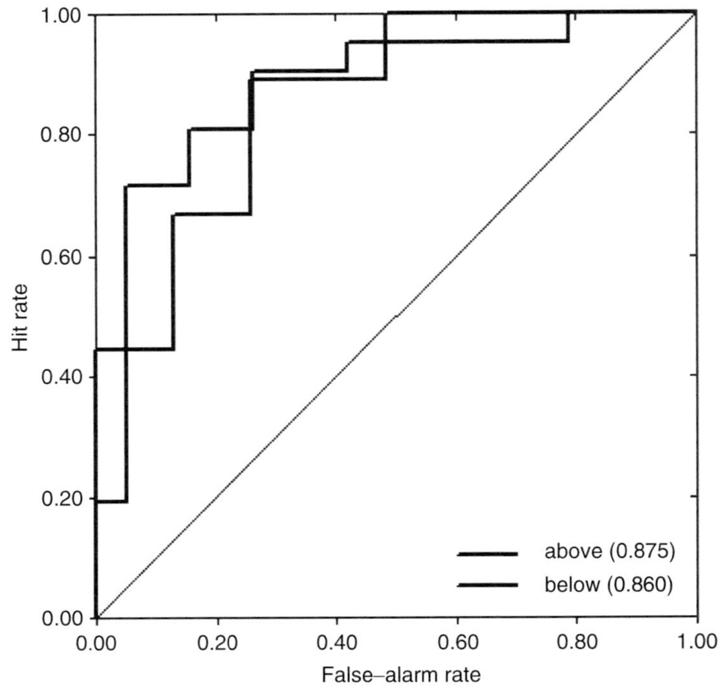

Fig. 4 The Relative Operating Characteristics (ROC) for above and below normal probability forecast of frequency of cyclonic disturbances (CDs) over the BOB

4.2 Verification of the Probability Forecast

For the probability forecast the normal frequency of CDs over the BOB during the post-monsoon season is considered to be between lower tercile (3) and the upper tercile (4). Above normal indicates frequency above 4 and below normal indicates frequency lower than 3 in case of BOB cyclonic disturbances in post-monsoon season. As discussed above the different verification scores considered for the verification of probability forecast is HS, HSS and the ROC. The HS and HSS are calculated during the whole period as given in Table 2 in column 'b', which is found to be 62.5 % and 43.8 %, respectively. As the ROC curve (above normal and below normal) lies above the diagonal line in Fig. 4 the probability forecast is found to be having much better forecast skill than climatology. The ROC curve for above normal and below normal forecast is shown in Fig. 4 with the ROC areas of above normal (below normal) is found to be 0.87 (0.86) respectively.

The probability forecast of CDs frequency in terms of below normal, normal and above normal probability during the test period from 2011 to 2013 is almost captured in the model. The observed frequency of CDs during 2011, 2012 and 2013 is found to be 1, 3 and 5 respectively, which is below normal (BN), normal (N) and above normal (AN), respectively. The below normal CD frequency during

2011 OND season is very well captured in six parameters PCR model with dominant 62 % probability of 'BN' category followed by 23 % of 'N' and 14 % of 'AN'. Similarly, the slight above normal frequency of CDs during 2013 was also well captured with highest probability under the 'AN' category (38 %) followed by 34 % of 'N' and '28 %' of 'BN' category. The normal frequency during 2012 is also captured reasonably well with all the three categories very closely distributed with no dominant category.

5 Conclusions

A PCR model is developed for the seasonal forecast of frequency of CDs over the Bay of Bengal during October–December season by using six predictors. The performance of the PCR model is found to be highly significant both for the deterministic forecast and probability forecasts. The CC between forecast CD frequency and observed CD frequency is found to be highly significant with a value of about 0.7 during the whole period and RMSE and mean absolute error is found to be close to 1. The forecast mean bias is negligible and very close to zero indicating the average mean forecast and average mean observation is very close to each other. With respect to the verification of probability forecast the HS is found to be about 63 %. The ROC curves for above normal and below normal lies above the diagonal line and is found to be having much better forecast skill than climatology with the ROC areas for above normal and below normal being 0.87 (0.86) respectively. The below normal CD frequency during 2011 season is very well captured in six parameters model with dominant 62 % probability of below normal category was indicated. Similarly, the slight above normal frequency of CDs during 2013 was also well captured with highest probability (38 %) under the above normal category. The normal frequency during 2012 is also captured reasonably well with all the three categories very closely distributed.

Acknowledgements The author is very much grateful to the Director General of Meteorology for allowing him to participate in the National Conference on cyclone held in New Delhi in July 2014. Thanks are due to IRI, USA for making the CPT available to us, which was used in our study. Thanks are also due to the NCEP reanalysis Centre for providing the wind data used in the present study.

References

Balachandran, S. and Geetha, B. (2012). Statistical prediction of seasonal cyclonic activity over North Indian Ocean. *Mausam*, **63** (**1**), 17–28.

Chan, J.C, Shi, J.E. and Lam, C.M. (1998). Seasonal forecasting of tropical cyclone activity over the western North Pacific and the South China Sea. *Weather Forecasting*, **13**, 997–1004.

Chan, J.C, Jiu-En Shi and Kin Si Liu (2001). Improvements in the seasonal forecasting of tropical cyclone activity over the Western North Pacific. *Weather Forecasting*, **16**, 491–198.
Camargo, S.J, Barnston, A.G. and Zebiak, S.E. (2005). A statistical assessment of tropical cyclones in atmospheric general circulation models. *Tellus*, **57A**, 589–604.
Girishkumar M.S. and Ravichandran, M. (2012). The influences of ENSO on tropical cyclone activity in the Bay of Bengal during October–December. *Journal of Geophysical Research*, **117**, C02033, doi: 10.1029/2011JC007417.
Gray, W.M. (1984a). Atlantic seasonal hurricane frequency: Part I: El Nino and 30 mb quasi-biennial oscillation influences. *Monthly Weather Review*, **112**, 1649–1668.
Gray, W.M. (1984b). Atlantic seasonal hurricane frequency: Part II: Forecasting its variability. El Nino and 30 mb quasi-biennial oscillation influences. *Monthly Weather Review*, **112**, 1669–1683.
Kalnay, E, Kanamitsu, M, Kistler, R, Collins, W, Deaven, D, Gandin, L, Iredell, M, Saha, S, White, G, Woollen, J, Zhu, Y, Leetmaa, A, Reynolds, R, Chelliah, M, Ebisuzaki, W, Higgins, W, Janowiak, J, Mo, K.C, Ropelewski, C, Wang, J, Roy, J. and Joseph, D. (1996). The NCEP/NCAR 40- year reanalysis project. *Bulletin of the Americal Meteorological Society*, **77**, 437–471.
Kikuchi, K. and Wang, B. (2010). Formation of tropical cyclones in the northern Indian Ocean associated with two types of tropical intraseasonal oscillation modes. *Journal of the Meteorological Society of Japan*, **88**, 475–496, doi:10.2151/jmsj.2010-313.
Krishnakumar, G, Rajeevan, M. and Gadgil, A.S. (2007). Use of wind and temperature anomalies over the Indian Ocean in the estimation of cyclonic disturbance activity. *Vayu Mandal*, **33**, Jan–Dec 2007.
Michaelsen, J. (1987). Cross-validation in statistical climate forecast models. *Journal of Climate and Applied Meteorology*, **26**, 1589–1600.
Mohapatra, M. and Adhikary, S. (2010). Seasonal prediction of cyclonic disturbances over Bay of Bengal during summer monsoon season: Identification of potential predictors. *Mausam*, **61**, 469–486.
Mohapatra, M. and Adhikary, S (2011). Modulation of cyclonic disturbances over the north Indian Ocean by Madden – Julian oscillation. *Mausam*, **62** (**3**), 375–390.
Nicholls, N. (1979). A possible method for predicting seasonal tropical cyclone activity in the Australian region. *Monthly Weather Review*, **107**, 1221–1224.
Mohapatra, M, Mandal, G.S, Bandyopadhyay, B.K, Tyagi, A. and Mohanty, U.C. (2012). Classification of cyclone hazard prone districts of India. *Natural Hazards*, **63**, doi:10.1007/s11069-011-9891-8.
Mohapatra, M. (2015). Cyclone hazard proneness of districts of India. *Journal of Earth System Sciences*, **124**, 515–526, doi:10.1007/s12040-015-0556-y.
Pattanaik, D.R. (2005). Variability of oceanic and atmospheric conditions during active and inactive periods of storm over the Indian Region. *International Journal of Climatology*, **25**, 1523–1530.
Rajeevan, M, Guhathakurta, P. and Thapliyal, V. (2000). New models for long range forecasts of summer monsoon rain fall over Northwest and Peninsular India. *Meteorology and Atmospheric Physics*, **73**, 211–225.
Singh, O.P. and Pai, D.S. (1996). An oceanic model for the prediction of SW monsoon Rainfall Over India. *Mausam*, **47**, 91–98.
Swets, J.A, Dawes, R.M. and Monahan, J. (2000). Better decisions through science. *Scientific American*, **283**, 82–87.
Vitart, F, Anderson, J.L. and Stern, W.F. (1997). Simulation of the interannual variability of tropical storm frequency in an ensemble of GCM integrations. *Journal of Climate*, **10**, 745–760.
WMO (2002). Standardised Verification System (SVS) for Long-Range Forecasts (LRF). New attachment II-9 to the Manual on the GPDS (WMO-No.485), Volume 1. Available on the internet at http://www.wmo.ch/web/www/DPS/LRF-standardised-verif-sys-2002.doc.

Tropical Cyclogenesis Prediction in the North Indian Ocean During 2013 Using OSCAT Derived Surface Wind Observations

N. Jaiswal, C.M. Kishtawal, and P.K. Pal

1 Introduction

The Indian sub-continent is one of the most adversely affected TC active basins that experience on an average 4–5 TCs every year. In comparison to other TC basins, this region is the most vulnerable due to relatively dense coastal population, shallow bottom topography and coastal configuration. Though the TCs formed in this region are considered to be much weaker in intensity and smaller in size as compared to other region, yet the number of deaths in the region is highest in the globe (300,000 human deaths were estimated from TC associated storm surge in Bangladesh in 1970). Out of nine recorded cases of heavy loss of human lives (~40,000) by TCs during last 300 years, seven cases (77 %) occurred in Indian sub-continent (Frank and Husain 1971). To overcome such loss, the advance predictions of TC in terms of their genesis, track and intensity is highly important. These advance timely information can save the life of people and help in decision making for taking the preventive measures like evacuation during the TC landfall. The predictions of TC are generated based on the models, using the satellite observations and ground based radar networks when TC reaches close to the land (Mohapatra et al. 2013a). Due to the advancements in numerical prediction models and, high temporal and spatial satellite observations, during the last decades the track prediction accuracy has been improved drastically (Mohapatra et al. 2013b, 2015). However, the predictions of TC genesis and intensity are still challenging as in TC predictions (Mohapatra et al. 2013c), the requirements of accuracy in genesis, track and intensity are higher compared to normal numerical weather predictions. Skillful prediction of tropical cyclogenesis will be of great benefit for the affected

N. Jaiswal (✉) • C.M. Kishtawal • P.K. Pal
Atmospheric and Oceanic Sciences Group, EPSA, Space Applications Centre (ISRO), Ahmedabad, Gujarat 380015, India
e-mail: neeru@sac.isro.gov.in

© Capital Publishing Company 2017

M. Mohapatra et al. (eds.), *Tropical Cyclone Activity over the North Indian Ocean*,
DOI 10.1007/978-3-319-40576-6_14

communities as it may provide sufficient time for planning and preparations (e.g. Brunet 2010). Due to unavailability of conventional observations, satellite data are being employed and found to be useful tool to study and understand tropical cyclogenesis (Liu et al. 1995; Zehr 1998; Katsaros et al. 2001; Sharp et al. 2002; Li et al. 2003; Chelton et al. 2004; Wang et al. 2007, 2008). The microwave scatterometers onboard polar-orbiting satellites have helped to study the early stages of cyclogenesis (Sharp et al. 2002). Sea-winds scatterometer, aboard the QuikScat satellite that infers surface wind speed and direction, has encouraged various studies regarding early identification of tropical disturbances (Sharp et al. 2002; Li et al. 2003). The wind pattern matching based technique using scatterometer derived surface wind have been discussed by Jaiswal and Kishtawal (2011) and Jaiswal et al. 2013. The above technique depends on the availability of coverage of the cyclonic disturbance during the satellite overpass. In the present study, the cyclogenesis prediction of the systems formed during year 2013 has been discussed using the OSCAT derived surface wind vector based cyclogenesis prediction algorithm. The brief methodology is provided in Sect. 2. The results are discussed in the Sect. 3 and the conclusions of the study are given in Sect. 4.

2 Data Used and Methodology

2.1 Data Used

The data used in this study is the surface wind observations which have been obtained from the scatterometers viz., QuikSCAT and OSCAT. The QuikSCAT derived wind fields have been used to prepare the database of the wind patterns during the cyclogenesis stage of the system, while the OSCAT observations have been used to provide the real time predictions. In the preparation of database, the cyclogenesis stage has been taken as 1–4 days prior to the declaration of TC as a tropical storm ($\geq$34 knots) by Joint Typhoon Warning Centre (JTWC). QuikSCAT data for the cyclogenesis stage of TCs formed in the North Indian Ocean during the year 2000–2009 have been obtained from 'www.ssmi.com'. The OSCAT derived wind data was obtained from the Oceansat-2 data portal of National Remote Sensing Centre (NRSC), India. This data can be accessed through NRSC website 'www.nrsc.gov.in'. The study area used in the present study is the North Indian Ocean basin (0°–30° N and 50°–100° E). The summary of TCs formed during year 2013 in the North Indian Ocean has been given in Table 1 and their tracks are depicted in Fig. 1.

Table 1 Summary of TC formed during year 2013 in North Indian Ocean

TC Name	Life duration	Maximum wind speed (knots)
Viyaru	11–16 May 2013	50
Phailin	09–12 October 2013	140
Helen	19–22 November 2013	60
Lehar	23–28 November 2013	75
Madi	6–12 December 2013	70

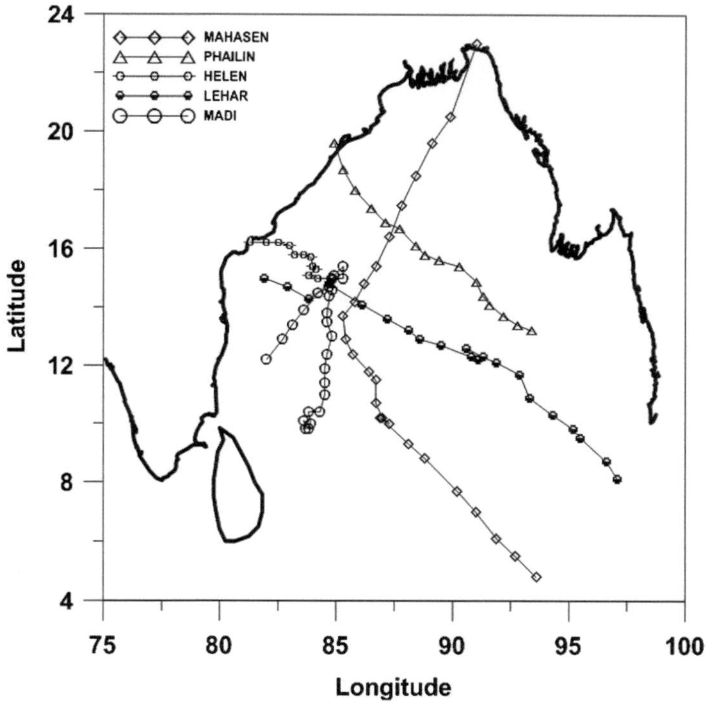

Fig. 1 Track of TCs formed in the North Indian Ocean during year 2013

2.2 *Methodology*

The technique used in this work is inspired by the pattern matching approach. It is based on the assumption that there is some degree of similarity in low-level wind circulation among the low pressure systems that could develop into TCs (the developing systems) at later stages, which can be utilised to distinguish them from non-developing systems. This similarity of wind patterns is measured quantitatively using a vector (defined as complex numbers) pattern correlation coefficient, termed as the 'matching index'.

For the pattern matching based technique, the first requirement is to have a rich reference database. Wind pattern of low-level circulations during the cyclogenesis period, observed by scatterometer onboard QuikSCAT satellite have been used to prepare a reference database. Total numbers of 251 QuikSCAT scenes, corresponding to the early stages (0–96 h before the declaration of system as tropical storm by JTWC) of TCs (intensity $\geq$ 34 knots) formed in the North Indian Ocean basin during the period 2000–2008 were found to be available. The scenes that covered the circulation partially (< 60 %) were discarded. In our earlier work we used vorticity values for fast and automatic identification of such patterns (Jaiswal and Kishtawal 2011). In some situations, vorticity may exist, without a well defined cyclonic circulation which is a pre-requisite for cyclogenesis (Katsaros et al. 2001). Thus in the subsequent work, a different approach was used to select the representative samples of circulation from the QuikSCAT data that develop into TC (Jaiswal et al. 2013). In this approach a synthetic TC template was generated using the following function (Chan and Williams 1987):

$$V(r) = V_{\mathrm{m}}\left(\frac{r}{R_{\mathrm{m}}}\right)\exp\left[\frac{1}{b}1 - \left(\frac{r}{R_{\mathrm{m}}}\right)^{\mathrm{b}}\right] \tag{1}$$

where V is a tangential velocity at radius r; V_m is the maximum value of tangential velocity; R_m is the radius at which V_m occurs and b is shape parameter. These synthetic TC templates were matched with a randomly selected one of the QuikScat scene, to identify the cyclonic wind circulation. The vector wind fields of the synthetic TCs were matched to the QuikScat winds, using the following expression in a vector block wise manner.

$$cc = \frac{\frac{1}{N}\sum_{i=1}^{N}\left(A_i - \overline{A}\right) \times \left(B_i - \overline{B}\right)}{\sqrt{\frac{1}{N}\sum_{i}^{N}\left(\left|A_i - \overline{A}\right|\right)^2 \times \sqrt{\frac{1}{N}\sum_{i}^{N}\left(\left|B_i - \overline{B}\right|\right)^2}}} \tag{2}$$

where $\overline{A}$ and $\overline{B}$ represents the mean value of the complex vectors A and B, respectively. N is the dimension of vector A (or B) and, A and B are the complex numbers formed using the wind vectors (for example $A = (u + iv)$). The above approach of matching wind fields yields high values ($>> 0.5$), if cyclonic circulations are detected in the QuikScat scenes. It can be noticed that this approach emphasises more on cyclonic circulations, almost irrespective of the structural parameters (e.g. b, R_{m} and V_{m}) of synthetic TCs. Therefore, the values of V_{m}, R_{m} and b were fixed as 20 knots and 100 m and 0.5, respectively, and this single template was used for all the matches to form the database. Even though there may be large variability in cyclonic circulation in QSCAT data, single template was found sufficient to detect a pattern automatically without manual interface. Template generated by Eq. 2 yields high correlations whenever there was a cyclonic circulation in QuikSCAT data, irrespective of the intensity of circulation. Above template of cyclonic wind circulation was matched to each of the above QuikSCAT derived wind fields around the location, where JTWC reported the formation of TC. A block of size 31 × 31 (7.75° longitude and 7.75° latitude) around the point, that was matched to the above template, was selected and the wind vectors in this region were stored as the reference scene in database. The formed database was assumed to be the representative of the wind vector signatures of various cyclogenesis stages, which turned into TCs of different intensities.

A cyclogenesis in a given wind field of a scatterometer scene, was detected by matching its wind vectors with all the reference scenes, in database using the Block Matching Algorithm (BMA). BMA is a commonly used algorithm for searching similarity of patterns in the images, and signal processing applications (Lu and Liou 1997). In the present work, BMA is used to find the similarity between the wind vectors of two scenes. The algorithm automatically detects the cyclonic circulation in the region of interest, and also computes its similarity with the archived scenes in terms of 'matching index'. First the region of interest (ROI) covering the North Indian basin (40°–100° E/0°–30° N) was selected. For the prediction of possibility of cyclogenesis, ROI (analysing scene) was matched with all the archived scenes using vector BMA. A block (analysing block) was selected in an analysing scene, and matched to a scene in the database (reference block). For the matching purpose, the vectors should be of similar dimensions thus, the size of an analysing block was taken similar to the size of reference block (31 × 31). The data points in the blocks represent the wind vector with two components (u and v). A block was excluded from the analysis, if it contains the undefined values (land area or uncovered pass) in more than 50 % of the block area. In this way, the two vectors A and B were constructed, with wind components associated to the analysing and reference blocks, respectively. The matching between the vector fields was determined by considering them as complex variables, (for example $A = (u + iv)$). The matching index (c) between the two sets of such complex numbers was computed as follows:

$$c = 1 - \frac{\frac{1}{N}\sum_{i=1}^{N}(A_i - B_i)^2}{\sqrt{\frac{1}{N}\sum_{i}^{N}\left(\left|A_i - \overline{A}\right|\right)^2} \times \sqrt{\frac{1}{N}\sum_{i}^{N}\left(\left|B_i - \overline{B}\right|\right)^2}} \tag{3}$$

where $\overline{A}$ and $\overline{B}$ represents the mean value of the complex vectors A and B respectively. N is the dimension of vector A (or B), which is the number of valid pairs between two scenes to be matched. The analysing block was slided over the whole ROI and at each shifts, the 'c' between the data of the two blocks was computed. The matching was performed with all the archived scenes in the database. The maximum value of matching index (c_{max}) was determined and compared with the threshold value.

Before using the above technique of wind pattern matching, the threshold used for tropical cyclogenesis prediction was optimised, as a smaller value of c_{max} would result in undesirable false alarms and a high value of c_{max} may miss some potential genesis events. The non-developing systems generally result in small values of c_{max}, while for the developing systems, the value of c_{max} was found to be higher than a certain threshold. If 'c_{max}' was found to exceed the threshold limit, the system was identified as developing. However, no assessment can be made if the matching index was less than the threshold limit. The optimum value of threshold c_{max} for matching index was determined using the QuikSCAT data of TC active months of the years 2007–2009. Winds of both the passes (ascending and descending) for all the days during the years 2007–2009 have been matched with archived scenes in a database and c_{max} was computed. By minimising the false alarm rate (FAR) and maximising the probability of detection (POD) the optimum value of threshold was obtained as 0.6 for predicting tropical cyclogenesis.

3 Results and Discussion

Based on the value of matching index obtained using the above discussed wind pattern matching approach, the cyclogenesis was predicted in the real-time during the TC active season of the study area i.e. during April–June and October–December 2013 using the OSCAT derived winds. OSCAT provides daily two passes over Arabian Sea and two passes over the Bay of Bengal. For the better representation, the results of Arabian Sea and Bay of Bengal have been plotted separately. The time series of matching indices over the Arabian Sea region for the months of April–June (91 days i.e. 182 OSCAT passes) and October–December (92 days i.e. 184 OSCAT passes) have been shown in the Figs. 2a and 2b, respectively. The threshold limit (0.6) has been shown with the bold horizontal (red) line. The values exceeding the threshold line represent the prediction of cyclogenesis. During the year 2013, no TC was formed in the Arabian Sea, however, few low pressure systems were formed that could not develop further. Few of these low pressure systems were predicted as a cyclogenesis, and these false cases can be seen in the Figs. 2a and 2b.

The time series of matching indices obtained in the OSCAT passes of each day over the Bay of Bengal region for the months April–June and October–December have been shown in the Figs. 2c and 2d, respectively. The days on which, the TCs formed have been marked with the broken vertical lines. It can be seen that all the five TCs have been predicted before their formation into tropical storm. However, there were few false predictions that could not turn into TC.

The first named TC in the North Indian Ocean in the year 2013 was TC Viyaru. TC Viyaru was designated as cyclonic storm on 11 May 2013 by Regional Specialised Meteorological Centre, New Delhi. The present technique predicted the cyclogenesis of this system on 6 May 06Z i.e. 4 days in advance as the matching index was obtained as 0.61, which was exceeding the threshold value (0.6). The OSCAT pass on 6 May 06Z, showing the low-level circulation corresponding to the genesis of TC Viyaru (marked with a box) has been shown in the Fig. 3a.

The TC Phailin formed on 9 October 2013 and turned into very severe tropical TC. It made landfall near the Odisha–Andhra coast on October 12 as a very severe TC. The cyclogenesis prediction of this system was done in 1 day advance i.e. on

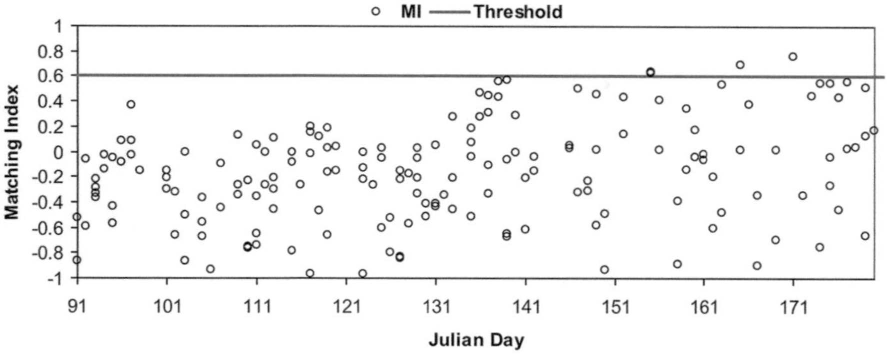

Fig. 2a Matching index obtained during April–June 2013 in the Arabian Sea

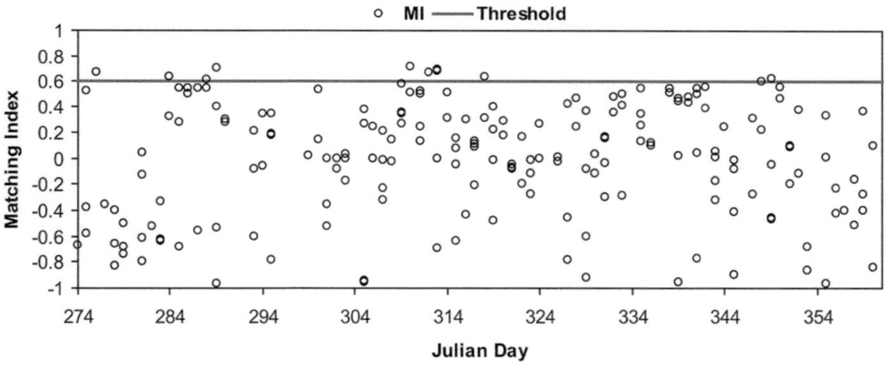

Fig. 2b Matching index obtained during October–December 2013 in the Arabian Sea

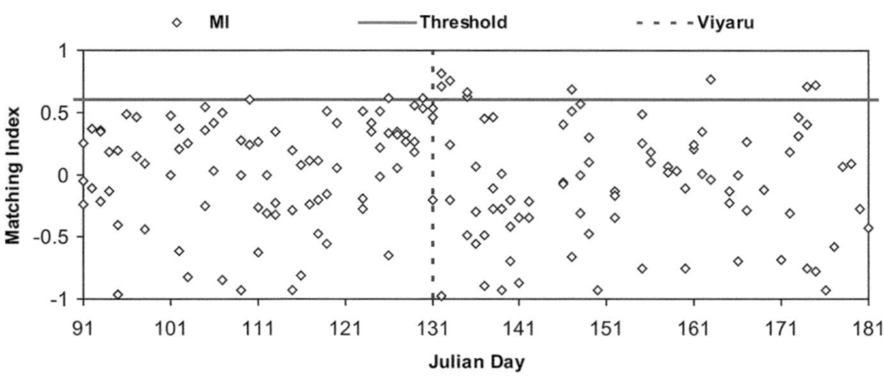

Fig. 2c Matching index obtained during April–June 2013 in the Bay of Bengal

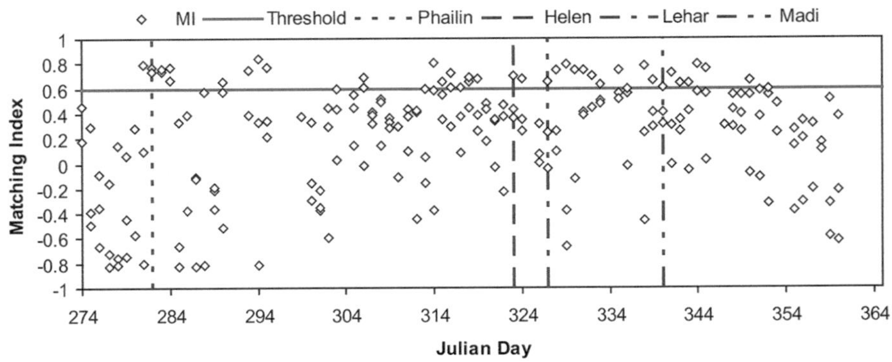

Fig. 2d Matching index obtained during October–December 2013 in the Bay of Bengal

8 October, as the matching index was found as 0.79. The system was lying over the land before that day and thus can not be captured by OSCAT winds. The OSCAT pass on 8 October 16Z, showing the low-level circulation corresponding to the genesis of TC Phailin has been shown in the Fig. 3b.

The TC Helen was the next TC that formed on 19 November 2013 in the study region and made landfall on 23 November at the Andhra coast. The earliest signature of the formation of this system was predicted by the present model before 9 days, ahead of its formation i.e. on 10 November, as the matching index was 0.8. The OSCAT pass on 10 November 16Z, showing the low-level circulation corresponding to the genesis of TC Helen has been shown in the Fig. 3c.

After TC Helen, TC Lehar formed on 23 November 2013 in the Bay of Bengal. The earliest detection of the formation of this system was predicted on that day itself using the above approach. The OSCAT pass on 23 November 16Z, showing the low-level circulation corresponding to the genesis of TC Lehar has been shown in the Fig. 3d. The TC Lehar turned into the very severe TC category and made landfall on 28 November as a depression.

TC Madi which formed on 6 December in the Bay of Bengal, was predicted 6 days ahead by the present model i.e. on 30 November, as the matching index was obtained as 0.63. The TC Madi has very unique track as it was moving northerly after its formation till 15.7° N and then re-curved southwestwards to Tamil Nadu coast. The OSCAT pass on 29 November 16Z, showing the low-level circulation corresponding to the genesis of TC Madi is shown in the Fig. 3e.

The above results show that the cyclogenesis of all the five TCs formed during the year 2013 in the North Indian Ocean viz., Viyaru, Phailin, Helen, Lehar and Madi have been predicted before 4, 1, 9, 1 and 6 days in advance, respectively. The mean lead time of cyclogenesis prediction has been computed as 78 h based on the

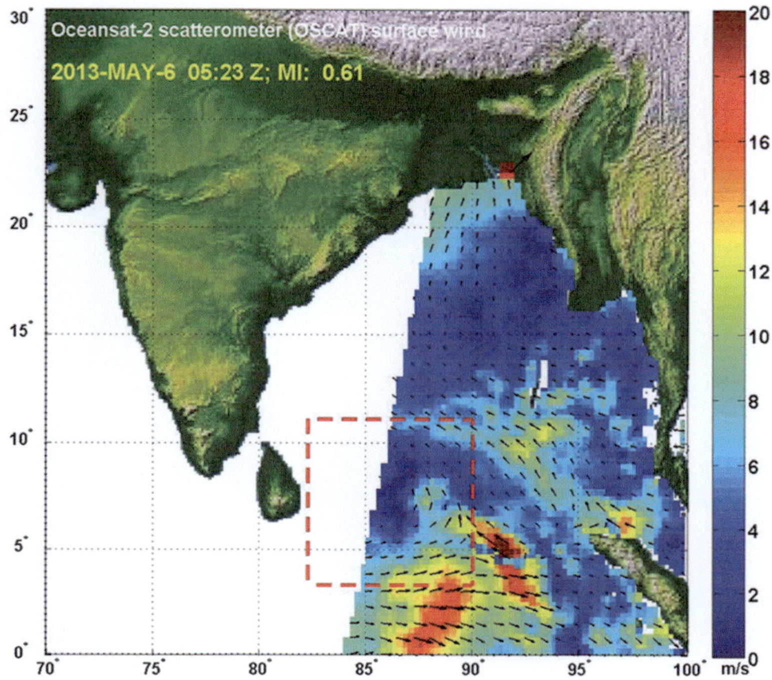

Fig. 3a Earliest detection of cyclogenesis of TC Viyaru (6 May 2013, 0530Z)

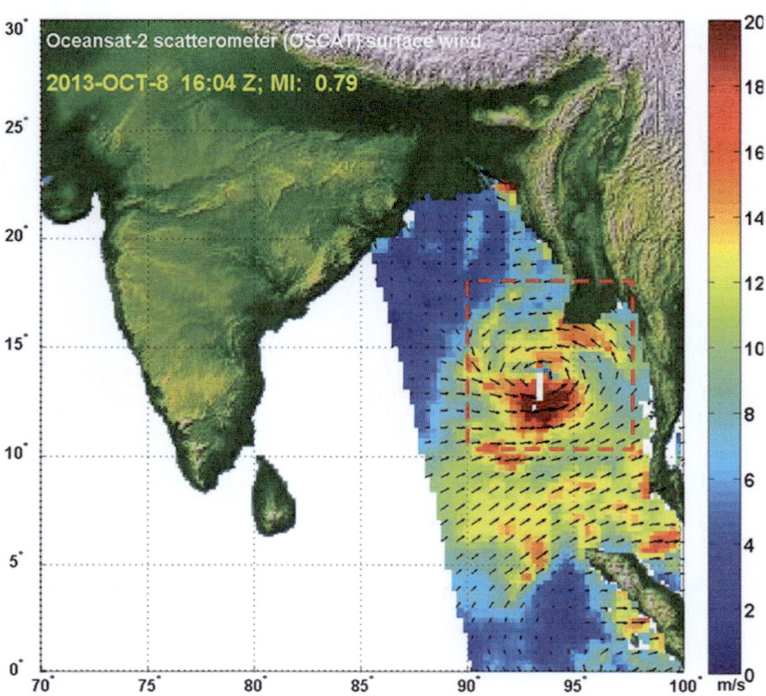

Fig. 3b Earliest detection of cyclogenesis of TC Phailin (8 October 2013, 1600 Z)

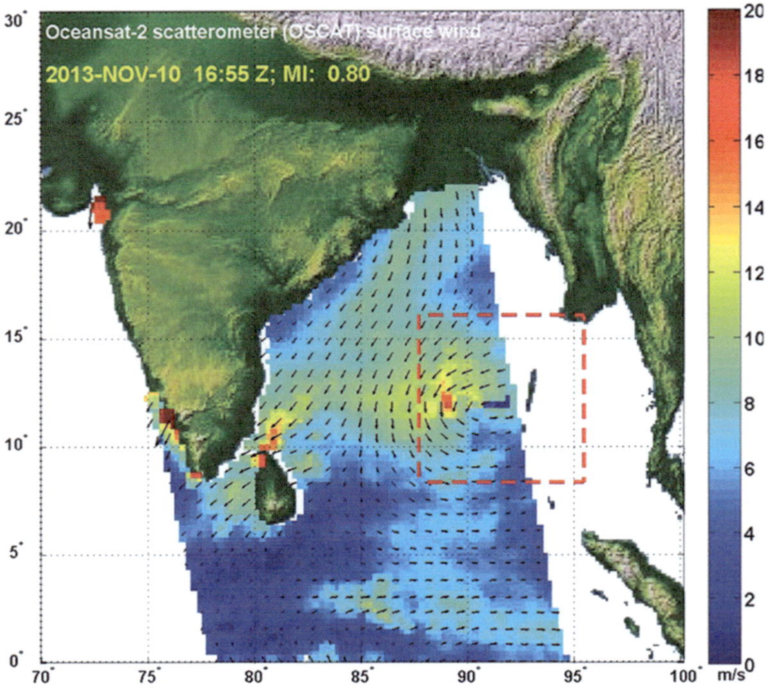

Fig. 3c Earliest detection of cyclogenesis of TC Helen (10 November 2013, 1700 Z)

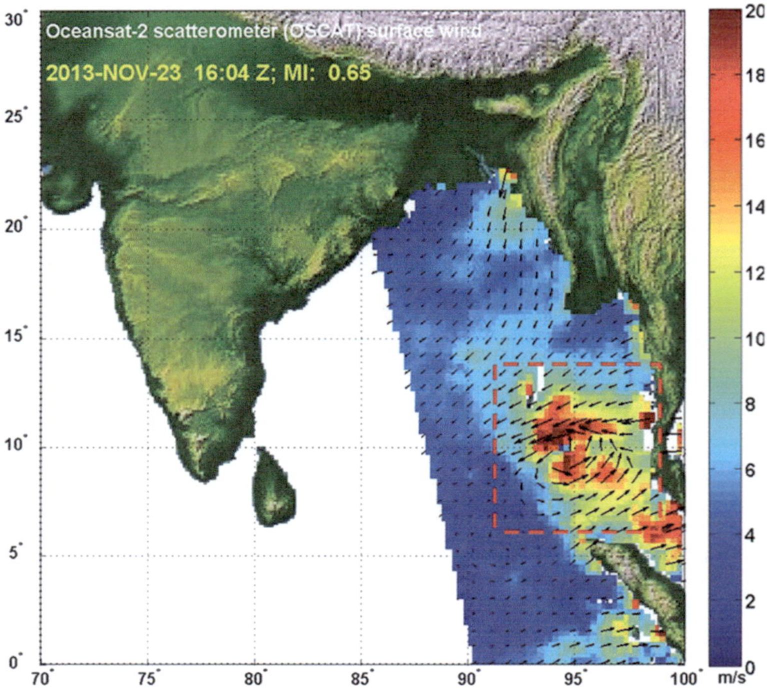

Fig. 3d Earliest detection of cyclogenesis of TC Lehar (23 November 2013, 1600Z)

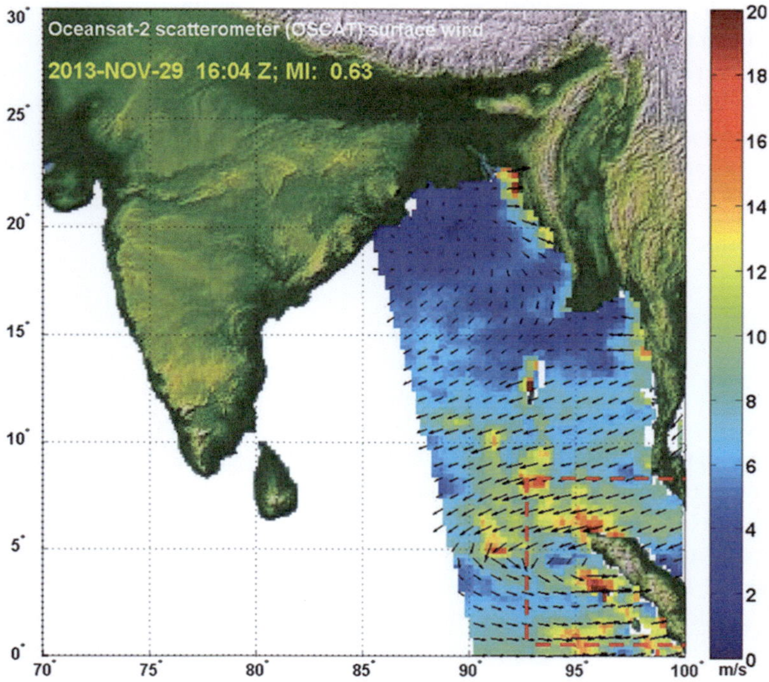

Fig. 3e Earlierst detection of cyclogenesis of TC Madi (29 November 2013, 1600Z)

Table 2 Four cell contingency table for cyclogenesis prediction

	Observed = Yes	Observed = No
Forecast = Yes	Hits = 5	False = 31
Forecast = No	Miss = 0	Correct negative = 849

above 5 cases. The four-cell contingency Table has been formed and given in the Table 2 to estimate the accuracy of the technique. The probability of detection (POD), FAR, and the accuracy of the technique has been determined using the following expressions.

$$\text{Probability detection (POD)} = \frac{X}{X+Y} \tag{4}$$

$$\text{Falscalarm rate (FAR)} = \frac{Z}{Z+W} \tag{5}$$

$$\text{Accuracy} = \frac{X+W}{X+Y+Z+W} \tag{6}$$

where, X is the number of hits, Y is the number misses, Z is the number of false detection and W is the number of correct negative detection.

As all the five systems have been predicted by this technique, the value of hits (X) is 5 and miss (Y) is 0. Thus the probability of detection (POD) is computed as 1. There were also few false alarms during the year 2013, which have been shown in the Figs. 2a, 2b, 2c, and 2d. It can be seen from the Table 2 that there were total 31 scenes that turn out as a false cases. The number of false cases (Z) is 31, and the correct negative detection (W) is 849. Thus the value of FAR was computed as 0.035 (i.e. 3.5 %). The accuracy of the technique has been obtained as 96.5 %.

4 Conclusions

The pattern matching based technique to detect the cyclogenesis using Indian satellite Ocensat-2 onboard scatterometer derived surface wind vectors have been discussed in this study. The main feature of this technique is that it does not depend on many data sources and do not need much computational resources. The algorithm is fast (30–45 s in the single processor computer) and generates the results just after the acquisition of satellite data. The only limitation of the technique is the coverage of the developing system within the satellite swath. The cyclogenesis of five tropical TCs that developed in the year 2013 were predicted in the real-time, well in advance before there formation into tropical storm category using the OSCAT observations. The prediction lead time of cyclogenesis of TC Viyaru, Phailin, Helen, Lehar and Madi was 4, 1, 8, 1 and 6 days, respectively using the above approach. The mean prediction lead time of the technique was estimated as 78 h. The probability of detection was computed as 100 %, whereas the FAR was estimated as 3.5 %. The accuracy of the technique for the year 2013 is found as 96.5 %. The performance of the technique can be further improved by including other parameters e.g. the wind shear values and oceanic conditions.

Acknowledgements The authors are grateful to Remote Sensing Systems for providing the QuikScat satellite data (www.ssmi.com). The authors would like to thank Joint Typhoon Warning Center (http://www.usno.navy.mil) for providing the real-time TC updates and the archive for TC warnings issued in the real-time. Authors thank Mr. Aman Waheed Khan for the proof reading of manuscript.

References

Brunet, G. (2010). Collaboration of the weather and climate communities to advance sub-seasonal to seasonal prediction. *Bulletin of the American Meteorological Society*, **91**, 1397–1406.

Chan, J.C.L. and Williams, R.T. (1987). Analytical and numerical studies of the beta-effect in tropical motion. *Journal of Atmospheric Sciences*, **44**, 1257–1265.

Chelton, D.B, Schlax, M.G, Freilich, M.H. and Milliff, R.F. (2004). Satellite measurements reveal persistent small-scale features in ocean winds. *Science*, **303**, 978–983.

Frank, N.L. and Husain, S.A. (1971). The Deadliest TC in History. *Bulletin of the American Meteorological Society*, **52**, 438–445.

Jaiswal, N. and Kishtawal, C.M. (2011). Prediction of tropical cyclogenesis using scatterometer data. *IEEE Transaction on Geoscience and Remote Sensing*, **49**, 4904–4909.

Jaiswal, N, Kishtawal, C.M. and Pal, P.K. (2013). Prediction of tropical cyclogenesis in North Indian Ocean using Oceansat-2 scatterometer (OSCAT) winds. *Meteorology and Atmospheric Physics*, ***119***, 137–149.

Katsaros, K.B, Forde, E.B, Chang, P. and Liu, W.T. (2001). QuikSCAT's SeaWinds facilitates early identification of tropical depressions in 1999 hurricane season. *Geophysical Research Letters*, **28**, 1043–1046.

Li, T, Fu, B, Ge, X, Wang, B. and Peng, M. (2003). Satellite data analysis and numerical simulation of TC formation. *Geophysical Research Letters*, **30**, 2122.

Liu, G, Curry, J.A. and Clayson, C.A. (1995). Study of tropical cyclogenesis using satellite data. *Meteorology and Atmospheric Physics*, **56**, 111–123.

Lu, J. and Liou, M.L. (1997). A simple and efficient search algorithm for block matching motion estimation. *Circuits and Systems for Video Technology, IEEE transactions on*, **7**, 429–433.

Mohapatra, M, Nayak, D.P, Sharma, R.P. and Bandyopadhyay, B.K. (2013a). Evaluation of official tropical cyclone track forecast over north Indian Ocean issued by India Meteorological Department. *Journal of Earth System Sciences,* **122**, 433–451, doi:10.1007/s12040-013-0291-1

Mohapatra, M, Bandyopadhyay, B.K. and Nayak, D.P. (2013b). Evaluation of operational tropical cyclone intensity forecasts over north Indian Ocean issued by India Meteorological Department. *Natural Hazards*, **68**, 589–601, doi:10.1007/s11069-013-0624-z

Mohapatra, M, Sikka, D.R, Bandyopadhyay, B.K. and Tyagi, A. (2013c). Outcomes and challenges of Forecast Demonstration Project (FDP) on landfalling cyclones over Bay of Bengal. *Mausam,* **61**, 1–15.

Mohapatra, M, Nayak, D.P, Sharma, M, Sharma, R.P. and Bandyopadhyay, B.K. (2015). Evaluation of official tropical cyclone landfall forecast issued by India Meteorological Department. *Journal of Earth System Sciences*, **124**, 861–874, doi:10.1007/s12040-015-0581-x

Sharp, R.J, Bourassa, M.A. and Brien, J.J.O. (2002). Early detection of tropical TCs using SeaWinds-derived vorticity. *Bulletin of the American Meteorological Society*, **83**, 879–889.

Wang, L, Lau, K.H, Fung, C.H. and Gan, J.P. (2007). The relative vorticity of ocean surface winds from the QuikSCAT satellite and its effects on the geneses of tropical TCs in the South China Sea. *Tellus (Ser. A)*, **59**, 562–569.

Wang, L, Lau, K.H, Zhang, Q.H. and Fung, C.H. (2008). Observation of nondeveloping and developing tropical disturbances over the South China Sea using SSM/I satellite. *Geophysical Research Letters*, **35**, 1–5.

Zehr, R. M. (1998). Satellite diagnosis of Tropical Cyclones. Proceedings of 3rd Conference on Satellite Meteorology and Oceanography, Anaheim, CA, pp. 241–246.

The Influence of Madden-Julian Oscillation on the Bay of Bengal Tropical Cyclogenesis During the Year 2013

P.C.S. Rao, S.A. Nair, and M. Khole

1 Introduction

The aim of the study is to correlate the role of the Madden-Julian Oscillation (MJO) with the genesis of the cyclonic storms that occurred during the year 2013 in the Indian Ocean/Bay of Bengal (BOB) region. The MJO is the major fluctuation in tropical weather on weekly to monthly timescales. This eastward propagating tropical disturbance was first identified by Madden and Julian (1971).The MJO can be characterised as an eastward moving 'pulse' of cloud and rainfall near the equator that typically recurs every 30–60 days and has a global wave number one structure with a spatial wavelength of roughly 12,000–20,000 km (Nakazawa 1986). The MJO is characterised by an eastward progression of large regions of both enhanced and suppressed tropical rainfall, observed mainly over the Indian and Pacific Ocean. This oscillation is associated with eastward-moving convection with an enhanced convective phase followed by a non-convective (suppressed convection) phase at any one location. The low-level wind fields associated with MJO are characterised by fluctuations between easterly and westerly phases and can be observed in several other parameters such as surface pressure, upper tropospheric wind, and proxies for deep convection such as outgoing long-wave radiation (OLR).

As per a study carried out by Krishnamohan et al. (2012), 118 cyclones were identified which satisfy the intensity criteria for NIO basin from 1979 to 2008.

P.C.S. Rao (✉) • S.A. Nair • M. Khole
India Meteorological Department, Pune, India
e-mail: pcsraoimd@gmail.com

© Capital Publishing Company 2017

M. Mohapatra et al. (eds.), *Tropical Cyclone Activity over the North Indian Ocean*,
DOI 10.1007/978-3-319-40576-6_15

2 Data and Methodology

The real-time multivariate MJO index (RMM) developed by Wheeler and Hendon (2004) available at http://cawcr.gov.au/staff/mwheeler/maproom/RMM/ is used in this study. This index is developed based on the first two empirical orthogonal functions (EOF) of the combined fields of equatorially averaged 850 hPa zonal wind, 200 hPa zonal wind, and OLR. The two leading principal components corresponding to the two EOFs, known as RMM1 and RMM2, respectively is used to calculate the phase and amplitude of eastward MJO propagation. When the index is within the central circle i.e. amplitude < 1 the MJO is considered weak, implying it is difficult to discern using the RMM methods. Outside this circle, the index is stronger and will usually move in an anti-clockwise direction as the MJO moves from west to east.

The MJO days can be clustered into eight phases (P1, P2, P3 ... P8, hereafter) depending on the phase-angle variations in the phase space created by the RMM1 and RMM2 indices (Wheeler and Hendon 2004). Each phase represents the geographical location of the enhanced convection of the MJO with respect to the phase space. On this, the phases P1, P2, P3 and P4 represent the convective phase of MJO where the equatorial Indian Ocean is dominated by convection. Similarly, the phases P5, P6, P7 and P8 represent the period when convection is suppressed in the equatorial Indian Ocean. The genesis dates of cyclones are clustered into these eight groups based on the MJO phase in which a cyclone genesis date resides. The genesis date of a cyclone is defined as the first day of the cyclone reported in the Best Tracks dataset reported by IMD (2011).

3 Results

As per IMD definition, a tropical cyclone is a rotational low pressure system in tropics when the central pressure falls by 5–6 hPa from the surrounding and maximum sustained wind speed reaches 34 knots (about 62 kmph). The year 2013 witnessed 10 cyclonic disturbances developed over North Indian Ocean. Out of the 10 cyclonic disturbances, 3 were Very Severe Cyclonic Storms (VSCSs) namely Phailin over the BOB (08–14 October 2013), Lehar (23–-28 November 2013), Madi (06–13 December 2013), one Severe Cyclonic Storm (SCS) namely Helen over BOB (19–23 November 2013), one Cyclonic Storm (CS) namely Viyaru (10–16 May 2013), 3 depressions in BOB, one deep depression over Arabian Sea and one land depression. Although the long period average of occurrence of cyclones in the entire North Indian Ocean is 5.5, there were 5 cyclones over the BOB and no cyclone over the Arabian Sea in the year 2013. Although the year 2010 witnessed 5 Cyclonic Storms in the North Indian Ocean (NIO) region but 5 cyclones occurring in BOB happened for the first time after 1987.

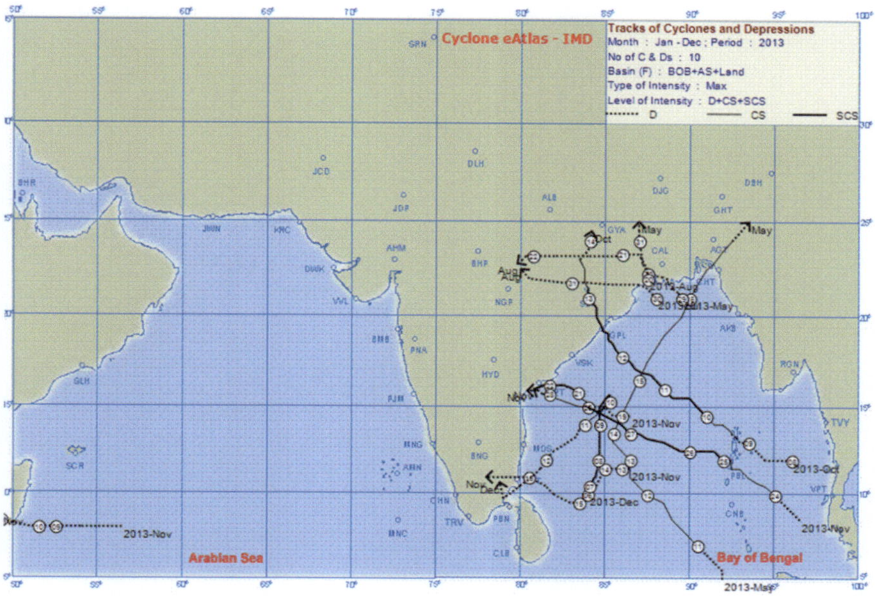

Fig. 1 Tracks of cyclones and depressions formed in the year 2013 in NIO region

Figure 1 represents the tracks of cyclones and depressions formed in the year 2013 in NIO region. It is seen that out of 10 disturbances, 2 developed during pre-monsoon, 2 during monsoon and 6 during post-monsoon season. Post-monsoon season was very active, especially over the BOB with the formation of three very severe cyclonic storms and one severe cyclonic storm.

A phase space diagram is created using the RMM1 and RMM2 indices to represent the different phases of MJO. The phase space diagram along with the distribution of cyclones is given by Fig. 2. In Fig. 2, the circle in the center of phase space with radius 1 represents days with MJO amplitude 1 and less. The black dots represent the date of cyclogenesis. Any day falling inside this circle is considered as a weak MJO day. From the phase-space diagram, it can be seen that cyclogenesis dates are clustered on the convective phase of MJO (P3, P4 and P6). In 3 cases i.e. during CS 'Viyaru', VSCS 'Phailin' and VSCS 'Madi', cyclogenesis occurred in the enhanced convection phases of MJO (in phase 4, 6, 3 and amplitude greater than 1) and on 2 occasions viz. VSCS 'Lehar' and SCS 'Helen' the cyclogenesis occurred in the suppressed convection phases of MJO (amplitude less than 1 over phase 4).

It endorses the earlier findings of Mohapatra and Adhikary (2011). The spatial distributions of cyclones are found to be highly influenced by the synoptic conditions associated with the MJO categories. The large-scale circulation patterns and associated changes in environmental parameters such as 850 hPa relative vorticity and vertical wind shear are important in modulating the tropical cyclone activity over the ocean basins (Chand and Walsh 2009; Belanger et al. 2010). Parameters

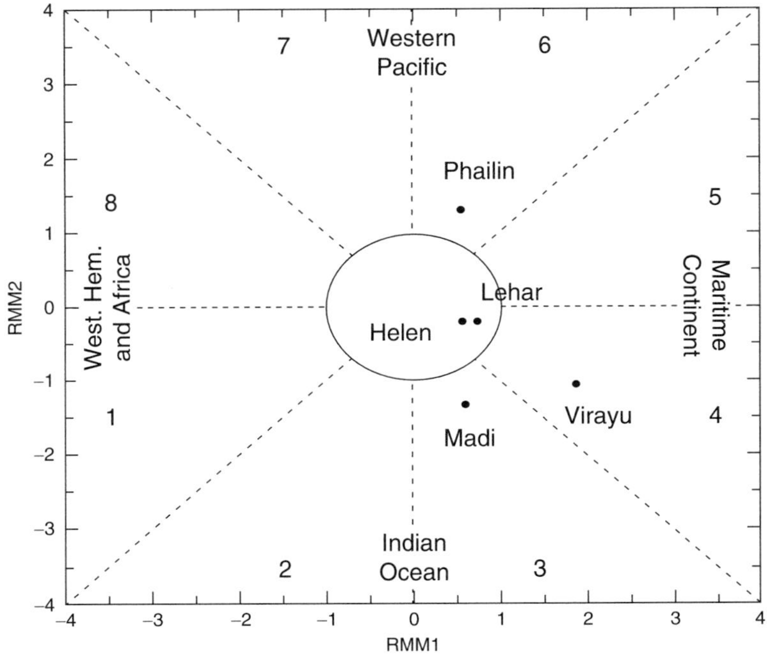

Fig. 2 Phase space diagram along with the distribution of cyclones from Bureau of Meteorology Australia

such as OLR, 850 hPa wind flow, 850 hPa vorticity, play a significant role in understanding an impact of the MJO on the cyclogenesis. The OLR anomaly patterns for the storms are given in Fig. 3. During the cyclogenesis of CS Virayu, SCS Helen and VSCS Lehar the MJOs convective envelope (negative OLR anomalies) were in phase 4 whereas during the cyclogenesis of VSCS Phailin it was in phase 6 and during the cyclogenesis of Madi the MJOs convective envelope was in phase 3. During the life span of CS Virayu from 11 to 16 May and VSCS Phailin from 08 to 14 October, 3 days had MJO amplitude greater than 1 and 3 days it was less than 1. Though VSCS Phailin was in phase 6 which is the suppressed phase of convection in the Indian Ocean, other environmental and atmospheric conditions like upper level divergence, increased low-level convergence along with low-level relative vorticity, decreased vertical wind shear of horizontal wind to about 10–20 knots and oceanic conditions like sea surface temperature of about 28–29 °C and ocean thermal energy of about (60–80 KJ/cm^2) helped in its intensification.

During the life span of SCS Helen from 19 to 23 November and VSCS Lehar from 23 to 28 November all the days had MJO amplitude less than 1 and during the life span of VSCS Madi from 6 to 13 December all the days witnessed MJO greater than 1.

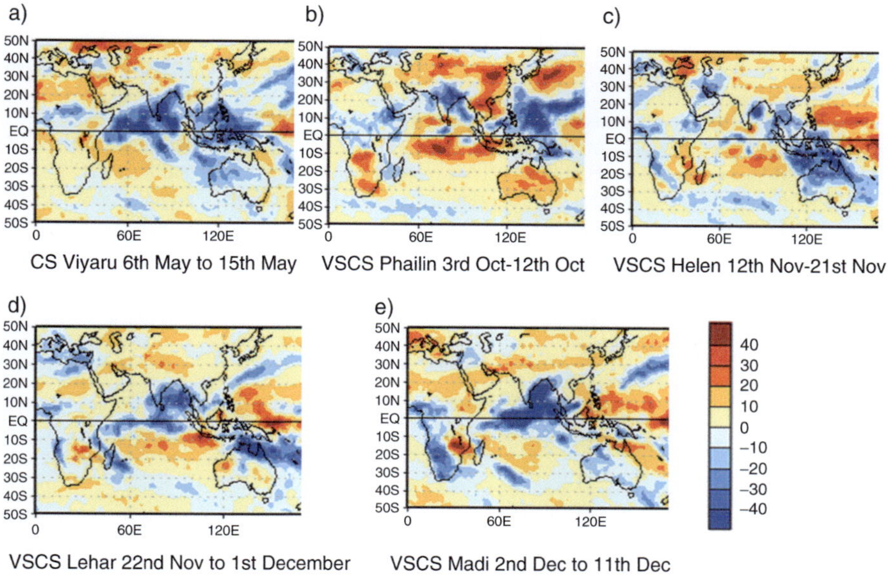

Fig. 3 (**a–e**) Ten day average OLR anomalies from Climate Prediction Centre (CPC), USA for (**a**) cyclonic storm (CS, Viyaru (6–15 May 2013), (**a**) cyclonic storm (CS), Viyaru (6–15 May 2013), (**b**) very severe cyclonic storm (VSCS), Phailin (3–12 October 2013), (**c**) severe cyclonic storm (SCS), Helen (12–21 November 2013). (**d**) VSCS, Lehar (22 Nov–01 Dec. 2013), (**e**) VSCS, Madi (2–11 December 2013)

4 Conclusions

Out of the 5 cyclones formed over the BOB in 2013, 3 were formed in the phase 4 and one each in phase 3 and phase 6 of MJO. It is seen that the number of cyclogenesis days are more in the convective phases (P1, P2, P3 and P4) compared with the suppressed convection phases.

References

Belanger, J.I, Curry, J.A. and Webster, P.J. (2010). Predictability of north Atlantic tropical cyclone activity on intraseasonal time scales. *Monthly Weather Review*, **138**, 4362–4374.

Chand, S.S. and Walsh, K.J.E. (2009). Tropical cyclone activity in the Fiji region: Spatial patterns and relationship to large-scale circulation. *Journal of Climate*, **22**, 3877–3893.

IMD (2011). Track of storms and depressions over Bay of Bengal and Arabian Sea: Cyclone e-Atlas, Cyclone Warning and Research Centre, Regional Meteorological Centre, IMD, Chennai.

Krishnamohan, K.S, Mohanakumar, K. and Joseph, P.V. (2012). The influence of Madden-Julian Oscillation in the genesis of North Indian Ocean tropical cyclones. *Theoritical and Applied Climatology*. Doi:10.1007/s00704-011-0582-x.

Madden, R.A. and Julian, P.R. (1971). Detection of a 40–50 day oscillation inthe zonal wind in the tropical Pacific. *Journal of Atmospheric Sciences*, **28**, 702–708.

Mohapatra, M. and Adhikary, S. (2011). Modulation of cyclonic disturbances over the north Indian Ocean by Madden-Julian oscillation, *Mausam*, **62**, 375–390.

Nakazawa, T. (1986). Intraseasonal variations of OLR in the tropics during the FGGE year. *Journal of Meteorological Society Japan*, **64**, 17–34.

Wheeler, M.C. and Hendon, H.H. (2004). An all-season real-time multivariate MJO index: Development of an index for monitoring and prediction. *Monthly Weather Review*, **132**, 1917–1932.

Relation of Frequency of Tropical Cyclones Over North Indian Ocean and North West Pacific Ocean with Sea Surface Temperature Anomaly Over Nino 3.4 Region and Indian Ocean Dipole

R. Chand and C. Singh

1 Introduction

Tropical cyclones (TCs) are one of the most devastating weather phenomena that cause large number of human casualties and loss of property. It is well known that the TCs over North Indian Ocean (NIO) have caused the maximum loss of human lives. The death toll in Bangladesh cyclone of November 1970 has been estimated to be about 300,000. The tropical depression (Maximum Sustained Surface Wind Speed (MSSWS) 17–33 knots or more) and TCs MSSWS 34 knots or more form over warm ocean surfaces, with SST more than 26.0 °C, low magnitudes of vertical wind shear and large magnitudes of low-level relative vorticity, coriolis force and mid-tropospheric level relative humidity (Gray 1968). Gray (1979), using 20 years of data has shown that the global tropics produce about 80 TCs in a year out of which only 6 form over NIO. Most of the TCs form over NIO during October–December (primary peak season) and March–May (secondary peak season). The period of May-November is the peak season for tropical depressions (TDs) over the NorthWest Pacific Ocean (NWPO).

The genesis of TDs is governed by several coupled ocean atmospheric phenomena in addition to Gray parameters. The El-Niño is manifested as an anomalous warming of the eastern and central tropical Pacific Ocean and La-Niña refers to an anomalous cooling of the eastern and central tropical Pacific Ocean. The SST anomaly over Nino 3.4 region (170.0°–119.0° W and equator to $\pm$ 5.0° N & S) is commonly used to represent El Nino/La Nina conditions- El Nino, if the departure from normal is $\geq$ 0.5 °C and La Nina, if the departure from normal is $\leq$ –0.5 °C.

R. Chand (✉) • C. Singh
India Meteorological Department, Mausam Bhawan, Lodi Road, New Delhi 110003, India
e-mail: r58.chand@gmail.com

© Capital Publishing Company 2017

M. Mohapatra et al. (eds.), *Tropical Cyclone Activity over the North Indian Ocean*,
DOI 10.1007/978-3-319-40576-6_16

Sea Surface Temperature (SST) plays a significant role in the formation and intensification of TCs (Palmen 1949; Miller 1958). The El-Nino Southern Oscillation (ENSO) events have significant impacts on the hurricane intensity (Landsea et al. 1999) and genesis locations (Elsner and Kara 1999) of the TCs. Gray has concluded in his paper that there is a tendency towards less (or more) TCs over the Atlantic basin during the El-Niño (La-Niña) years (Gray 1984). During past few decades, many studies have focused on the relationship between Western North Pacific (WNP) TCs and the ENSO (Chia and Ropelewski 2002; Wang and Chan 2002; Camargo and Sobel 2005; Chan 2007; Huang and Xu 2010; Kim et al. 2011). For example, Wang and Chan (2002) found that a southeastward (northwestward) shift in the positions of TCs genesis in the WNP occurs during El-Niño (La Niña) years which in turn may favour (suppress) the development of intense TCs. Camargo and Sobel (2005) revealed that the accumulated cyclone energy is positively correlated with ENSO indices, suggesting that stronger (weaker) and longer-lasting (shorter-lived) TCs tend to form during an El-Niño (La Niña) event. Following this, Camargo et al. (2007a, b) investigated the impacts of ENSO on the tracks and genesis locations of TCs. Making use of a genesis potential index (Emanuel and Nolan 2004), suggested that vorticity and relative humidity play an important role in the eastward shift in the mean genesis location of TCs over WNP. Chan (2007) also pointed out that interannual variations in intense typhoons over WNP are unlikely to be determined by local SST. But, TCs are related to changes in planetary-scale atmospheric circulation (vorticity and wind shear) and thermodynamic structure (moist static energy) associated with the El-Niño phenomenon. More recently, Huang and Xu (2010) attributed the increase in the number of super typhoons (STYs) in El-Niño years to changes in SST, the monsoon trough, and vertical wind shear. These studies focussed mainly on the effect of ENSO on intense TCs (Chan 2007; Huang and Xu 2010) or considered all TCs as a whole regardless of their intensity (Chia and Ropelewski 2002; Wang and Chan 2002; Camargo et al. 2007b; Kim et al. 2011). Thus, the effect of ENSO on weaker typhoons or TDs is unclear, and the connection between ENSO and TCs with different intensities remains uncertain. In addition, less effort on studying the TC–ENSO relationship during the ENSO transition phase has been found. Frank and Young (2007) suggested that the variance in factors that control the formation of TCs can ultimately result in storms with different degrees of intensity. Thus, the impact of ENSO on TCs with different intensities is also expected to vary. Kessler (1990); Kinter et al. (2002); Zhou and Chan (2007); Wang et al. (1999) indicated that the strongest ENSO signal occurs in the sub-surface as a result of wind stresses driven by ENSO. However, few previous investigations have used upper ocean heat content (OHC) to study the impact of ENSO on TCs. A recent study by Wada and Chan (2008) suggested that a decrease in OHC in the WNP is related to the passage of TCs, though they did not deal deeply into the relationship between OHC, ENSO and TCs. Wada and Chan (2008) described in his paper the role of OHC, to examine the impact of ENSO on TCs with different intensities.

In 1997, the year of the strongest El-Niño of the century, the WNP had 23 typhoons (the second largest number of any year on record), an unprecedented number of very intense TCs (11 became STYs) and a large eastward displacement of the genesis location. That year, two TCs (Oliwa and Paka) formed in the Central North Pacific (east of the data line) and entered the WNP before reaching STY intensity (Lander and Guard 2001). Wang and Chan (2002) also noted that the mean lifetime of TCs tends to increase during strong El-Niño events, as more TCs form in the southeast quadrant of the WNP, thus experiencing a longer traveling time (westward and northward) before encountering the continent or colder mid-latitude water. This is one possible explanation for the effect, the tendency towards more intense typhoons in El-Niño years. A few previous studies have briefly documented a relationship between TC intensity and ENSO. The main ones of which we are aware are those of Pudov and Petrichenko (1998, 2001), who found an increase in the intensity of TCs in El-Niño years. Camargo (2005) concluded in his paper that the shifts towards both longer lifetimes and greater intensities in El-Niño years are due to the eastward shift in genesis location, which any one can naturally expect to lead to longer tracks over a warm ocean surface. While this seems plausible, particularly in the case of the lifetime effect, it has not been proven. The intensity effect in particular could also be due to some extent, to other influences of ENSO on the mean regional climate of the NWPO.

Many studies have documented for NWPO, but very few are seen for NIO. On an average about 5–6 TCs with MSSWS 34 knots or more form over this basin every year (Singh 2001) in two cyclone seasons, such as pre-monsoon season (March–May) and post-monsoon season (October–December). Frequency of formation of TDs and their intensification have been studied for NIO by Srivastav et al. (2000). The average frequency of TDs and storms over NIO and NWPO during post satellite era (1960–2013) is discussed in detail in Sect. 3 of present paper. The favourable region of cyclogenesis in the NIO oscillates in both time and space. Ramesh Kumar and Sankar (2010) studied the impact of global warming on the cyclonic storms over NIO, it states that the frequency of storms and severe storms do not show a dramatic rise despite increase over the SST in the Bay of Bengal (BOB) during 1951–2007 compared to the 1901–1951, this study further states that there is large decrease in the mid-tropospheric humidity over the BOB during the period 1951–2007 and the atmospheric parameters such as low-level vorticity, mid-tropospheric humidity and vertical wind shear, all play an important role in the genesis and intensification of storms over this basin.

The effect of the Indian Ocean Dipole (IOD) on the formation of TDs over NIO and NWPO is also studied in this paper. IOD is defined by the difference in SST between Arabian Sea in western Indian Ocean and south of Indonesia in the eastern Indian Ocean. In positive phase, the greater-than-average SST and greater precipitation in the western Indian Ocean region, with a corresponding cooling of waters in the eastern Indian ocean which generally tends to cause droughts in adjoining areas of Indonesia and Australia. The negative phase of the IOD brings about the opposite conditions. The IOD is commonly measured by an index that is the difference between SST in the western (50.0°–70.0° E and 10.0° S–10.0° N) and eastern (90.0°–110.0° E and 10.0°–0.0° S) equatorial Indian Ocean. The index is called the Dipole Mode Index (DMI). Sumesh and Ramesh Kumar (2013) have suggested that the cyclogenesis over the NIO is a complex phenomena, as it is influenced by several coupled ocean atmospheric phenomena such as El-Nino, El-Nino Modki, IOD and MJO. Saha and Wasimi (2013) have analysed 30 years' data (1976–2006) which reveals that individual IOD pole (Western or Eastern) has peak negative correlation (−0.5 and −0.4, respectively) with the seasonal occurrences of TCs in Australia. Mainly, the TC occurrences in western and eastern sub-regions are correlated with IOD index of both the poles. Northern sub-region is more or less free from the influence of IOD index.

2 Data and Methodology

In present study, the tropical disturbances (TDs) formed over NIO and NWPO under RSMCs New Delhi and Tokyo have been considered, as shown in area 3 and 4 in Fig. 1. The annual frequency of TDs and storms over NIO and NWPO are taken from Regional Specialized Meteorological Centre (RSMC), New Delhi and RSMC Tokyo (website http://en.wikipedia.org/wiki/2012_Pacific) respectively. Month-wise SST anomaly of Nino 3.4 region is obtained from the website http://www.cpc.ncep.noaa.gov/products. Indian Ocean Dipole (IOD) years is obtained from the website http://www.bom.gov.au/climate/IOD/.

In present study, statistical methods are used to find the relation of annual frequency of TDs and storms over NIO and NWPO with SST anomaly in Nino

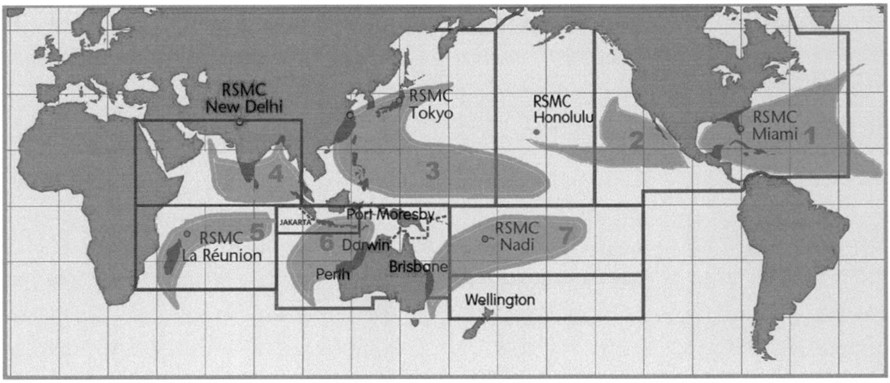

Fig. 1 The present study areas is shown as area 3 and 4 in above figure

3.4 region which is the most dominant mode in year-to-year climate variations in the tropics, and its impact on depressions and TCs activity. The relation is also studied with the annual IOD conditions as it also affects the formation and intensification of TCs. The student's *t*-test at a confidence level of 90 % or more has been used to compare means of two independent groups of above and below average frequency of depressions and TCs with SST anomaly in Nino 3.4 region and annual IOD status to determine if they are significantly different from one another. Similarly, the year with warm and cold anomaly over Nino 3.4 region have been compared with frequency of depressions and storms in corresponding years and the difference in mean frequency with respect to cold and warm anomaly has been tested for significance and analysed. The similar comparison has been carried out with respect to + ve and – ve IOD years.

3 Discussion

In IMD the classification for analysing intensity over the NIO is as follows; it is called a super cyclonic storm (SuCS), if MSSWS is 120 knots and above, a very severe cyclonic storm (VSCS) if MSSWS 64–119 knots, a severe cyclonic storm (SCS) if it is 48–63 knots, a cyclonic storm (CS) if it is 34–47 knots, a deep depression (DD) if it is 28–33 knots and depression (D) it is 17–27 knots. For NWPO, the TDs are of four categories namely as STY, typhoon, tropical cyclone and tropical depression. The STY group includes TCs that reach at least 114 knots (Cat 4 and 5 in the Saffir–Simpson Hurricane Scale), and representing the most intense TCs. The typhoon group comprises moderately strong TCs with a MSSWS between 64 and 114 knots (Cat 1–3 in the Saffir–Simpson Hurricane Scale). The TCs with a MSSWS of 34–63 knots is classified as the tropical cyclone and system with MSSWS between 17 and 33 knots are tropical depression.

The last 54 years (1960–2013) data shows the average annual frequency of tropical depressions or more over NIO is 6.2 with SD 2, CS 3.75 with SD 1.42 and SCS 1.22 with SD 1.19. Similarly, the average annual frequency of Tropical Storms over NWPO is 34 with SD 7, Tropical cyclone 27 with SD 5, Typhoon 16 with SD 4 and STY is 4 with SD 2.

3.1 Genesis and Intensification of TDs in NIO with Respect to El-Nino/La-Nina Conditions

SST anomaly in Nino 3.4 region is analysed with respect to above normal, normal and below normal frequency of depression and CS in the NIO region and Student's *t* test is carried out to identify statistically significant differences. Table 1 presents the number of years with above, below and normal frequency of depression and CS

Table 1 Number of years with above, below and normal frequency of depressions and CS over NIO along with SST anomaly in Nino 3.4 region

	Depression (SST anomaly)			CS (SST anomaly)		
	Average anomaly (°C)	SD (°C)	No.	Average anomaly (°C)	SD (°C)	No.
Above average frequency years	0.29	0.75	10	0.55	0.43	4
Below average years	0.69	0.53	3	−0.05	0.66	4
Normal	−0.11	0.60	41	−0.05	0.66	48

Table 2 Average frequency of SCS over NIO during +ve IOD and neutral years

	Positive IOD Years	Neutral IOD years
Average	2	1.06
SD	1.25	1
No. of years	10	35

Table 3 No of years with above average, below average and normal frequencies of D, CS, SCS over the NIO corresponding to +ve IOD, −ve IOD and neutral IOD conditions

	Depression			Cyclonic storm			SCS	
Years	Above Avg > 8	Below Avg < 4	Normal 4–8	Above Avg > 5	Below Avg < 2	Normal 2–5	Above Avg > 2	Normal 0–2
+ve IOD	2 (20 %)	1 (33 %)	7 (17 %)	1 (20 %)	0	9 (19 %)	3 (37)	7 (15 %)
−ve IOD	3 (30 %)	0	5 (12 %)	0	0	7 (14 %)	1 (12 %)	7 (15 %)
Neutral	5 (50 %)	2 (66 %)	29 (70 %)	4 (80 %)	1	32 (67 %)	4 (50 %)	32 (70 %)

over NIO along with the SST anomaly in Nino 3.4 region. During the years with below average frequency of D, the Nino 3.4 SST anomaly satisfies the El Nino conditions. On the other hand, the Nino 3.4 SST anomaly satisfies El Nino conditions during the years when of above average frequency of CS over the NIO.

3.2 *Genesis and Intensification of TDs in NIO with Respect to IOD Conditions*

Analysis of frequency of depression, CS, SCS and VSCS in the NIO region with respect to positive, neutral and negative IOD conditions indicates significantly higher frequency of SCS over NIO during positive IOD years at 95 % level of confidence (Table 2). There is no significant difference for depression and CS frequency with respect to IOD conditions. Table 3 presents the number of years with above average, below average and normal frequencies of D, CS, SCS corresponding to positive, negative and neutral IOD conditions. In the case of years with above average frequency of depressions, 20 % are of positive IOD years, 30 %, negative IOD years and 50 % neutral IOD years. For below average frequency of depressions, 33 % are of positive IOD years and 66 % neutral IOD years. In the case of above average frequency of CS years, 20 % are of positive IOD years and 80 % neutral years. In case of above average frequency of SCS, 37 % are of positive IOD years, 12 % negative IOD years and 50 % neutral years.

Difference in frequency of SCS during positive and neutral IOD years is significant at 95 % level of confidence.

3.3 Genesis and Intensification of TDs Over the NWPO in Respect of El-Nino/La Nina Conditions

For NWPO the student *t*-test for frequency of disturbances with warm and cold SST anomalies years of Nino 3.4 region gives significant relation for Typhoon and STY with the SST anomaly. The SST anomaly is significantly positive (higher) during the years with above average frequency of typhoons and STYs as compared to years of normal frequency as well as below average frequency. However, no such correlation is found between SST anomaly over Nino 3.4 region with depressions and storms (Table 4). The SST anomaly is significantly positive (higher) for STY in case of above average frequency as compared to normal frequency as well as below average frequency in NWPO (Table 5).

Table 4 Average frequency of typhoons/super typhoons over NWPO during warm and cold anomaly years

	Warm anomaly years	Cold anomaly years
Average (Typhoon/ST)	17.5/5.05	14.09/3.48
SD (Typhoon/ST)	3.27/2.93	5.45/2.17
No. of years (Typhoon/ST)	20/20	23/23

Table 5 Mean SST anomaly over Nino 3.4 region during years with above and below average frequency of super typhoons over NWPO

	SST anomaly, Nino 3.4 region		
Frequency of years (ST)	Average	SD	No.
Above average	0.47	0.85	8
Below average	−0.62	0.61	7
Normal	−0.017	0.65	38

Table 6 Average frequency of super typhoons over the NWPO during +ve, −ve and neutral IOD years

Super typhoon	Positive IODs	Negative IODs	Neutral IODs
Average	5.7	2.44	2.16
SD	2.98	1.88	2.16
No. of years	10	9	35

3.4 Genesis and Intensification of TDs Over the NWPO in Respect of IOD Conditions

Student's *t*-test for frequency of disturbances over NWPO with annual IOD conditions yields that the frequency of STY is significantly higher during +ve IOD years compared to −ve IOD years as well as in neutral years. No comparison is found to be significant in case of depression, tropical storm and typhoon with annual IOD conditions (Table 6). Relation of frequency of TDs with annual IOD conditions over NWPO showed that the possibility of genesis of depressions and storms are higher in +ve IOD Years (Table 7). At the same time there seems no relation in frequency of storms and typhoons with IOD condition. However, in +ve IOD years frequency of STY is higher and in −ve IOD years frequency is less.

4 Conclusion

The present state of knowledge regarding TC activity in NIO and NWPO basins and the El-Niño phenomenon is reviewed in this study. The genesis of TCs is influenced by various factors such as El-Nino and IOD. Though there is no significant linear relationship between the number of TCs and El-Nino conditions in particular year, a nonlinear relation between El-Nino conditions and the number of TCs has been found in this study. On the basis of analysis of 54 years data of SST anomaly of Nino 3.4 region and annual IOD conditions and frequency of depression, CS and SCS over NIO and NWPO, the following conclusions are drawn:

1. During the years with below average frequency of depressions, the Nino 3.4 SST anomaly satisfies the El Nino conditions. On the other hand, the Nino 3.4 SST anomaly satisfies El Nino conditions during the years when of above average frequency of CS over the NIO.
2. Frequency of Typhoons and STYs is significantly higher over the NWPO during warm anomaly years in Nino 3.4 region. SST anomaly is significantly higher during years with higher frequency of STY.
3. The possibility of genesis of depressions and intensification into storms, typhoons and STYs over NWPO are higher in + IOD years and lower in −ve IOD years.

Table 7 Frequency of years with different IOD conditions associated with above average, below average and normal frequency of depression, storm, typhoon and super typhoon over NWPO

IOD years	Depression			Storm			Typhoon			Super typhoon		
	A	B	N	A	B	N	A	B	N	A	B	N
	Avg > 40	Avg < 27	27–40	Avg > 32	Avg < 22	22–32	Avg > 20	Avg < 11	11–20	Avg > 6	Avg < 2	2–6
+ve IOD years	3 (37 %)	2 (25 %)	6 (16 %)	1 (17 %)	0	9 (21 %)	1 (25 %)	1 (12 %)	8 (19 %)	4 (44 %)	1 (14 %)	5 (13 %)
−ve IOD years	1 (12 %)	1 (12 %)	6 (16 %)	1 (17 %)	2 (33 %)	6 (14 %)	1 (25 %)	2 (25 %)	6 (14 %)	0	4 (57 %)	5 (13 %)
Neutral years	4 (50 %)	5 (53 %)	26 (68 %)	4 (66 %)	4 (66 %)	27 (65 %)	2 (50 %)	5 (63 %)	28 (67 %)	5 (56 %)	2 (28 %)	28 (74 %)

Note: *A* above, *B* below and *N* normal

Acknowledgements We would like to express our sincere thanks to Dr. M. Mohapatra Head (RSMC and cyclone warning division) of IMD for giving constructive suggestions/comments for significant improvement of this paper.

References

Camargo, S.J. (2005). Western North Pacific Tropical Cyclone Intensity and ENSO. *American Meteorological Society*, **18**, 2996–3006.

Camargo, S.J. and Sobel, A.H. (2005). Western North Pacific tropical cyclone intensity and ENSO. *Journal of Climate*, **18**, 2996–3006.

Camargo, S.J, Robertson, A.W, Gaffney, S.J, Smyth, P. and Ghil, M. (2007a). "Cluster analysis of typhoon tracks", Part II: Large-scale circulation and ENSO. *Journal of Climate*, **20**, 3654–3676.

Camargo, S.J, Emanuel, K. A. and Sobel, A. H. (2007b). Use of a genesis potential index to diagnose ENSO effects on tropical cyclone genesis. *Journal of Climate*, **20**, 4819–4834.

Chan, J.C.L. (2007). Interannual variations of intense typhoon activity. *Tellus A*, **59**, 455–460.

Chia, H.H. and Ropelewski, C.F. (2002). The interannual variability in the genesis location of tropical cyclones in the northwest Pacific. *Journal of Climate*, **15**, 2934–2944.

Elsner, J. B. and Kara, A. B. (1999), Hurricanes of the North Atlantic: Climate and Society, New York, Oxford, Oxford University Press.

Emanuel, K.A. and Nolan, D. (2004). Tropical cyclone activity and the global climate system, Preprints, 26th Conf. on Hurricanes and Tropical Meteorology, Miami, FL, American Meteorological Society, CD-ROM, 10A.2.

Frank, W.M. and Young, G.S. (2007). The interannual variability of tropical cyclones. *Monthly Weather Review*, **135**, 3587–3598.

Gray, W.M. (1968). Global view of the origin of tropical disturbances and storms. *Monthly Weather Review*, **96**, 669–700.

Gray, W.M. (1979). Hurricanes: Their formation, structure and likely role in the tropical circulation. *In:* Shaw, D.B. (ed.) *Meteorology over the Tropical Oceans*. Royal Meteorological Society, James Glaisher House, Gvenville Place, Bracknell, Berkshire, RG 121 BX, 155–218.

Gray, W.M. (1984). Atlantic seasonal hurricane frequency. Part I: El Ni˜no and 30 mb quasi-biennial oscillation influences. *Monthly Weather Review*, **112**, 1649–1668.

Huang, F. and Xu, S. (2010). Super typhoon activity over the western North Pacific and its relationship with ENSO. *Journal of Ocean University of China*, **9**, 123–128.

Kessler, W.S. (1990). Observations of long Rossby waves in the northern tropical Pacific. *Journal of Geophysical Research*, **95**, 5183–5217.

Kim, H.M, Webster, P.J. and Curry, J.A. (2011). Modulation of North Pacific tropical cyclone activity by three phases of ENSO. *Journal of Climate*, **24**, 1839–1849.

Kinter, J.L, Miyakoda, K. and Yang, S. (2002). Recent change in the connection from the Asian monsoon to ENSO. *Journal of Climate*, **15**, 1203–1215.

Lander, M.A. and Guard, C.P. (2001). Western North Pacific, North Indian Ocean, and Southern Hemisphere tropical cyclones of 1997. *Monthly Weather Review*, **129**, 3015–3036.

Landsea, C.W, Pielke, R.A, Jr, Mestas-Nuñez, A.M and Knaff, J.A. (1999). Atlantic Basin hurricanes: Indices of climatic changes. *Climatic Change*, **42**, 89–129.

Ramesh Kumar, M.R. and Sankar. S. (2010). Impact of global warming on cyclonic storms over north Indian Ocean. *Indian Journal of Marine Sciences*, **39**, 516–520.

Miller, B.I. (1958). On the maximum intensity of hurricanes. *Journal of Meteorology*, **15**, 184–195.

Palmen, E. (1949). On the formation and structure of tropical hurricanes. *Geophysics*, **3**, 26–38.

Pudov, V.D. and Petrichenko, S.A. (1998). Relationship between the evolution of tropical cyclones in the Northwestern Pacific and El Niño. *Oceanology*, **38**, 447–452.
Pudov, V.D. and Petrichenko, S.A. (2001). 1997–1998 El Niño and tropical cyclone genesis in the northwestern Pacific. *Izvestiya Atmospheric and Oceanic Physics,* **37**, 576–583.
Saha, K.K. and Wasimi, S.A. (2013). Interrelationship between Indian Ocean Dipole (IOD) and Australian Tropical Cyclones. *International Journal of Environmental Science and Development*, **4**, 647–651.
Singh, O.P. (2001). Long term trends in the frequency of monsoonal cyclonic disturbances over the north Indian ocean. *Mausam*, **52**, 655–658.
Srivastav, A.K, Sinha Ray, K.C. and De, U.S. (2000). Trends in the frequency of cyclonic disturbances and their intensification over Indian seas. *Mausam*, **512**, 113–118.
Sumesh, K.G. and Ramesh Kumar, M.R. (2013). Tropical cyclones over north Indian Ocean during El-Niño Modoki years. *Natural Hazards*, **68**, 1057–1074.
Wada, A. and Chan, J.C.L. (2008). Relationship between typhoon activity and upper ocean heat content. *Geophysical Research Letters*, **36**, L17603, doi:10.1029/2008GL035129.
Wang, B. and Chan, J.C.L. (2002). How strong ENSO affect tropical cyclone activity over the western North Pacific. *Journal of Climate*, **15**, 1643–1658.
Wang, B, Wu, R. and Lukas, R. (1999). Roles of the western North Pacific wind variation in thermocline adjustment and ENSO phase transition. *Journal of Meteorological Society Japan*, **77**, 1–16.
Zhou, W. and Chan, J.C.L. (2007). ENSO and South China Sea summer monsoon onset. *International Journal of Climatology*, **27**, 157–167.

Governing Factors Associated with Intensification of TC-A Diagnostic Study of VSCS PHAILIN and LEHAR

G.K. Das and G.C. Debnath

1 Introduction

Tropical cyclones (TCs) are synoptic scale intense low pressure systems which form over the warm tropical oceans characterised by strong cyclonic winds and organised convection with heavy rainfall. The TC causes enormous damage to life and property at the time of crossing the coast and subsequent movement over the land. The TCs can impact over a wide area with its strong winds, heavy rains and storm surges. Gray (1979) prepared a detailed climatology of the TCs in the global ocean basins. He observed that annually about 80 TCs form globally of which half to two thirds reach hurricane strength (maximum sustained winds greater than 33 m/s). East coast of India is frequently affected by TCs that form over Bay of Bengal (Tyagi et al. 2010). During the period 1891–2010, about 134 severe TCs crossed the east coast of India (IMD 2011; Atlas of storm track). Anthes (1982) has extensively studied the three dimensional structure of TC. He observed that the surface pressure is lowest at the centre of the TC and increases outward. The wind speed reaches its maximum value at nearly 40–80 km from the centre beyond which it decreases. The central parts are warmer than the surroundings and the temperature anomaly could be more than 10° K at the upper troposphere. The TC grows upto a height of 10–15 km. Intense TCs frequently develop an eye, which is a cloud free region at the centre of the storm characterised by the presence of subsidence.

Mandal (1991) studied year-to-year fluctuations in the frequency of cyclonic disturbances. He concluded that there is no trend or periodicity in the series. The decadal frequency, however, suggested a decreasing trend since 1950s. Mohapatra et al. (2013b) has also shown decreasing trend in TCs, severe TCs and very severe TCs over north Indian Ocean during satellite era (1961–2010). The influence of

G.K. Das • G.C. Debnath (✉)
Area TC warning Centre, Regional Meteorological Centre, Kolkata 700027, India
e-mail: gcdebnath@yahoo.com

© Capital Publishing Company 2017
M. Mohapatra et al. (eds.), *Tropical Cyclone Activity over the North Indian Ocean*,
DOI 10.1007/978-3-319-40576-6_17

greenhouse gas-induced warming, sea-surface temperature anomalies and phenomena such as the El Niño–Southern Oscillation (ENSO) on annual TC activity have been investigated by (Gray 1984; Chan 1995). Patwardhan and Bhalme (2001) concluded the decreasing trend of cyclonic disturbance over Indian and adjacent ocean based on (1891–1998) data.

TC intensity is defined by the maximum mean (sustained) wind speed experienced around the centre of the TC. Many authors have studied TC movement and intensity in details taking various parameters as an input. Majority of studies on TC intensification have concentrated on the role of sea surface temperature (SST). Muthuchami and Sridharan (2008) have studied the intensification and movement of cyclonic storm in the Bay of Bengal during post-monsoon season. They concluded that the orientation of isotherms of SST of Bay of Bengal influenced the direction of motion of TC. They have found that wind shear is responsible for weakening of TC. Muthuchami (2000) concluded that during post-monsoon season in the Bay of Bengal cyclonic storms have a particular annual behaviour in direction of motion in each year, i.e., the storms in a year tend to move in a particular direction. Krishna Rao (1997) remarked that during the formative stage when the low-level convergence zone is superimposed by upper level divergence it gets intensified. Das et al. (2012) has shown that the TC moves towards and cross near the station having relatively steeper decrease of geopotential height upto mid-tropospheric level followed by increase in geopotential height. Schade (2000) concluded that the effect of SST on the intensity of TCs can be separated into two distinct contributions: one from the large-scale SST field that is in equilibrium with the atmosphere and another one from a local reduction of the SST under the eye of the TC due to surface winds of the TC. Kotal and Roy Bhowmik (2013) used upper level divergence, lower level relative vorticity and relative humidity and vertical winds shear to find out the Rapid Intensification Index (RII) for TC over the Bay of Bengal.

Holland (1997) suggested that the role of the Ocean is to initially contribute a favourable environment for the development and then to provide the additional energy required for the intensification into a high intensity TC. The interaction of TC with SST can be considered as a feedback mechanism for the intensification of the TC. As explained by Ginis (1995), SST feedback mechanism is favourable for TC intensification in the initial development stages. As TC intensifies, the increasing wind speed enhances the evaporation rate and the increased moisture supply in-turn increases the latent heat energy released and causes further intensification of the TC.

The intensification of a TC after its genesis is greatly dependent on both thermodynamic and dynamic parameters besides regional influences of individual basins. Therefore, in this paper an attempt has been made to find out the role of Oceanic and atmospheric parameters for intensification/weakening of VSCS PHAILIN and LEHAR with a view to improve the understanding about forecasting of TC intensity.

2 Data and Methodology

2.1 Brief Life History of VSCS PHAILIN and LEHAR

Very severe cyclonic storm (VSCS) PHAILIN (8–14 October 2013), developed from a remnant of cyclonic circulation from the South China Sea, was seen as a low pressure area over Tenasserim coast on 6 October 2013 (RSMC, New Delhi 2014) . It lay over north Andaman Sea as a well marked low pressure area on 7th and concentrated into a depression on 8 October near latitude 12.0° N and longitude 96.0° E. Moving west-northwestwards, it intensified into a deep depression on 9th morning and further into cyclonic storm (CS), 'PHAILIN' in the same evening. Moving northwestwards, it further intensified into a severe cyclonic storm (SCS) in the morning and into a VSCS in the forenoon of 10th October over east central Bay of Bengal. It crossed Odisha and adjoining north Andhra Pradesh coast near Gopalpur (Odisha) around 2230 h IST of 12 October 2013 with a sustained maximum surface wind speed of 215 kmph. After landfall it moved in northerly to north westerly direction and reached upto Bihar as a depression. The track of the TC and the typical satellite imagery are shown Fig. 1a.

VSCS LEHAR (23–28 November 2013) developed from remnant cyclonic circulation from South China Sea which emerged into south Andaman Sea on 23 November 2013 evening when it lay as depression located about 550 km south-southeast of Port Blair (Fig. 1b). It intensified into a deep depression in the same night and into a cyclonic storm (CS), 'LEHAR' in the early morning of 24th. It further intensified into a severe cyclonic storm and crossed Andaman & Nicobar Island near Port Blair in the morning of 25th with a wind speed of about 110–120 kmph. On 25th it emerged into southeast Bay of Bengal and moved west-northwestward, intensified into a very severe cyclonic storm in the early hours of 26th near southeast Bay of Bengal. It maintained its VSCS intensity with a maximum wind speed reaching upto 140–150 gusting to 165 kmph till noon of 27th. Afterwards, it rapidly weakened into a deep depression and crossed Andhra Pradesh coast close to south of Machilipatnam around 1400 h IST of 28 November as depression (RSMC, New Delhi 2014).

The two VSCS PHAILIN and LEHAR have been analysed by addressing (i) Sea surface temperature (°C), (ii) vertical wind shear (VWS) of horizontal wind (knot) (iii) lower tropospheric absolute vorticity (AV) (10^{-5} per s), (iv) lower level convergence (CON) (10^{-5} per s) and (v) upper level divergence (DIV) (10^{-5} per s).

The SST and VWS data associated with the two TC have been obtained from Regional Specialized Meteorological Centre (RSMC), IMD New Delhi bulletin issued on real time basis whereas AV and convergence/divergence data have been obtained from ARPEG 1.5 (FRANCE) model using synergy system installed at RMC (ACWC) Kolkata.

The intensity forecast at IMD has been issued from deep depression stage onwards for 12, 24, 36, 48, 60, 72, 84, 96, 108 and 120 h forecast periods. It has

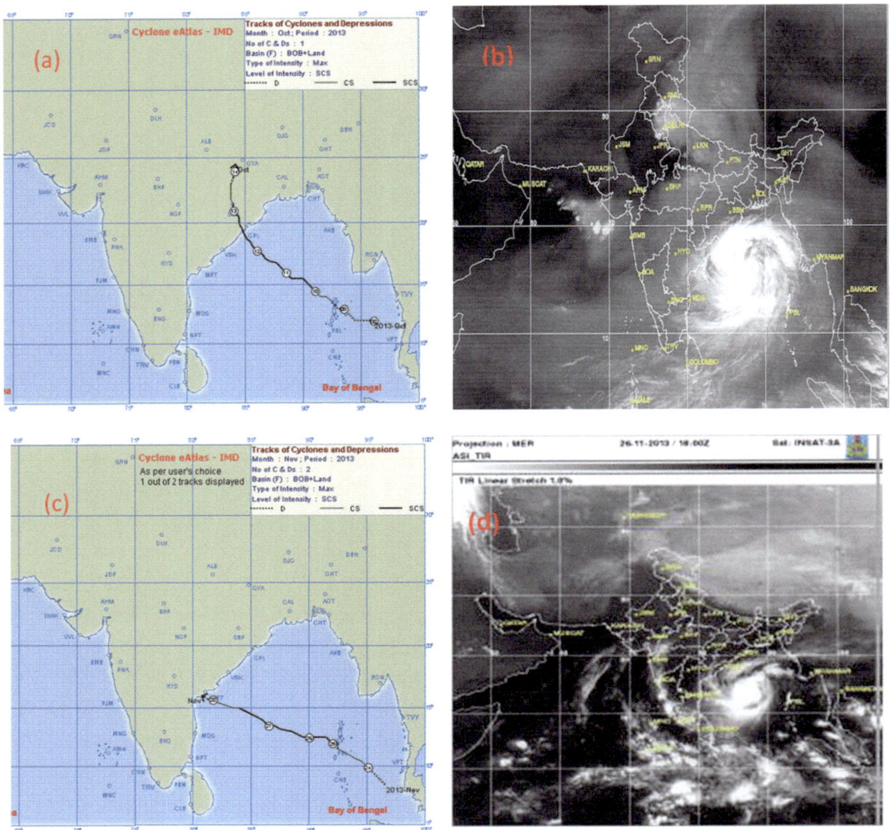

Fig. 1 (**a & b**): Track and satellite imagery of VSCS PHAILIN and (**c & d**) track and satellite imagery of VSCS LEHAR

been issued every 3 h during the life period. Following methods are used by IMD for intensity forecasting of TCs over the NIO (Mohapatra et al. 2013a, b).

1. Statistical Techniques: (a) Analogue, (b) Persistance and (c) Climatology.
2. Synoptic Technique: Empirical techniques.
3. Satellite Technique: Empirical technique.
4. Radar Technique: Empirical technique.
5. (v) NWP Models
   - Individual models (global and regional)
   - IMDGFS (382, 574), ARPEG (Meteo-France), ECMWF, JMA, UKMO, NCEP WRF (IMD, IITD, IAF), HWRF (IMD)
6. Dynamical Statistical Model (Statistical TC Intensity Prediction – SCIP).

A statistical–dynamical model for TC intensity prediction (SCIP) has been implemented for real time forecasting of 12-h intensity upto 72 h. The model parameters are derived based on model analysis fields of past CS. The parameters selected as predictors are: Initial storm intensity, Intensity changes during past 12 h, Storm motion speed, Initial storm latitude position, VWS averaged along the storm track, Vorticity at 850 hPa, Divergence at 200 hPa and SST. For the real-time forecasting, model parameters are derived based on the forecast fields of ECMWF-GFS model. The method is found to be promising for the operational use. In the synoptic method, prevailing environmental conditions like VWS, SST, Ocean thermal energy, low-level inflow, upper level outflow etc. are considered for intensification. The development of characteristic features in satellite and radar observations is also taken into consideration for predicting the intensity. Although, the synoptic, statistical and satellite/radar guidance help in short range intensity forecast (upto 12/24 h), the NWP guidance is mainly used for 24–120 h forecasts. Hence, the RSMC, New Delhi intensity forecast results from a manually analysed forecasting process, which relies on output from several NWP models as mentioned above. Finally, consensus forecast that gather all or part of the numerical forecast intensities along with synoptic and statistical guidance are utilised to issue forecast (Mohapatra et al. 2013a, b).

3 Results and Discussion

3.1 SST and Intensification of PHAILIN and LEHAR

The SST has historically been mentioned as a significant factor for the development and intensification of TCs. The maximum intensity of a TC is known to have a direct relationship with SST (Palmen 1948). Figure 2 shows the time series plot of SST during movement/intensification of the two VSCS PHAILIN and LEHAR. It is seen that in case of VSCS PHAILIN the SST value was much higher than the threshold value of 26.5 °C. It is also seen from Fig. 2a that warmer SST of the order of about 28.5 °C persisted over Bay of Bengal from genesis stage. Moreover, the value increased to 30 °C, 24 h before its land fall over Odisha coast near to Gopalpur on 12 October.

The distribution of SST value for LEHAR indicates that like PHAILIN, it also maintained higher value (28.5 °C) than the normal from genesis till 0600 UTC of 27th and fell sharply thereafter reducing the value to 24 °C before landfall as depression over Andhra Pradesh coast near Machilipatnam (Fig. 2b). Thus it is observed that SST played a very vital role for intensifying and decaying of the above two TCs.

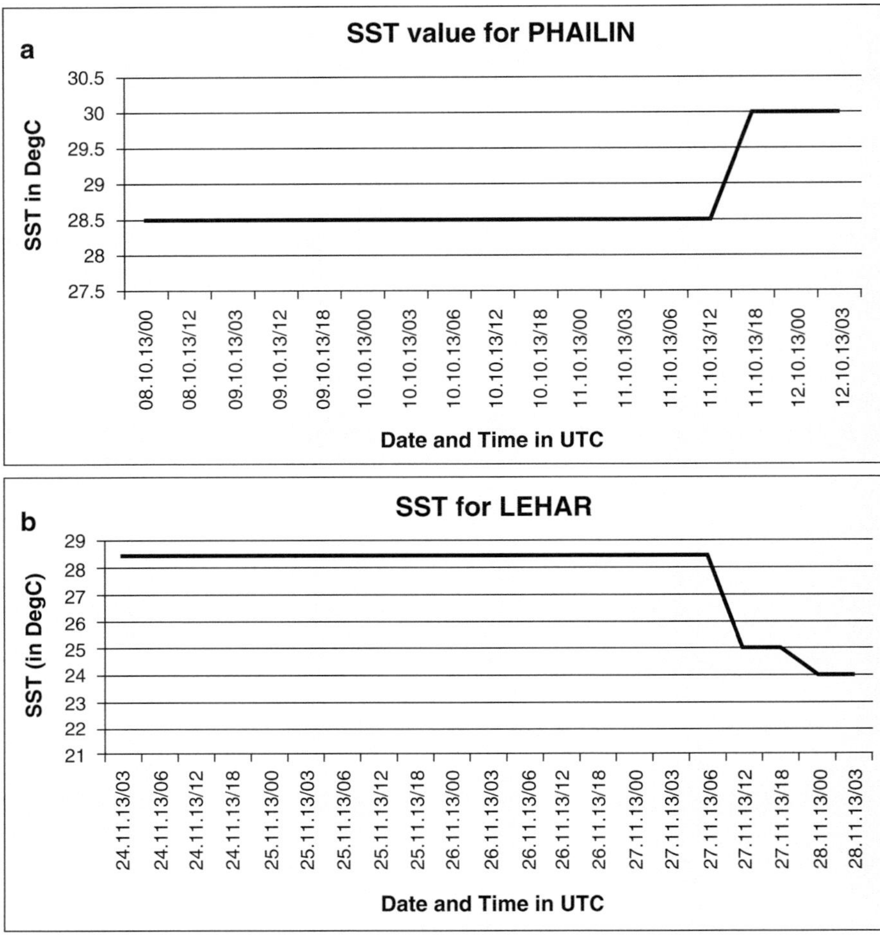

Fig. 2 (**a** and **b**) Sea surface temperature (SST) condition during VSCS PHAILIN and LEHAR

3.2 Analysis of VWS

The VWS is one of the important parameters for intensification of low pressure systems. High wind shear causes the structure of the eyewall region to become highly asymmetric throughout the depth of the storm. The asymmetries in the upper troposphere, where the storm circulation is weaker, become sufficiently so strong that air with high values of potential vorticity and equivalent potential temperature are mixed outward rather than into the eye. This allows the shear to ventilate the eye resulting in a loss of the warm core at upper levels, which causes the central pressure to rise, weakening the entire storm.

It has been seen that for a weak system, the VWS between 850 and 500 hPa are considered for determination of its intensity whereas for a strong system like VSCS, the VWS shear between 200 and 850 hPa are considered

VWS based on satellite observation is defined as follows.

VWS = 150–300 hPa layer mean minus 700–925 hPa layer mean horizontal wind VWS is classified as weak (favourable for development): If VWS is 5–10 knots.

It is moderate (unfavourable for weak system, or neutral for mature TC), if it is 10–20 knots and high (unfavourable) it VWS is $>$ 20 knots.

To study the role of VWS, the zonal vertical VWS between 850 and 200 hPa levels was obtained for each day of the two storms. It is found that the system weakened rapidly when it entered into high VWS region particularly when they entered into upper level westerly regime.

Figures 3a, b present the zonal vertical VWS computed between 850 and 200 hPa levels during the storm period. It can be seen that in the case of VSCS PHAILIN the VWS started from the range of 10–20 knots (low moderate) then reduced to 10–15 during its movement and further reduced to 5–10 knots (low) during middle of its path upto landfall (Fig. 3a). In case of LEHAR it started at high range (15–25 knots) became moderate (10–20 knots) during its movement and increased again to its high range before 24 h of its landfall (Fig. 3b). Thus low VWS in the storm region rapidly enhanced the intensity of the storm PHAILIN and high VWS in the storm region reduced the intensity of LEHAR.

3.3 AV and Intensity

AV indicates the measurement of rotation of the air parcel and high value of rotation is associated with intensification of the system. The distribution of AV over the central region under consideration at 850 hPa level is shown in Fig. 4. The AV in case of PHAILIN started initially from 15×10^{-5} per s increased significantly during its course of movement and reached to 46×10^{-5} per s before its landfall (Fig. 4a) and hence may be considered an important factor for its intensification. In case of LEHAR it started initially at moderate value of 25×10^{-5} per s reached to a level of around 35×10^{-5} per s and then decreased sharply to less than 20×10^{-5} per s and LEHAR weakened to depression before its landfall (Fig. 4b).

3.4 Analysis of Convergence

The strong lower level convergence along with upper level divergence is associated with any strong low pressure system. The weakening of lower level convergence is an indication of weakening of low pressure system. Figure 5a, b indicates the low level (925 hPa) convergence associated with the systems. The convergence started

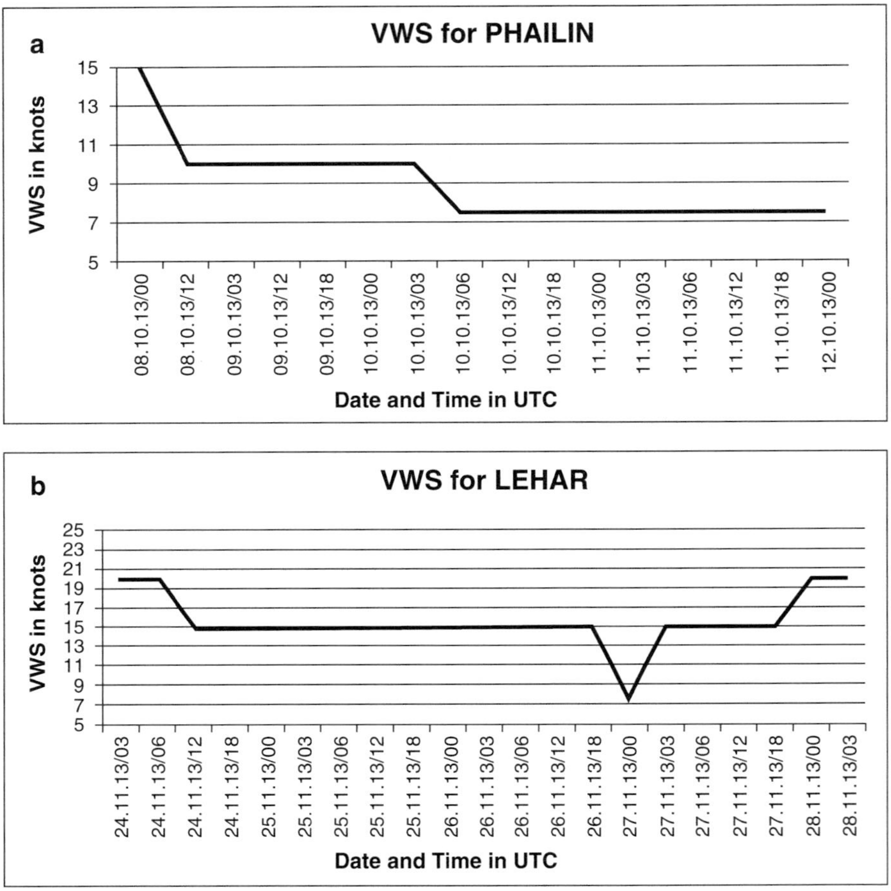

Fig. 3 (**a** and **b**) Vertical wind shear (VWS)during VSCS PHAILIN and LEHAR

initially at moderate value of 6×10^{-5} per s with PHAILIN decreases to 4×10^{-5} per s in the middle and increased sharply to 12×10^{-5} per s unit before its landfall (Fig. 5a) whereas in case of LEHAR it started with moderate value of 6×10^{-5} per s and reduced 2×10^{-5} per s at the time of landfall and associated with the weakening of the system before its landfall (Fig. 5b).

4 Conclusions

The results of the observational study leads to the following conclusions:

1. SST has reached to highest value of 30 °C at the time of landfall from the initial value of 28.5 °C in PHAILIN whereas it was reverse for LEHAR, recording 24 °C at the time of landfall from the starting value 28.5 °C.

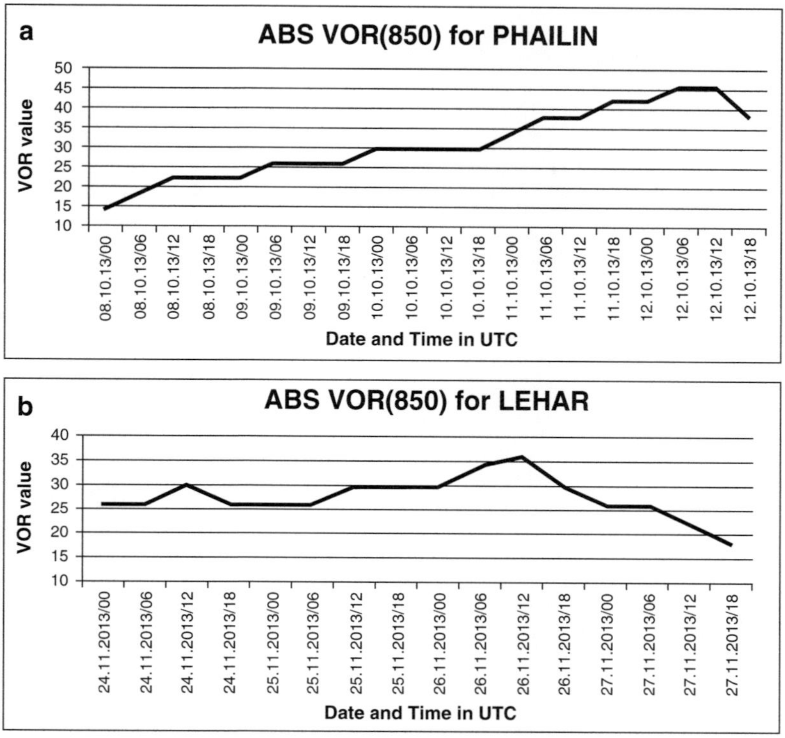

Fig. 4 (**a** and **b**) Absolute vorticity (AV) at 850 hPa level ($\times 10^{-5}$ per s) during VSCS PHAILIN and LEHAR

2. Low VWS in the storm region rapidly enhanced the intensity of the storm PHAILIN before landfall and high VWS in the storm region decreased the intensity of LEHAR at the time of landfall.
3. PHAILIN maintained increasing trend of AV throughout its life cycle reaching the highest value (46×10^{-5} per s) at the time of land fall but in case of LEHAR the value reached to the lowest (26×10^{-5} per s) before landfall.
4. Low level convergence sharply increased to 12×10^{-5} per s at landfall from the initial value of 6×10^{-5} per s in case of PHAILIN whereas for LEHAR it decreased to 2×10^{-5} per s at landfall from the initial value of 6×10^{-5} per s.

References

Anthes, R.A. (1982). Tropical Cyclones: Their Evolution, structure and effects. *Meteorological Monographs, American Meteorological Society*.

Chan, J.C.L. (1995). Prediction of annual TC activity over the Western North Pacific and the South China Sea. *International Journal of Climatology*, **15**, 1011–1019.

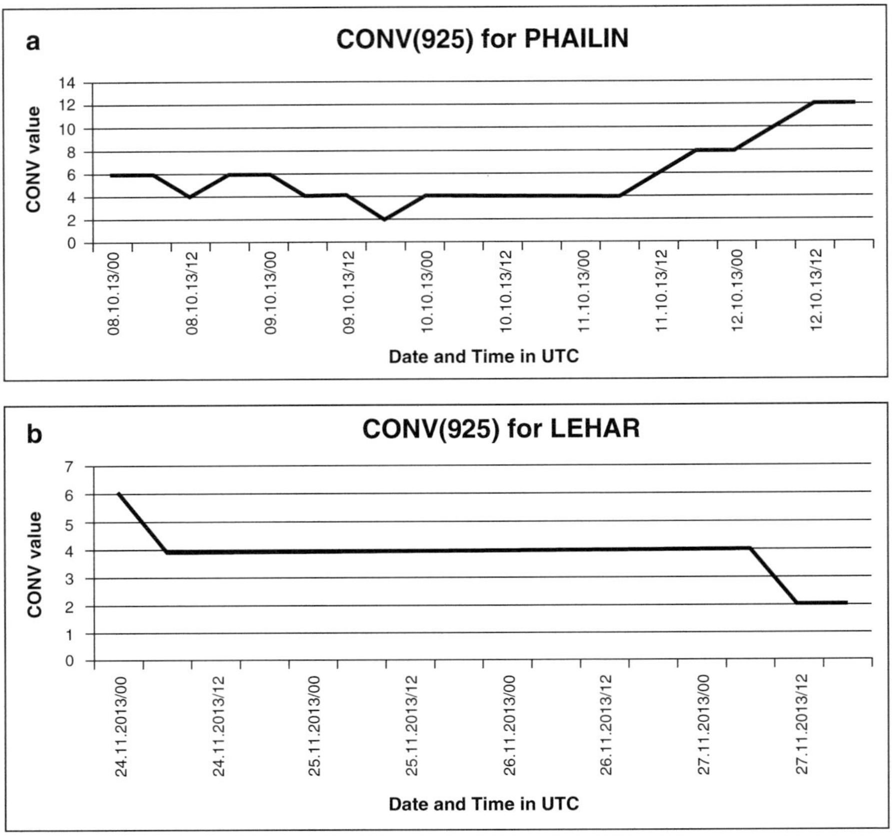

Fig. 5 (**a** and **b**) Convergence (Conv) at 925 hPa level ($\times 10^{-5}$ per s) during VSCS PHAILIN and LEHAR

Das, G.K, Midya, S.K, Debnath, G.C. and Roy, S.N. (2012). The relationship between geopotential height and movement & landfall of TC over the Bay of Bengal. *Mausam*, **63**, 469–474.

Ginis, I. (1995). Interaction of TCs with the ocean. Global Perspectives on TCs. *In*: Elsberry, R.L. (ed.) Tech. Doc. WMO/TD No. 693, Chapter 5, World Meteorological Organization, Geneva, Switzerland, pp. 198–260.

Gray, W.M. (1979). Hurricanes: Their formation structure and likely role in the tropical circulation. *In*: Shaw D.B. (ed.) Meteorology over the Tropical Oceans. Royal Meteorological Society, pp. 155–218.

Gray, W.M. (1984). Atlantic seasonal hurricane frequency. Part I: El Niño and 30 mb quasi-biennial oscillation influences. *Monthly Weather Review*, **112**, 1649–1668.

IMD (2011). Cyclone e-Atlas-IMD Tracks of Cyclones and Depressions over North Indian Ocean (www.rmcchennaieatlas.tn.nic.in).

Holland, G.J. (1997). The maximum potential intensity of TCs. *Journal of the Atmospheric Sciences*, **54**, 2519–2541.

Kotal, S.D. and Roy Bhowmik, S.K. (2013). Large-scale characteristics of rapidly intensifying TCs over the Bay of Bengal and a Rapid Intensification (RI) index. *Mausam*, **64**, 13–24.

Krishna Rao, A.V.R. (1997). TCs-synoptic methods of forecasting. *Mausam*, **48**, 239–256.

Mandal, G.S. (1991). TCs and their forecasting and Warning System in the north Indian Ocean. WMO/TD No.430, Report No. TCP-28.

Mohapatra, M, Bandyopadhyay, B.K, Nayak, D.P. (2013a). Evaluation of operational tropical cyclone intensity forecasts over north Indian Ocean issued by India Meteorological Department. *Natural Hazards*, 68, 589–601, doi:10.1007/s11069-013-0624-z

Mohapatra, M, Bandyopadhyay, B.K. and Tyagi, A. (2013b). Constraction and Quality of Best Track Parameters for study of Climate Change Impact on Tropical Cyclones over the North Indian Ocean during satellite era. published in Monitoring and Prtediction of Tropical Cyclones in the North Indian Ocean, Copublished by Springer & Capital Publishing Company, pp. 3–15.

Muthuchami, A. (2000). Origin and movement of cyclonic storms in the Bay of Bengal. *Mausam*, **51**, 225–230.

Muthuchami, A. and Sridharan, S. (2008). Intensification and movement of cyclonic storms in the Bay of Bengal during post monsoon season. *Mausam*, **59**, 51–68.

Palmen, E. (1948). On the formation and structure of tropical hurricanes. *Geophysica*, **3**, 26–38.

Patwardhan, S.K. and Bhalme, H.N. (2001). A study of cyclonic disturbances in India and the adjacent ocean. *International Journal of Climatology*, **21**, 527–534.

RSMC, New Delhi, 2014, Report Cyclonic Disturbances over north Indian Ocean during 2013. RSMC, New Delhi.

Schade, L.R. (2000). TC intensity and sea surface temperature. *Journal of Atmospheric Sciences*, **57**, 3122–3130.

Tyagi, A, Mohapatra, M, Bandyopadhyay, B.K. and Naresh Kumar (2010). Inter-annual variation of frequency of cyclonic disturbances landfalling over WMO/ESCAP Panel Member Countries. WMO/TD-No. 1541 on 1st WMO International conference on Indian Ocean Tropical Cyclones and Climate Change, Muscat, Sultanate of Oman 08–11 March, 2009, WWRP-2010/2.

Part IV
NWP Modelling for Tropical Cyclone Forecast

Numerical Simulations with WRF to Study the Impact of Sea Surface Temperature on the Evolution of Tropical Cyclones Over Bay of Bengal

C.V. Srinivas, G.M. Mohan, D.V. Bhaskar Rao, R. Baskaran, and B. Venkatraman

1 Introduction

Tropical cyclones (TCs) belong to the class of severe weather systems associated with strong ocean-atmospheric coupling. They are highly disastrous weather phenomena that cause damage to the life and physical infrastructure in tropical maritime countries. The Indian coastal lands, especially the east coast, are highly vulnerable to the TCs in the post-monsoon season of October to December. Weak vertical wind shear, high sea surface temperature (SST) (>26.5 °C), pre-existing disturbance like tropical easterly waves are some of the favorable conditions (Anthes 1982; Gray 1968) favouring the development and sustenance of TCs in this season in the North Indian Ocean (NIO). SST is a crucial influential parameter for the formation and development of TCs. A threshold SST value of 26.5 °C was defined by Gray (1968) for the genesis and further development of TCs. A TC is characterised with an outward diverging motion in the upper atmosphere and converging motion at the surface. Higher upper air divergence facilitates further deepening through enhanced convergence at the lower levels. The upper ocean provides heat energy to the overlying atmospheric boundary layer and for the deepening process by enhancing the convection. Earlier studies have shown that TCs experience sudden intensification when they entered oceanic areas with higher SSTs. There are several observational and modelling studies which explained the upper-ocean response to TCs (e.g. Price 1981; Price et al. 1994; Shay and Elsberry 1987; Sanford et al. 1987; Bender and Ginis 2000; Shay et al. 2000). Among a

C.V. Srinivas (✉) • G.M. Mohan • R. Baskaran • B. Venkatraman
Radiological Safety and Environment Group, Indira Gandhi Centre for Atomic Research, 603102 Kalpakkam, India
e-mail: cvsri@igcar.gov.in

D.V.B. Rao
Department of Meteorology and Oceanography, Andhra University, Visakhapatnam, India

© Capital Publishing Company 2017

M. Mohapatra et al. (eds.), *Tropical Cyclone Activity over the North Indian Ocean*,
DOI 10.1007/978-3-319-40576-6_18

number of factors, the TC intensification is partly controlled by the surface heat and moisture fluxes that feed the energy to the storm and the dissipation by the roughness to the winds at the sea surface (Chen et al. 2013). The upper ocean heat content and surface temperature are influential in the ocean–atmosphere interaction and the supply of energy through sensible and latent heat fluxes for the development and sustenance TCs. Emanuel (2005) and Webster et al. (2005) suggested that the rising SSTs in the Atlantic Ocean are related to the increase in the hurricane activity. A number of studies have shown that the strong surface wind in a TC drives sea surface waves and underlying ocean currents and enhance upper-ocean turbulent mixing, cools the SST, and result in a cold wake behind (Price 1981), which, in turn, provides a negative feedback on TC intensity (Schade and Emanuel 1999; Chan et al. 2001). It has been reported that warmer SST associated with large ocean heat content causes TC intensification (Hong et al. 2000; Shay et al. 2000; Bright et al. 2002), whereas negative SST anomalies associated with cold-core eddies or TC-induced cold wake weakens TC systems (Walker et al. 2005). A few workers (Knutson and Tuleya 2004; Michaels et al. 2006) using computer models have shown that increase in SST due to green house warming leads to increase in hurricane intensity. Michaels et al. (2006) reported that the strong relationship of TC intensity with SST is not clear at the upper range of SSTs. Though SST is known to influence the intensity of TCs, it is still not clear how it influences the movement of the storms. It is necessary to investigate the air–sea interaction processes associated with SST parameter and upper ocean heat content by analysis of the air–sea fluxes through numerical experiments. In this study, an attempt is made to understand the role of SST on the movement and intensity of the TCs by performing numerical simulations with a mesoscale atmospheric model.

2 Data and Methodology

Weather Research and Forecast (WRF) mesoscale atmospheric model with Eulerian mass dynamical core developed and sourced by the National Center for Atmospheric Research (NCAR), USA is used to study the impact of SST parameter on the evolution of TCs. It is a non-hydrostatic model with terrain following vertical coordinate, accurate numerics, higher order mass-conservation equations and dynamical nesting (Skamarock et al. 2008) to simulate various scales of atmospheric motions. A two-way interactive double nested (27 km, 9 km resolutions) domain configuration with 40 vertical levels is used for the present study (Fig. 1). The outer domain covers the NIO region with 27 km resolution and 173×162 grids. The inner domain has 9-km resolution with 292×229 grids covering the Bay of Bengal (BOB) and adjoining coastal land areas. The data for the terrain elevation, land-use and soil types for the 27 and 9 km resolution domains are taken from the U.S. Geological Survey (USGS) topography data at $10'$ and $5'$ resolutions. The case of a very severe TC Thane-2011 is chosen for the study. The TC Thane formed as a depression on 25 December 2011 in the southeast Bay of Bengal (BOB),

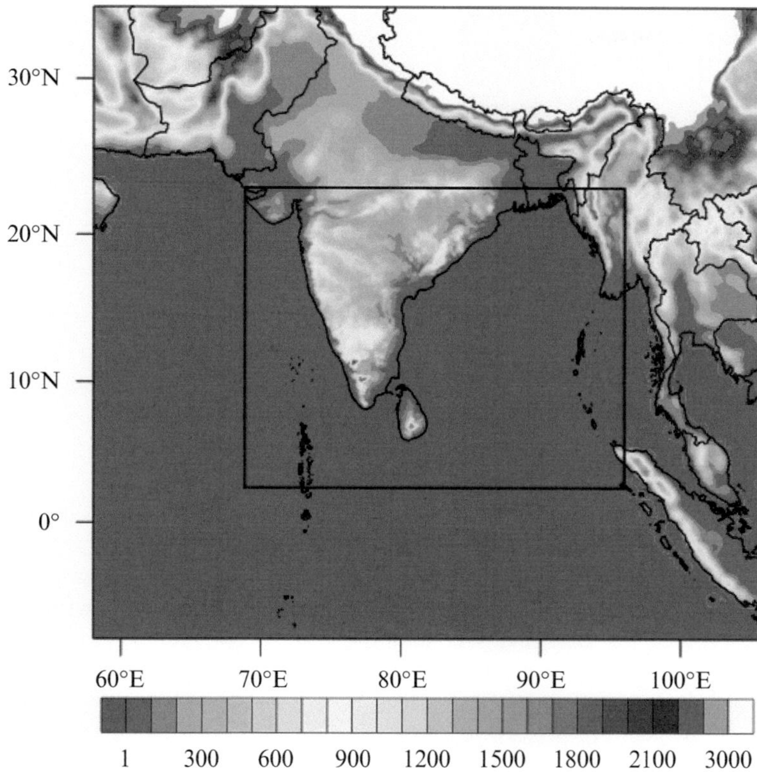

Fig. 1 Domains used for simulation of TC Thane. Shades give the terrain height

progressively intensified as a very severe TC, moved in a north-north-northwestward direction and crossed North Tamilnadu and Puducherry coast on 30 December 2011 (IMD 2012).

The model is initialised at 00 UTC 27 Dec 2011 and integrated up to 96 h. The US-National Centre for Environmental Prediction (NCEP) Global Forecasting System (GFS) analysis and 3-h forecasts available at 50-km resolution are employed for the initial and boundary conditions of Advanced Research WRF (ARW) model. No vortex initialisation and assimilation of observations are performed. For studying the impact of SST a series of experiments are performed in two groups. In the control simulation the SST is taken from NCEP real-time global SST analysis (http://polar.ncep.noaa.gov/sst/rtg_high_res/) for defining SST initial and boundary conditions. In the subsequent experiments SST is altered in steps of +1, +2, +3, −1, −2, −3 °C from the actual analysis SST at each grid point in the domains. All the experiments are performed with updation of SST at 3-h interval. In the second group of experiments the SST is altered as in the first group in steps of +1, +2, +3, −1, −2, −3 °C but over a limited region (300 × 300 km) around the TC and with updating of SST at 3-h interval. The names of the experiments along with the domain average SST used in each experiment are given in Table 1. India Meteorological Department (IMD) best track estimates for

Table 1 The details of the SST used in different experiments

Experiment name	SST data (°C)	SST used (°C)
CTL	NCEP analysis	~26.85
SST + 1	NCEP + 1	~27.85
SST + 2	NCEP + 2	~28.85
SST + 3	NCEP + 3	~29.85
SST − 1	NCEP-1	~25.85
SST − 2	NCEP-2	~24.85
SST − 3	NCEP-3	~23.85

intensity and vector track positions are used for comparisons of simulated intensity and track parameters from each of the experiments.

3 Results and Discussion

The changes in the movement of the system and the variation in the central sea level pressure (CSLP), maximum tangential winds with respect to variation of SST parameter are analysed from various experiments. The experiments without alteration of SST in the first and second groups are considered as control (CTL) experiments in the respective groups. In both the groups of simulations it has been found that alteration of SST leads to large variations in both intensification as well as movement of the storm. The results are discussed below.

3.1 Experiments with Alteration of SST Over Simulation Domain

In the first group of experiments SST is changed from analysis value over the modelling domain by applying increments or decrements in steps of 1, 2 and 3 °C. The boundary conditions for SST are also updated from the NCEP real time SST data. The average SST from analysis corresponding to the model initial time (00UTC 27 December 2011) over the model inner domain is about 26.85 °C. With alteration of SST over the analysis (Table 1), considerable changes are noted in the simulated track positions and intensity parameters. It has been found that when the SST is increased from the analysis value the track of the storm deviates to the north of the track in the CTL (Fig. 2a). However, when SST is decreased from the analysis value, it is seen that the track deviates to the south of the track in CTL. Higher track deviations are obtained with larger SST changes (Fig. 2b). It has been found that the experiment SST-2 with reduction of SST by −2 °C provides minimum track deviations of all the experiments. The track errors (Table 2) in the control run (CTL) are found to be 34, 79, 99, 181 km at 12, 24, 48, 72 h forecasts, respectively. Of all the simulations the experiment SST-2

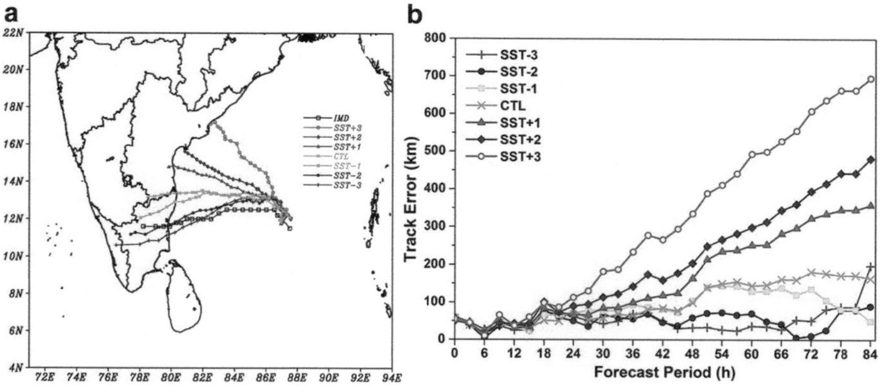

Fig. 2 (**a**) Simulated tracks of TC Thane and (**b**) time series of track errors from experiments using different SSTs over simulation domain

produced minimum track errors which are noted to be 39, 50, 59, 11 km at 12, 24, 48 72 h forecasts, respectively. However, the simulation SST + 3 with an increase of SST by 3 °C produced maximum track errors (112, 334, 608 km at 24, 48, 72 h forecasts).

When the SST is increased from analysis value (SST ~26.85 °C) in steps of 1, 2, 3 °C it is found that the central pressure reduces and maximum tangential winds increase progressively leading to an increase in intensity (Fig. 3). However, when the SST is reduced from analysis value in steps of 1, 2, 3 °C it is found that the pressure drop reduces and maximum winds decrease progressively leading to weakening of the storm. The experiment (SST + 2) with an increase of SST by 2 °C from analysis produced minimum errors in central sea level pressure (CSLP) and maximum sustained winds(MSW) (Tables 3 and 4). Thus, the use of SSTs lower than the analysis SST of ~26.85 °C, reduce the track errors appreciably, however, they increase the errors in maximum winds and central sea level pressure leading to underestimation of intensity of the storm. On the contrary, the use of SSTs higher than the actual analysis SST of ~26.85 °C, reduced the intensity errors but increased the track errors.

In the experiment with reduction of SST by 2 °C from analysis (Fig. 3), the track errors are reduced by 4.5 %, 36.7 %, 40.6 % at 24, 48, 72 h predictions, respectively. This experiment also simulated the timing of landfall (00 UTC 30 December 2011) with least error of −1 h (early landfall) and with least position error of 14 km (Table 5).

Table 2 Errors in vector track position from experiments using different SSTs

	Track error (in km)						
Forecast period (h)	SST − 3	SST − 2	SST − 1	SST	SST + 1	SST + 2	SST + 3
0	58.97	53.89	59.98	56.45	59.98	49.48	54.23
12	25.22	38.89	32.68	34.33	31.36	43.01	37.28
24	61.33	50.02	70.82	79.14	66.61	90.33	112.72
36	68.49	55.40	93.56	63.78	99.25	142.14	234.05
48	32.40	58.92	104.42	99.12	164.14	204.47	334.86
60	36.81	70.38	129.66	144.62	251.98	298.97	493.62
72	50.86	10.75	136.10	181.66	322.85	394.95	608.41
84	197.74	89.65	49.44	163.71	357.87	481.64	695.66

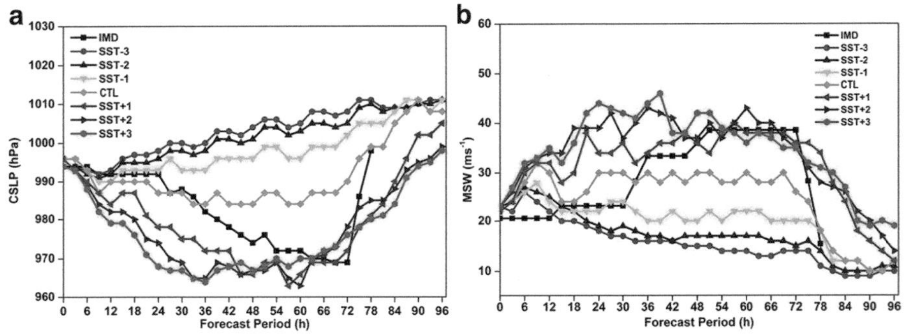

Fig. 3 Simulated central sea level pressure (**a**) and maximum sustained winds (**b**) of TC Thane in experiments with different SST values

Table 3 Errors in central sea level pressure from experiments using different SSTs

	CSLP error (in hPa)						
Forecast period (h)	SST − 3	SST − 2	SST − 1	SST	SST + 1	SST + 2	SST + 3
0	0	0	2	2	2	0	0
12	1	0	−2	−2	−8	−10	−13
24	6	4	1	−5	−14	−18	−24
36	18	16	11	2	−10	−17	−18
48	30	27	22	10	−8	−7	−7
60	33	31	24	12	−6	−9	−2
72	39	36	33	21	3	9	7
84	9	9	8	5	−10	−12	−16

Table 4 Errors in maximum surface winds from experiments using different SSTs

	MSW error (in m/s)						
Forecast period (h)	SST − 3	SST − 2	SST − 1	SST	SST + 1	SST + 2	SST + 3
0	2	2	1	1	1	2	2
12	1	4	3	9	11	13	14
24	−5	−4	−1	7	11	16	21
36	−17	−16	−13	−3	1	10	11
48	−21	−19	−16	−6	0	1	6
60	−25	−22	−17	−9	−1	4	−3
72	−25	−24	−19	−13	−3	−4	−4
84	−4	−3	−1	−1	11	13	14

Table 5 Errors in position and time of land fall from experiments using different SSTs

Experiment	Landfall point error (km) (experiment–actual)	Landfall time error (h) (experiment–actual)
SST + 3	603.00	8
SST + 2	397.13	12
SST + 1	301.66	8
SST	168.12	−2
SST − 1	122.80	−2
SST − 2	14.69	−1
SST − 3	68.33	−1

3.2 *Experiments with Alteration of SST Over the Region of TC*

In the second group of experiments the SST alteration was done over an area of 300 × 300 km of the T region. It is noticed that the SST variation over a limited ocean region around the TC has a smaller effect on the motion of the TC but has a greater impact on the intensification of the system. As in the first group of experiments higher SSTs (>26.85 °C) led to north/northwestward drift of the system leading to large track errors (Fig. 4a, b). However, lower SSTs (<26.85 °C) did not result in southwestward drift as in first group. The smaller effect of SST changes on the motion of the TC can be identified from the relatively lesser track errors (maximum track error ~400 km for SST + 3) in this group (Fig. 4b) as compared to the track errors (maximum track error ~600 km for SST + 3) in the first group. This shows that the variation of SST over a limited ocean region does not cause drastic influence on the dynamics of the system. As in the first group, use of higher SSTs (>26.85 °C) led to large intensity changes (Fig. 4c, d) with maximum impact given by SST + 3 °C in terms of reduction in central pressure and enhancement of MSW. However, the MSW increased upto 46 m/s and CSLP dropped upto 950 hPa which are much larger values relative to the first group. This suggests that the intensity increments in the case of limited area SST variation are much larger than

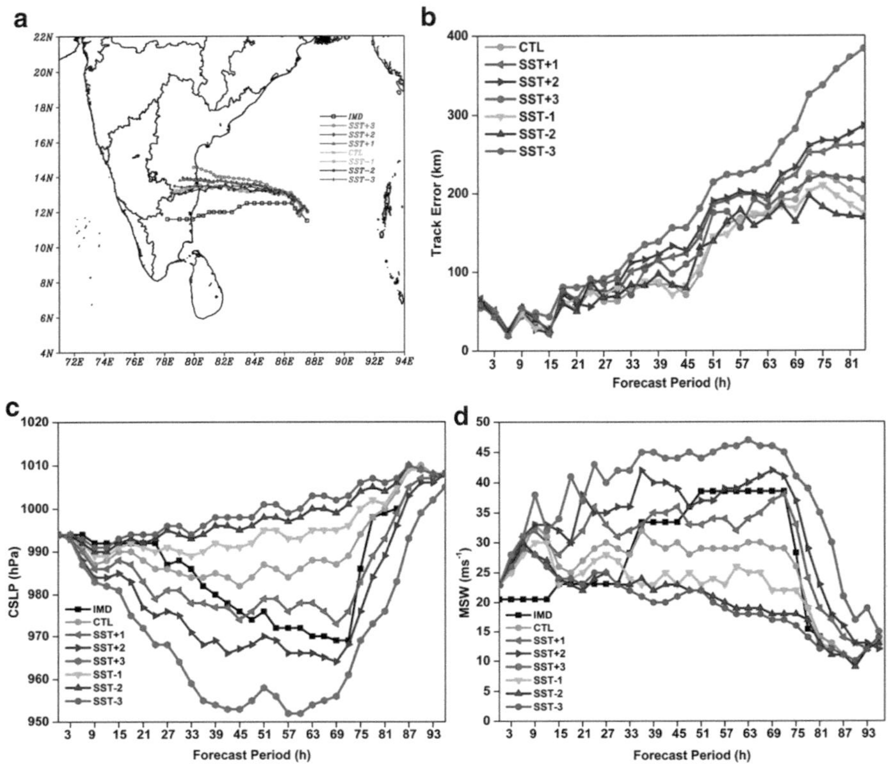

Fig. 4 Simulated tracks (**a**), track error (**b**), central sea level pressure (**c**) and maximum sustained winds (**d**) of TC Thane in experiments with different SST values for the experiments using SST change over a 300 × 300 km area

the intensity changes obtained due to domain wide SST variation. In TCs the processes of convergence of moist air, its upward motion and warming of the atmosphere occur more rigorously in the inner and outer core regions. These processes would more rigorously influence the thermodynamics of the atmosphere in a limited region of the TC, leading to enhancements in convection and further intensification. Further, the SST changes in TCs would also occur more rigorously in the region of the TC due to TC-induced wind and wave forcing and ensuing ocean mixing. Thus the results of limited area SST changes discussed above on the motion and intensification of the TCs assume importance.

3.3 Changes in Air–Sea Fluxes

The SST parameter is likely to influence the development of the storm through the transport of energy at the Ocean–Atmosphere interface. Upward fluxes of sensible,

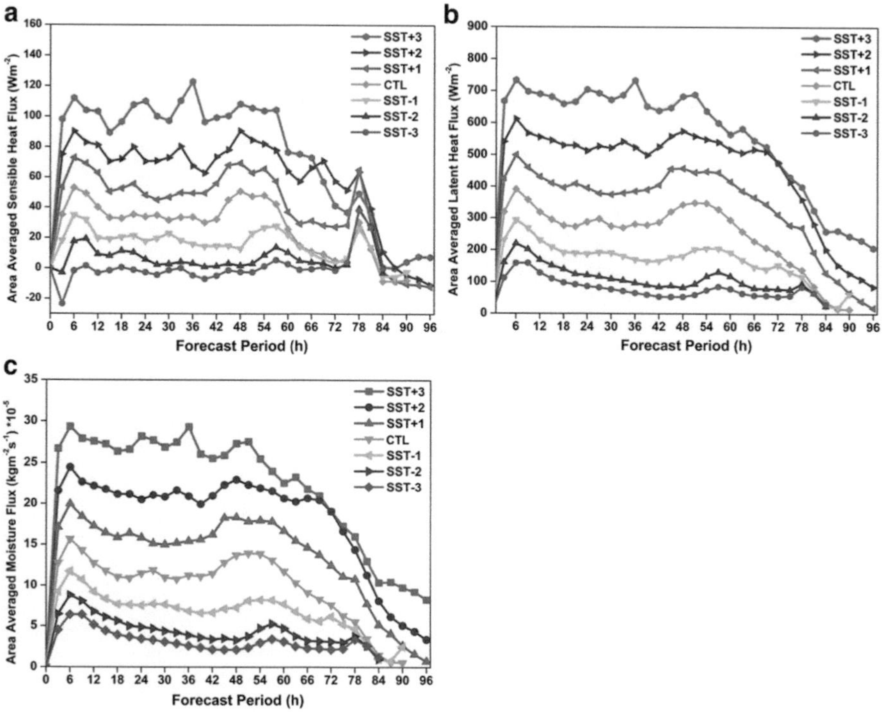

Fig. 5 Time series of area averaged (4 × 4° *rectangular* area around the cyclonic centre) fluxes (**a**) Sensible heat flux (Watts/m^2), (**b**) Latent heat flux (Watts/m^2) and (**c**) Moisture flux (Kg/m^2/s)

latent heat and moisture averaged over an area of 4° × 4° (~400 × 400 km) around the storm are analysed from different experiments to examine the changes in the fluxes due to changes in SST parameter. The time series of the surface fluxes from the first group of experiments with alteration of SST are presented in Fig. 5 It is found that as the SST increases all the three fluxes are enhanced throughout the life cycle of the storm, and the experiment with a rise of 3 °C in SST yielded highest fluxes. The larger upward fluxes would feed more moisture to the system and would enhance the warming of the atmosphere by release of latent heat leading to development of further convection, convergence and overall intensification.

These results suggest that the transport of energy to the TCs increases with higher SST values and thereby leads to intensity amplification and explains the mechanism for the air–sea interaction through SST parameter and its impact on the TC development.

The dynamical structure of tropical TCs consists of convergence and cyclonic motion at lower levels and divergence and anti-cyclonic motion at the upper atmosphere. The thermodynamical structure of the TCs gets modified due to changes in release of large amounts of latent heat flux associated with strong convection. The modification in temperature structure leads to vertical wind shear. Convection results in raising air motion and compensating sinking motion which transports heat and momentum between different vertical layers. The coupling of thermodynamical and dynamical quantities at various levels in the atmosphere is thus very important for the prediction of TCs in baroclinic flow. Some of the recent studies (Chan 2005; Wu and Wang 2000) have indicated that potential vorticity, which represents a coupling of the dynamic and thermodynamic variations can be used to study the TC movement. Wu and Wang (2000) have shown that the gradient towards the center of positive potential vorticity tendency (PVT) strongly coincides with the direction of the TC and so can be used as a tool for prediction of the movement. For the purpose of understanding the changes in the dynamics of the system resulting from changes in air–sea fluxes, a preliminary analysis is performed by computing different terms of vorticity equation from CTL, SST + 2, SST − 2 experiments from the first group. The equation for the tendency of the relative vorticity in the Cartesian coordinate system (*x*,*y*,*z*) can be written as

$$\underbrace{\frac{\Delta\xi_r}{\Delta t}}_{A} = \underbrace{-\left(u\frac{\xi_r}{\Delta x} + v\frac{\Delta\xi_r}{\Delta y}\right)}_{B} \underbrace{- w\frac{\Delta\xi_r}{\Delta z}}_{C} \underbrace{- v\frac{\Delta f_c}{\Delta y}}_{D} \underbrace{+ f_c\frac{\Delta w}{\Delta z}}_{E} \underbrace{+ \xi_r\frac{\Delta w}{\Delta z}}_{F} + \underbrace{\left(\frac{\Delta u}{\Delta z}\cdot\frac{\Delta w}{\Delta y} - \frac{\Delta v}{\Delta z}\cdot\frac{\Delta w}{\Delta x}\right)}_{G}$$

where u, v and w are zonal, meridional and vertical wind components, ξ_r is the relative vorticity and f_c is the Coriolis parameter. Here, A is the Relative vorticity tendency (VORTED), B is the Horizontal advection (HADV), C is the vertical advection (VADV), D is the Beta term (BETA), E is the Stretching term 1 (STR1), F is the stretching term 2 (STR2), and G is the tilting term (TILT). The various terms (A–G) of the vorticity tendency equation are computed at every 6 h interval starting at 12 h of the model integration. The vorticity analysis is done for the cases CTL, SST + 2 and SST − 2. Figure 6 shows the vorticity variations in experiments with different SST values.

It is seen that in the control run the horizontal advection, and tilting terms mainly contributed for the vorticity tendency. In the advanced stages (36–66 h) the tilting and advection terms are almost equal in magnitude and have opposing effects. It has been found that notable differences arise in the advection, tilting and stretching terms with variation of SST. In the experiment (SST + 2) with SST enhanced by 2 °C, the contribution of vorticity from the above terms increases to the positive side. In this case (SST + 2) the horizontal advection of vorticity especially increased from 36 to 72 h over other terms. The maximum vorticity tendency occurred at the lowest CSLP i.e. during peak intensification of the system. From the earlier discussions it has been found that this case (SST + 2) also simulated least intensity errors. It is seen that all the three terms (advection, stretching term2 and tilting term) show positive contributions during the peak intensification of the TC. The positive changes in advection, tilting and stretching terms seem to be responsible for the northward drift of the cyclonic systems with SST + 2 °C. In the experiment (SST − 2) with reduction of SST by 2 °C all the above terms are very small and

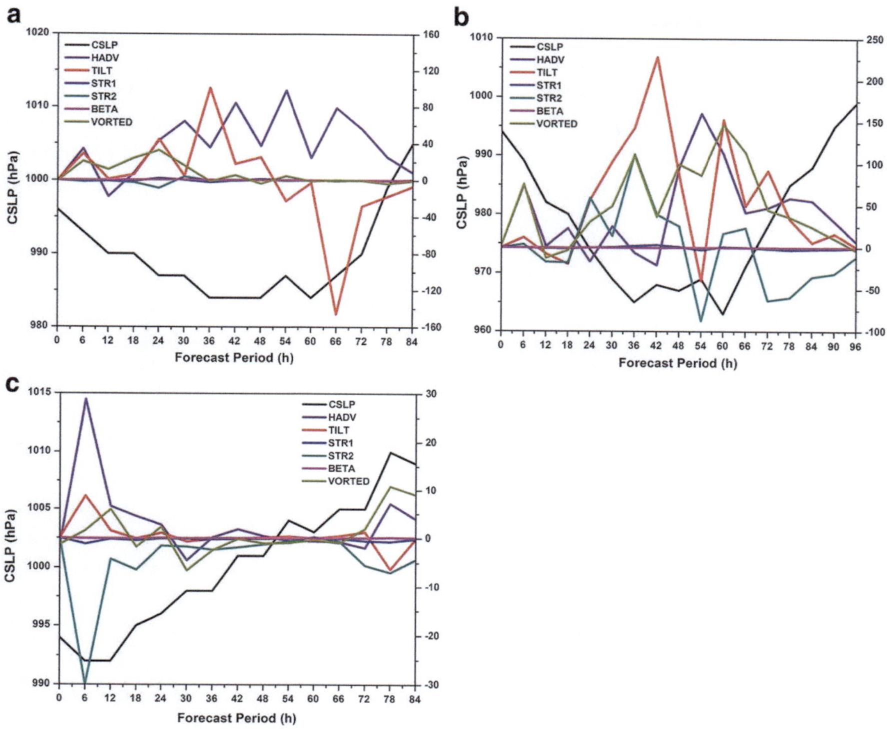

Fig. 6 Time series of various terms in vorticity equation (HADV, advection; TILT, tilting term; STR1, Stretching term1; STR2, Stretching term2; BETA, beta term) for (**a**) control experiment, (**b**) SST + 2 °C and (**c**) with SST − 2 °C experiments. The *left side* vertical axis is for CSLP and the *right side* vertical axis is for magnitude of vorticity terms

become nearly equal to the beta term but have a negative contribution. The negative changes in the vorticity terms seem to be responsible for the southward drift of the storm in the case of SST − 2. The results of the above experiments provide a mechanism for the air–sea interaction through SST parameter and its impact on the TC development. Among various predicted parameters the movement of the TC (or track) is more important as the final landfall position is often given more attention than the storm intensity in disaster mitigation (Mohapatra et al. 2015). From the perspective of track prediction the SST change by −2 °C simulated closer track to the observed track. This experiment (SST − 2) also produced the landfall position close to observations. Considering the intensity parameter, the SST change by +2 °C produced CSLP and winds closer to observations. For both intensity and track estimates the experiment (SST − 1) with reduction of SST by 1 °C produced minimum track as well as intensity errors.

4 Conclusions

An attempt is made to understand the role of SST on the evolution of TCs by performing numerical simulations with ARW mesoscale model. The case of a very severe TC Thane-2011 is simulated with a high resolution (9 km). Preliminary results indicate that SST has a large impact on the movement and intensification of the TCs. Significant differences are noticed in predictions for both track and intensity estimates when SST is altered over the analysis. While lower SSTs (<26.85 °C) led to southward deviation of the track higher SSTs (>26.85 °C) resulted in northward deviation of the track from the control experiment. The experiment with reduction of SST by −2 °C simulated minimum track errors. It has been found that SST changes influence the intensity of the storm, with positive (negative) increments of SST leading to enhancement (reduction) in pressure drop and maximum tangential winds. The experiment (SST + 2) with an increase of SST by 2 °C from analysis (~26.85 °C) produced intensity estimates close to observations. Results suggest SST changes produce opposite impacts on the movement and intensification of the storm. Analysis of average upward fluxes of sensible, latent heat and moisture in the region of TC indicated that SST changes alter the transport of heat and moisture to the TC. All the three fluxes are enhanced with higher SST values leading to enhancement of energy transport and amplification of intensity. A preliminary analysis of vorticity equation revealed that the SST changes lead to variation in the contribution by components advection, stretching and tilting terms leading to changes in the motion of the TC.

Acknowledgement Authors sincerely thank Shri S.A.V. Satyamurty Director, EIRSG, for their encouragement and support in carrying out the study. The WRF model is sourced from NCAR, USA. The GFS analysis and forecasts and SST analysis data are obtained from NCEP, USA. Authors wish to thank the anonymous reviewers for their technical comments which helped to improve the manuscript.

References

Anthes, R.A. (1982). Tropical cyclones: Their Evolution, Structure and Effects, American Meteorological Society, Science Press, Ephrata, 208.

Bender, M.A. and Ginis, I. (2000). Real-case simulations of hurricane-ocean interaction using a high-resolution coupled model: Effects on hurricane intensity. *Monthly Weather Review*, **128**, 917–946.

Bright, R.J, Xie, L. and Pietrafesa, L.J. (2002). Evidence of the Gulf Stream's influence on TC intensity. *Geophysical Research Letters*, **29**, 1801.

Chan, J.C. (2005). The Physics of TC motion. *Annual Review of Fluid Mechanics*, **37**, 99–128.

Chan, J.C.L, Duan, Y. and Shay, L.K. (2001). TC intensity change from a simple ocean–atmosphere coupled model. *Journal of Atmospheric Sciences*, **58**, 154–172.

Chen, S.S, Zhao, W, Donelan, M.A. and Tolman, H.L. (2013). Directional wind–wave oupling in fully coupled atmosphere–wave–ocean models: Results from CBLAST-Hurricane. *Journal of Atmospheric Sciences*, **70**, 3198–3215.

Emanuel, K. (2005). Increasing destructiveness of tropical cyclones over the past 30 years. *Nature*, **436**, 686–688.

Gray, W.M. (1968). Global view of the origin of tropical disturbances and storms. *Monthly Weather Review*, **96**, 669–700.

Hong, X, Chang, S.W, Raman, S, Shay, L.K. and Hodur, R. (2000). The interaction between Hurricane Opal (1995) and a warm core ring in the Gulf of Mexico. *Monthly Weather Review*, **128**, 1347–1365.

India Meteorological Department (2012). Report on Cyclonic Disturbances over North Indian Ocean during 2011. RSMC – TC Report NO. 02/2012. Regional Specialised Meteorological Centre Tropical Cyclones, New Delhi, 181 pp.

Knutson, T.R. and Tuleya, R.E. (2004). Impact of CO_2 – Induced warming on simulated hurricane intensity and precipitation: Sensitivity to the choice of climate model and convective parameterization. *Journal of Climate*, 17, 3477–3495.

Michaels, P.J, Knappenberger, P.C. and Davis, R.E. (2006). Sea-surface temperatures and tropical cyclones in the Atlantic basin. *Geophysical Research Letters*, **33**, L09708.

Mohapatra, M, Nayak, D.P, Sharma, M, Sharma, R.P. and Bandyopadhyay, B.K. (2015). Evaluation of official tropical cyclone landfall forecast issued by India Meteorological Department. *Journal of Earth System Science*, **124**, 861–874, doi:10.1007/s12040-015-0581-x

Price, J.F. (1981). Upper ocean response to a hurricane. *Journal of Physical Oceanography*, **11**, 153–175.

Price, J.F, Sanford, T.B. and Forristall, G.Z. (1994). Forced stage response to a moving hurricane. *Journal of Physical Oceanography*, **24**, 233–260.

Sanford, T.B, Black, P.G, Haustein, J, Fenney, J.W, Forristall, G.Z. and Price, J.F. (1987). Ocean response to hurricanes. Part I: Observations. *Journal of Physical Oceanography*, **17**, 2065–2083.

Schade, L.R. and Emanuel, K.A. (1999). The ocean's effect on the intensity of tropical cyclones: Results from a simple coupled atmosphere–ocean model. *Journal of Atmospheric Sciences*, **56**, 642–651.

Shay, G, Goni, J. and Black, P.G. (2000). Effects of a warm oceanic feature on Hurricane Opal. *Monthly Weather Review*, **128**, 1366–1383.

Shay, L.K. and Elsberry, R.L. (1987). Near-inertial ocean current response to Hurricane Frederic. *Journal of Physical Oceanography*, **17**, 1249–1269.

Skamarock, W.C, Klemp, J.B, Dudhia, J, Gill, D.O. Barker, D.M. Dudha, M.G. Huang, X. Wang, W. and Powers, Y. (2008). A Description of the Advanced Research WRF Version 30. NCAR Technical Note NCAR/TN-475+STR, Mesocale and Microscale Meteorology Division, National Centre for Atmospheric Research, Boulder, CO, pp. 113.

Walker, N.D.R, Leben, R, Balasubramanian, S. (2005). Hurricane-forced upwelling and chlorophyll a enhancement within cold-core TCs in the Gulf of Mexico. *Geophysical Research Letters*, **32**, L18610.

Webster, P.J, Holland, G.J, Curry, J.A. and Chang, H.-R. (2005). Changes in TC number, duration, and intensity in a warming environment. *Science*, **309**, 1844–1846.

Wu, L. and Wang, B. (2000). A potential vorticity tendency diagnostic approach for TC motion. *Monthly Weather Review*, **128**, 1899–1911.

Performance of NCMRWF Model TC Track Forecasts During 2013

R. Ashrit, A. Ashish, K. Sharma, A. Dube, I. Rani, M. Dasgupta, G.R. Iyengar, and E.N. Rajagopal

1 Introduction

There are two tropical cyclone (TC) seasons over the North Indian Ocean (NIO), (including the Bay of Bengal (BOB) and the Arabian Sea (AS)), i.e. during the pre-monsoon months (April–early June) and the post-monsoon months (October–December) (Mohanty et al. 2010). Further the Indian subcontinent happens to be one of the world's highly vulnerable areas since the coastal population density is very high leading to an extensive damage to life and property. Therefore, forecasting of TC track and landfall location is critical for early warnings and mitigation of disaster. Track forecast errors over the NIO though improved significantly in recent years (Mohapatra et al. 2013, 2015) are still high relative to those over the Atlantic and Pacific Oceans. With advancements in computational power, development of better NWP models (both global and regional), the forecasting capability of meteorologists have greatly increased. Several meteorological centers like NCEP, UKMet office, ECMWF, JMA, JTWC etc give a real time forecast of TC tracks from their global NWP models (deterministic as well as Ensemble Prediction Systems (EPS)) (Hamill et al. 2011; Froude et al. 2007; Buckingham et al. 2010; Heming et al. 1995; Heming and Radford 1998). TC track prediction from an ensemble forecasting system besides providing a track from each ensemble member also provides the strike probability (Weber 2005). For the TCs of NIO, Mohapatra et al. (2013, 2015) provided a detailed verification of the official forecast tracks and its improvements in the recent past. This study provides a detailed verification of the NCMRWF NWP model forecasts of 2013 TC cases. Some of the earlier studies (Ashrit et al. 2014; Chourasia et al. 2013 and Mohandas and Ashrit 2014) focused

R. Ashrit (✉) • A. Ashish • K. Sharma • A. Dube • I. Rani • M. Dasgupta • G.R. Iyengar
E.N. Rajagopal
National Centre for Medium Range Weather Forecasting, Noida, India
e-mail: raghu@ncmrwf.gov.in

© Capital Publishing Company 2017

M. Mohapatra et al. (eds.), *Tropical Cyclone Activity over the North Indian Ocean*,
DOI 10.1007/978-3-319-40576-6_19

on the NCMRWF model TC forecasts and the impact of bogusing, assimilation and cumulus parameterisation etc. The present study is focused on the real time operational forecasts provided to India Meteorological Department (IMD). During May–December 2013, there were five TCs observed in the Bay of Bengal namely: Viyaru (May10–17), Phailin (October 4–14), Helen (November 19–23), Lehar (November 19–28) and Madi (December 6–13). This report summarises the performance of the real time prediction of these TC tracks by the NCMRWF Global Forecast Systems.

A brief description of the TC history is given in this section. It is followed by details on the modelling systems operational at NCMRWF and the TC tracking methodology in the Sect. 2. Verification results of the TC forecast tracks during 2013 are discussed in Sect. 3. The Summary of the results is discussed in Sect. 4.

1.1 Brief Description of the Tropical Cyclones During 2013

The first case is the cyclonic storm (CS) 'Viyaru' during 10–17 May 2013. It originated from an area of low pressure over the southern Bay of Bengal around 10 May 2013 which attained its peak intensity with winds of 85 km/h (50 mph) and a barometric pressure of 990 mb on 16 May 2013 (RSMC, New Delhi 2014). Shortly, thereafter Viyaru made landfall near Chittagong, Bangladesh. On May 17, it moved over the eastern Indian state of Nagaland. The second case is a very severe cyclonic storm (VSCS) 'Phailin' that lasted during 9–12 October 2013. On 4 October 2013, it started as a tropical depression within Gulf of Thailand and started tracking westwards into Bay of Bengal. On 9 October 2013, it was named 'Phailin' which intensified into VSCS on 10 October 2013 which tracked towards Odisha coast. It made landfall on 12 October 2013, near Gopalpur in Odisha coast at around 2130 IST (1700 UTC). It subsequently weakened over land as a result of frictional forces, before it was last noted on October 14, as it degenerated into a well-marked area of low pressure. Third case is of the severe cyclonic storm (SCS) 'Helen' that lasted during 19–23 November 2013. Emerging out of a well developed trough on 19 November 2013, 'Helen' intensified into a depression (D) and then into deep depression (DD) on same day and into CS and SCS on 21 November 2013 with peak intensity of 100 km/h (62 mph) with a central pressure of 990 mbar. It made landfall south of Machilipatnam, Andhra Pradesh and rapidly deteriorated into a deep depression. The VSCS 'Lehar' lasted during 23–28 November 2013. Emerging from a low pressure system in south China Sea that crossed to Andaman Sea on 22 November 2013, it intensified into D, DD and CS 'Lehar'. On 25 November, it gradually consolidated further and was upgraded to a SCS by the IMD. The following day, Lehar further intensified into a VSCS. It featured peak winds of 140 km/h (87 mph) and a central pressure of 982 mbar. Thereafter, Lehar rapidly weakened into a depression and made landfall near Machilipatnam on 28 November 2013. The VSCS 'Madi' that lasted during 6–13 December 2013 is perhaps the most challenging case for any track forecasting system. The system first tracked northwards and then tracked in the southwesterly direction to cross the

Tamil Nadu coast. The D formed on the 6 December 2013 near Sri Lanka. It further strengthened into a CS, and was named Madi. Later on same day it further intensified into SCS. It gradually tracked northwards and was upgraded to VSCS on 8 December 2013. The system weakened on 9th and 10th while kept tracking northwards. It started tracking south-westwards on 10 December and kept weakening. On 12 December 2013, the system crossed Tamil Nadu coast twice with the intensity of a depression; first, near Nagapattinam at around 1200 UTC and the near Tondi at around 1700 UTC (Regional Specilised Meteorological Centre, New Delhi 2014).

2 Global Forecast Model Details and Tracking of TCs

2.1 NCMRWF Global Analysis and Forecast Systems

The TC track forecasts are based on NGFS (T574L64) and the Unified Model (UM) of Met Office (NCUM) (Rajagopal et al. 2012) operational at NCMRWF. Additionally, the track and intensity forecasts based on the NCMRWF Global Ensemble Forecast System (GEFS; T190L28) are also provided to IMD on experimental basis. The NGFS (T574L64) and NGEFS (T190L28) forecasts are available out to 240 h while the NCUM forecasts are available up to 168 h. Table 1 lists some of the model configuration details for the two deterministic models (NGFS and NCUM). The ensemble system (NGEFS) is a 20 member ensemble system. The control for the ensemble system is obtained from the NGFS initial analysis. The NGEFS uses a set of 20 perturbed initial conditions to obtain 20 member forecasts. The perturbed initial conditions are obtained using the Ensemble Transform with Rescaling (ETR) method. The detailed description of the implementation of NGEFS at NCMRWF and its configuration is reported in (Ashrit et al. 2013).

2.2 The Use of Observations Over the Ocean in the Model Initial Conditions

Data coverage over the NIO has improved in recent years due to increased deployment of buoys and satellite (scatterometer) observations. These are assimilated in NCMRWF models NGFS and NCUM. Figure 1 shows the spatial coverage of surface and satellite observations over BOB for case of 'Helen' (18z20Nov 2013; *top panels*), 'Lehar' (06z24Nov 2013; *middle panels*) and 'Madi' (18z08Dec2013; *bottom panels*).

Table 1 NCMRWF realtime global forecast model configurations

	NGFS	NCUM
Horizontal resolution	T574 (~23 km near equator) with a Gaussian grid of 1760 × 880 points	N512 (~25 km at mid-latitudes) with a EW-NS grid of 1024 × 769 points
Vertical levels	Hybrid sigma-pressure (64 levels)	70 vertical levels
Model time step	2 min	10 min
Forecast length	10 days	10 days
Data assimilation	Grid point Statistical Interpolation – GSI	4D var Data assimilation system
Dynamics	Spectral, hybrid sigma-p, reduced Gaussian grids	Non-hydrostatic dynamics with deep atmosphere. Height vertical coordinates with levels transitioning from terrain following to height. Global latitude–longitude grids
Time integration	Leapfrog/semi-implicit	Semi-implicit integration with 3D semi-Lagrangian advection

2.3 *Tracking of TCs in the Global Model Analysis and Forecasts*

The TC forecast tracks are derived based on vertical weighted average of the maximum or minimum of several parameters in the vicinity of a vortex in the input first guess (latitude and longitude) and forecasts (Marchok 2002). Briefly, for TCs, seven parameters are tracked, including the relative vorticity maximum, geopotential height minimum and wind speed minimum at both 850 and 700 hPa, as well as the minimum in sea-level pressure. The locations based on these parameters are averaged together to provide an average position fix at each forecast hour. In order to avoid tracking weak and transient disturbances (either real or artifacts of model noise), two constraints have been added to the tracking criteria in order for a found disturbance to be reported as being a tracked storm viz: (1) the storm must live for at least 24 h within a forecast, and (2) the storm must maintain a closed Mean Sea Level Pressure (MSLP) contour, using a 2 mb contour interval.

Based on the above described tracking algorithm, the TCs tracks are obtained from the two deterministic models; NGFS and NCUM, as well as from each of the ensemble members of NGEFS. Based on the location of tracked TC at any given time in each of the member forecasts, the tracker also provided ‘strike probability’ map. It is the probability of finding a TC within 100 km of the mean track. Thus, from the ensemble system, (i) mean track, (ii) member tracks and (iii) strike probability, the three products of importance to TC are obtained.

The real time forecast tracks based on NGFS and NCUM along with the mean track and strike probabilities based on NGEFS are provided to IMD.

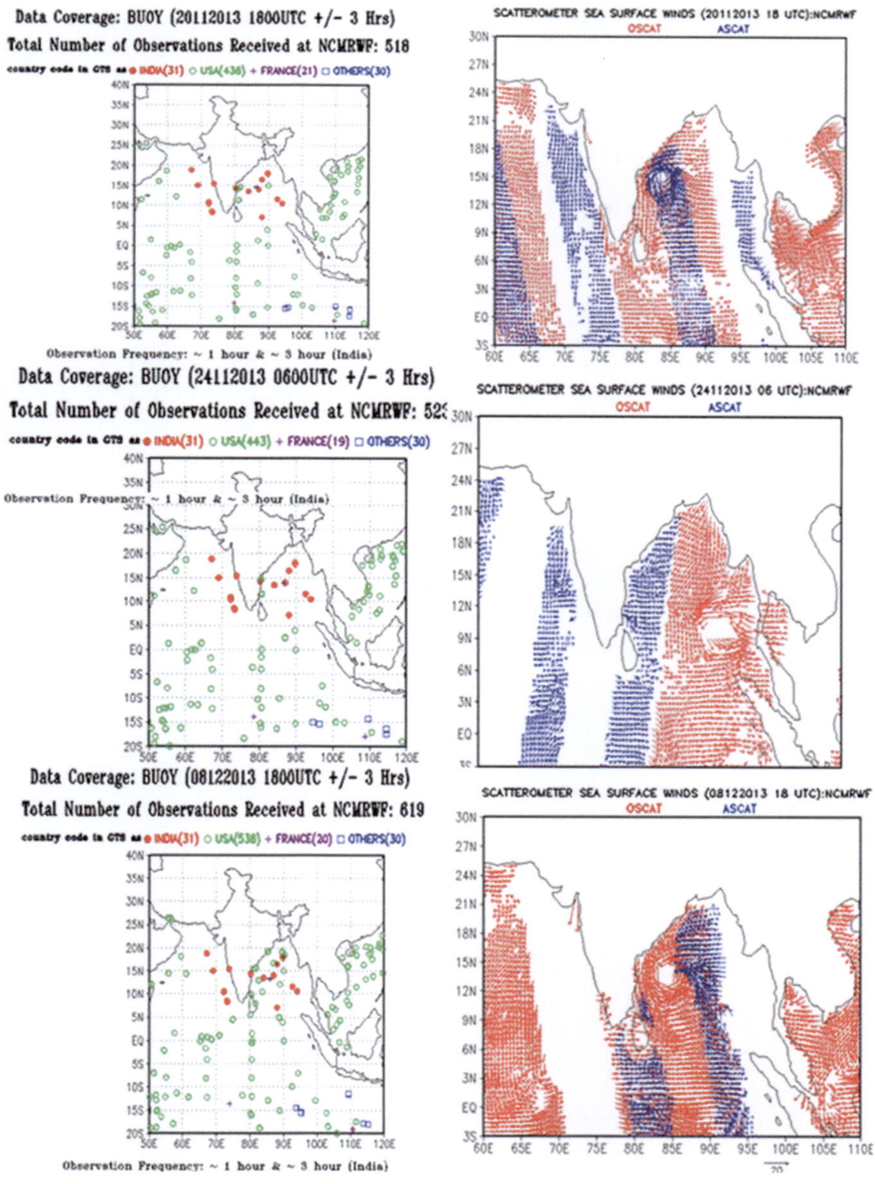

Fig. 1 The buoy data coverage (*left panels*) and the satellite scatterometer winds (*right panels*) over the Northern Indian Ocean (NIO) assimilated in NCMRWF model runs in respect of (i) TC Helen 20 Nov. 2013/18 UTC (ii) TC Lehar 24 Nov. 2013/06 UTC (iii) TC Madi 08 Dec. 2013/18 UTC

3 Verification of 2013 TC Track Forecasts

For verification of the TC forecast tracks, direct position error (DPE), along track error (ATE) and cross track error (CTE) are computed against the IMD reported best track data. Further, the forecast landfall position error and landfall time error are also computed.

3.1 'Viyaru' Forecast Tracks (11–16 May 2013)

The forecast tracks based on the 00 UTC initial conditions from 11 to 16 May 2013 were generated and communicated to the IMD in near-real time. Figure 2 shows the forecast tracks based on the 00 UTC 11 May 2013 initial conditions. The forecast based on the two deterministic models (NGFS and NCUM) indicate landfall over Myanmar. NGEFS tracks consistently show movement towards Myanmar. The ensemble average track and strike probability are also shown. Strike probability is the probability of a given location (grid point) being within a specified distance (~101 km) of an ensemble member track point. Strike probability is calculated both individually for each forecast hour and for the total accumulated probability up to 120 h forecast.

3.1.1 'Viyaru' Forecast Track Errors (11–16 May 2013)

Verification of forecast tracks is shown in Table 2. The average initial position error is lowest in the NGFS system at 52 km. The average forecast track error is lowest in NCUM in 24, 48 and 72 h forecasts. The average forecast track error in the NGEFS ensemble mean is consistently lower compared to NGFS at all lead times except at 120 h.

3.2 'Phailin' Forecast Tracks (9–12 October 2013)

Figure 3 shows the forecast tracks based on the 00 UTC 9 October 2013 initial conditions. Forecast positions based on NGFS and NGEFS are shown at 6-h interval while the forecast positions based on NCUM is shown at 24-h interval. The forecasts indicate landfall over Andhra Pradesh and Odisha border. The forecasts closely match with the observed track although with a time delay as can be seen by the time indicated in the Fig. 3.

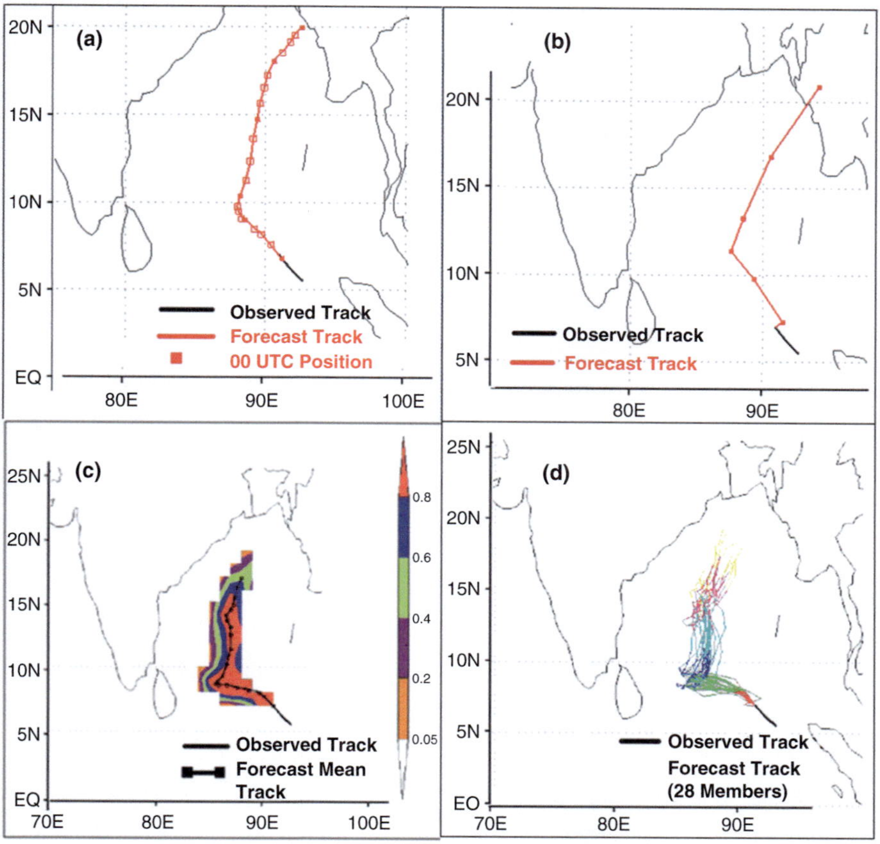

Fig. 2 Observed and forecast tracks based on 00UTC 11th May 2013 initial conditions in (**a**) NGFS (T574L64) (**b**) NCUM and NGEFS (**c**) strike probability and (**d**) ensemble member tracks

Table 2 Average forecast position errors (km) in the three modelling systems in respect of Viyaru (11–16 May 2013)

F/C hours	00	24	48	72	96	120
NGFS	52	205	383	542	584	324
NCUM	66	111	165	285	355	499
NGEFS	59	205	311	336	283	379

3.2.1 Forecast Track Errors (9–12 October 2013)

Forecast track errors are computed based on the TC positions from 00 UTC of 9–12 October 2013. The initial position errors in all three models are less than 50 km. The highest (lowest) initial position error of 45 km (26 km) is seen in NGFS (NGEFS) while the both NCUM model has initial error of 27 km. NCUM has the

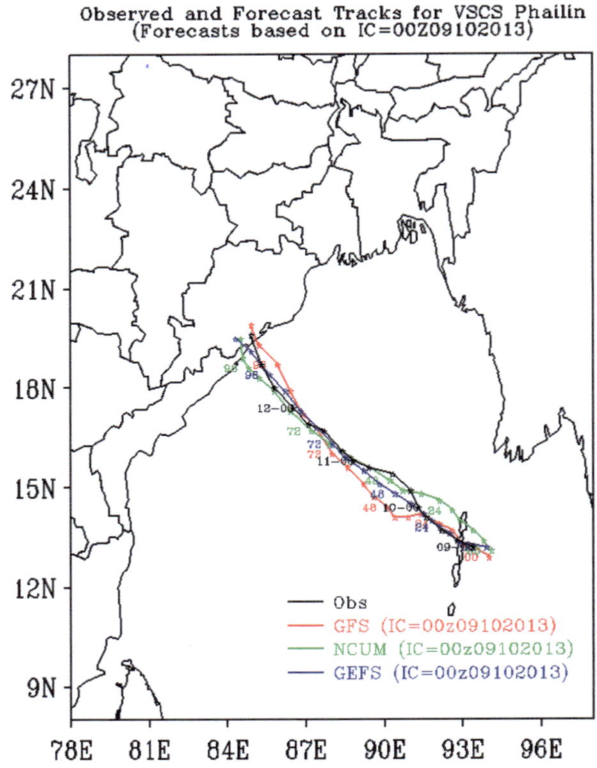

Fig. 3 Observed and forecast tracks of VSCS Phailin based on 9 October 2013

least position error at all lead times while NGEFS mean track consistently shows lower error than that of NGFS.

3.2.2 Landfall Position and Time Error

The IMD reported landfall is considered at 1500 UTC of 12 October 2013 at 19.1° N 85.0° E. Table 3 shows the landfall position and time errors based on all the available track forecasts. NCUM forecasts show least error in predicted landfall position and time with the exception of forecast based on 9 October 2013. NGEFS forecast can be considered next best. Similar to the forecast track errors discussed in the last section, NGEFS shows marginal improvement over the NGFS in predicting the landfall time and position.

Table 3 Forecast landfall position error for VSCS Phailin in the ESSO-NCMRWF global models (IMD reported landfall at 15 UTC of 12 October at location 19.1° N, 85.0° E)

	NGFS		NGEFS		NCUM	
Initial conditions	Position error (km)	Time error (h)	Position error (km)	Time error (h)	Position error (km)	Time error (h)
IC = 09102013	31	−15	10	−6	47	−15
IC = 10102013	84	−15	70	−12	11	−3
IC = 11102013	42	−9	33	−9	39	−3
IC = 12102013	115	−15	78	−21	69	−3

3.3 'Helen' Forecast Tracks (20–22 November 2013)

The observed and forecast tracks from NGFS, NCUM and NGEFS (mean and control) are presented based on 21 November 2013 in Fig. 4. The forecast positions are shown at 6-h interval.

The forecast tracks of all three models are much to the south of observed track. In NCUM the initial position is also much to the south of observed location. Forecast tracks based on 20 and 22 November 2013 (not shown) suggest NGEFS mean track closely compares with observed track. The NCUM and GFS track forecast based on 21 and 22 are also considerably improved compared to the tracks based on 20 Nov 2013.

3.3.1 Forecast Track Errors (20–22 November 2013)

Forecast track errors are computed based on 00 UTC TC positions from 20 to 22 November 2013 and average track errors are presented in Fig. 5a–c. The DPE, ATE and CTE are presented at 6-h interval up to 60 h. On one hand, positive (negative) values of ATE indicate that the movement of the TC in the forecasts is slower (faster) compared to the observations. On the other hand, positive (negative) values of CTE indicate that forecast track is right (left) of the observed track.

The initial position errors in NGFS and NGEFS models are less than 50 km. The highest (*lowest*) initial position error of 113 km (*14 km*) is seen in NCUM (N*GEFS*). NGEFS mean track shows least error at all lead times while NCUM shows highest average error at all lead times. NCUM and NGEFS mean tracks show relatively higher ATE up to 36 h varying from under 50 km to over 100 km. The prominently negative values in the CTE up to 54 h, shown in Fig. 5c indicate the forecast TCs tracks lie to the left of observed tracks. NGEFS mean track shows least CTE at all lead times.

Fig. 4 Observed and forecast tracks of SCS Helen based on 21 November 2013/00 UTC

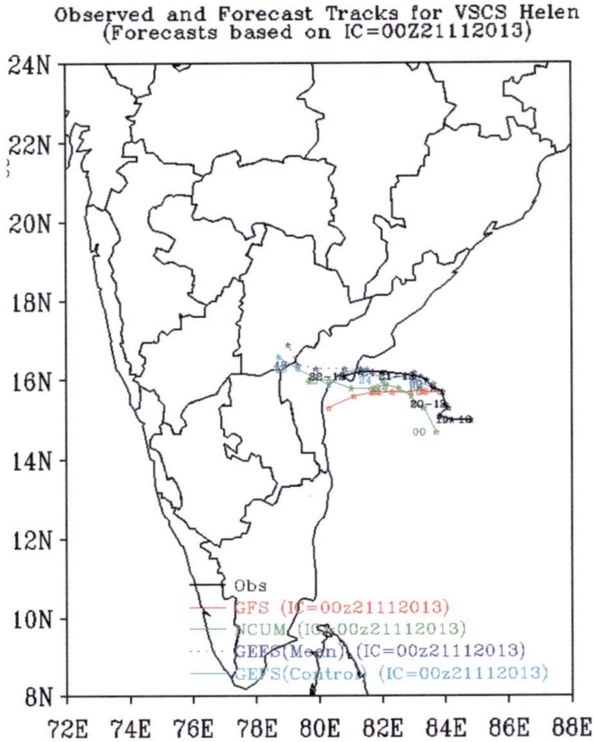

3.3.2 Error in Forecast Landfall Position and Time

The IMD reported that the SCS Helen crossed the coast between 0900 UTC of 22 November 2013 south of Machilipatnam in Andhra Pradesh at 16.1° N 81.3° E. Table 4 shows the landfall position and time errors based on all the available track forecasts. Forecasts show highest error in predicted landfall position based on 20th initial conditions. The predicted land fall time error varies from −3 to +9 h. Based on 21st and 22nd, NGEFS forecast shows least position error (48 and 24 km) and time error (+3 and −3 h). Both NCUM and NGFS have large position error.

3.4 'Lehar' Forecast Tracks (23–28 November 2013)

The observed and forecast tracks from NGFS, NCUM and NGEFS (mean and control) are presented based on initial conditions of 24–28 November 2013 in Fig. 6a–e. The forecast positions are shown at 6-h interval.

Forecasts based on 24, 25 and 26th initial conditions clearly suggest NGEFS mean track closely follows the observed track for most of the forecast period. The NCUM forecasts on the other hand show large deviation. NGFS forecasts too (to a

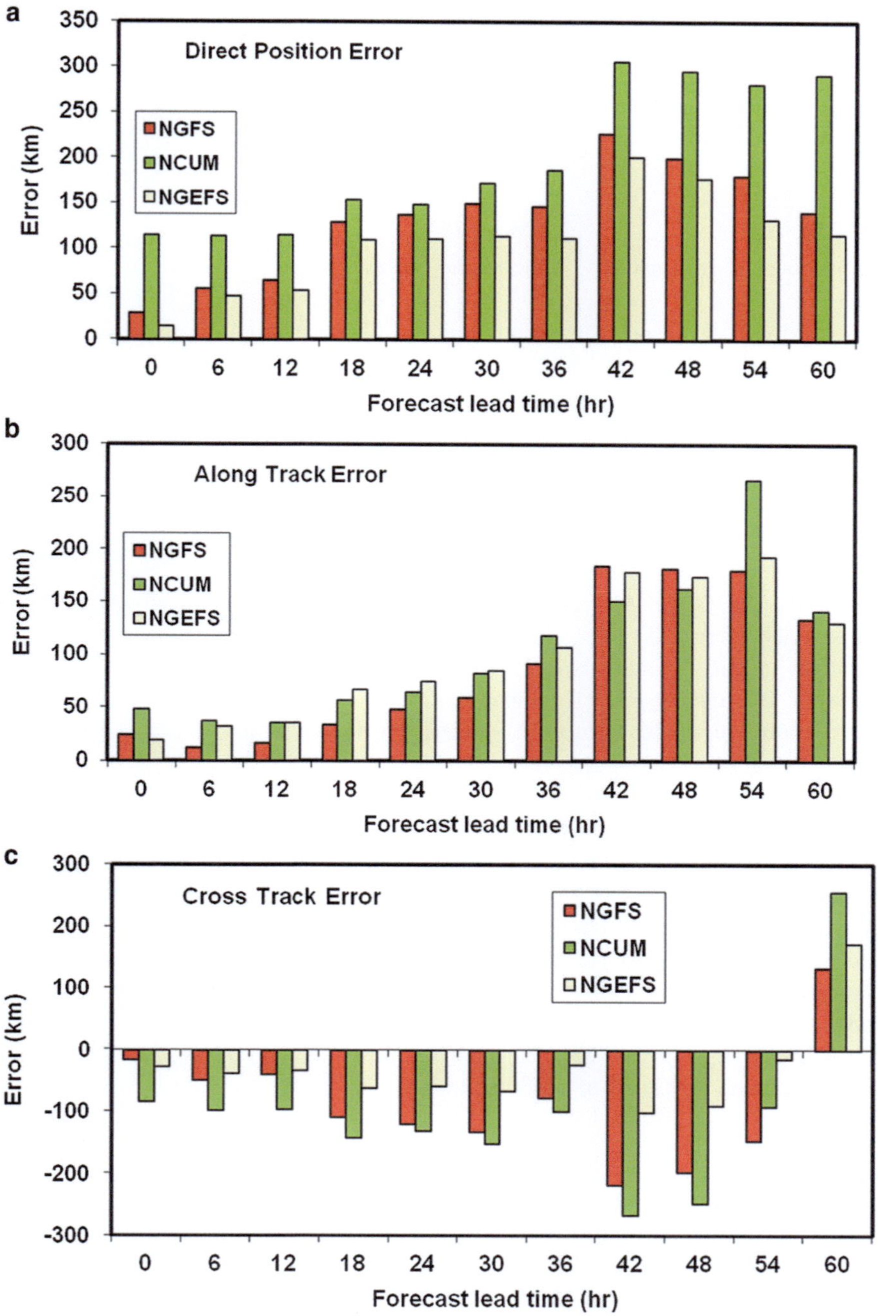

Fig. 5 Average forecast track errors for SCS 'Helen' expressed in terms of (**a**) Direct position error (**b**) Along track (*lag* or *lead*) error and (**c**) Cross track (*left* or *right*) error

Table 4 Forecast landfall position error for SCS Helen in the ESSO-NCMRWF global models (IMD reported landfall at 09 UTC of 22 November at location 16.1° N, 81.3° E)

	NGFS		NGEFS		NCUM	
Initial conditions	Position error (km)	Time error (h)	Position error (km)	Time error (h)	Position error (km)	Time error (h)
IC =20112013	147	+9	154	+9	267	−3
IC = 21112013	131	+9	48	+3	39	+9
IC = 22112013	39	+3	24	−3	101	−3

('+' indicates delayed landfall, '−' indicates early landfall)

lesser extent) show some deviation from the observed track. Forecasts based on 27th and 28th also show wide dispersion from the observed track.

3.4.1 Forecast Track Errors (24–28 November 2013)

Forecast track errors are computed based on the 00 UTC TC positions from 24 to 28 November 2013 and average track errors are presented in Fig. 7a–c. The direct position error (DPE), Along track error (ATE; time lag/lead in movement) and Cross track error (CT; left/right error) are presented at 6-h interval up to 60 h. Positive (*negative*) values of ATE indicate that the movement of the TC in the forecasts is slower (*faster*) compared to the observations. On the other hand positive (*negative*) values of CTE indicate that forecast track is right (*left*) of the observed track.

The initial position errors in NGFS and NGEFS models are less than 50 km. The highest (lowest) initial position error of 72 km (19 km) is seen in NCUM (NGEFS). Up to 48 h all three models have comparable track errors with marginally higher errors in NGEFS. Beyond 48 h, NGEFS and NGFS forecasts show comparable errors while NCUM track shows very large error. The ATE values range from about less than 10 km (NGEFS) in the first 24 h to about 200 km in NCUM up to 72 h forecasts. The predominantly positive values of CTE in NCUM shown in Fig. 7c indicate the forecast TCs tracks lie to the right of observed tracks. NGFS (NGEFS mean) tracks lie to the left up to 24 h (60 h). Subsequently, both NGFS and NGEFS mean tracks lie to the left of observed tracks.

3.4.2 Error in Forecast Landfall Position and Time

The IMD reported that the VSCS Lehar crossed the coast between 0900 UTC of 28 November 2013 south of Machilipatnam in Andhra Pradesh at 15.9° N 81.1° E. Table 5 shows the landfall position and time errors based on all the available track forecasts. Forecasts show highest error in predicted landfall position and time based on 24 November 2013 initial conditions. Highest landfall position error of about 500 km is seen in NCUM forecast based on 24 November 2013. The predicted

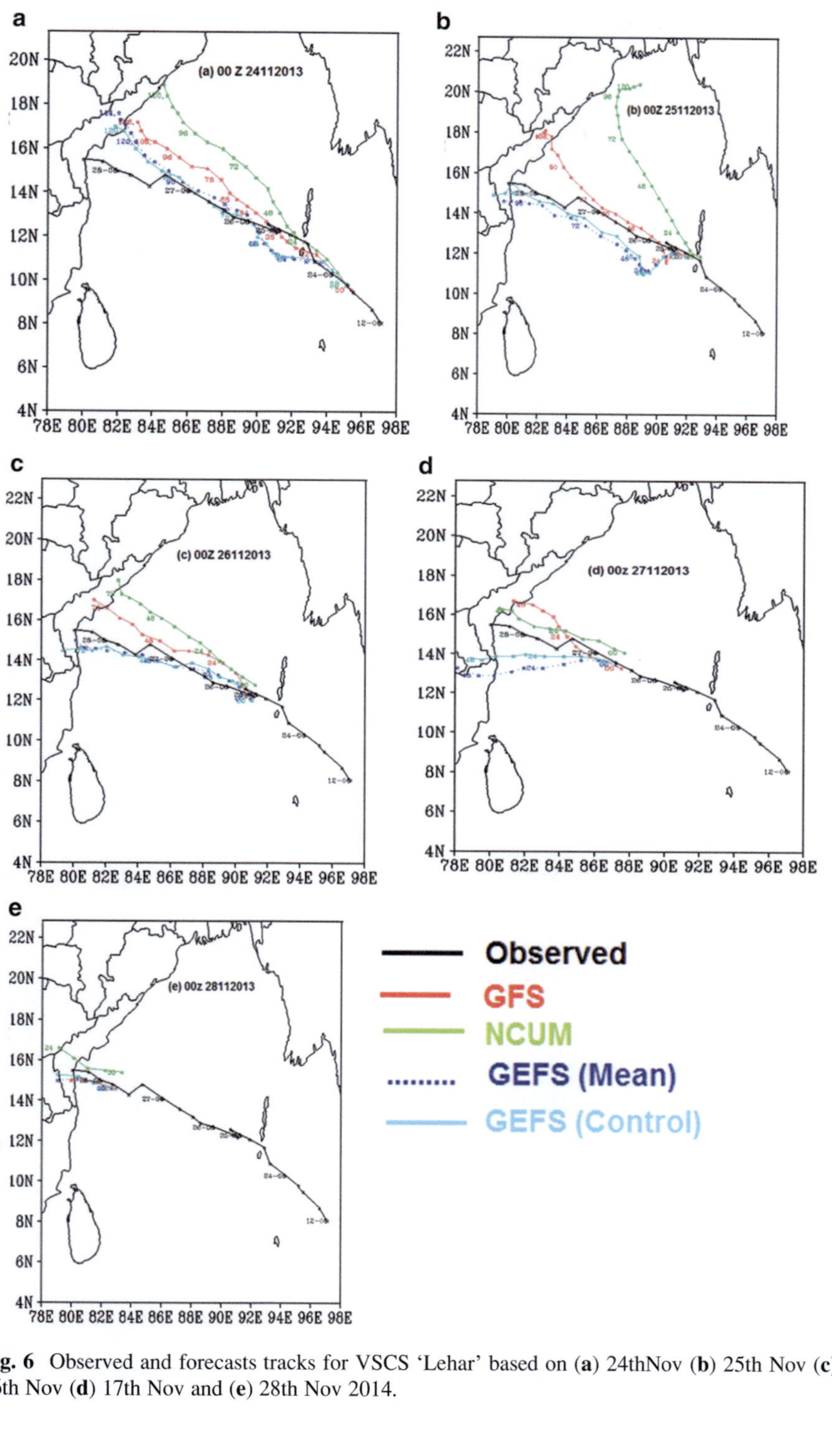

Fig. 6 Observed and forecasts tracks for VSCS 'Lehar' based on (**a**) 24thNov (**b**) 25th Nov (**c**) 26th Nov (**d**) 17th Nov and (**e**) 28th Nov 2014.

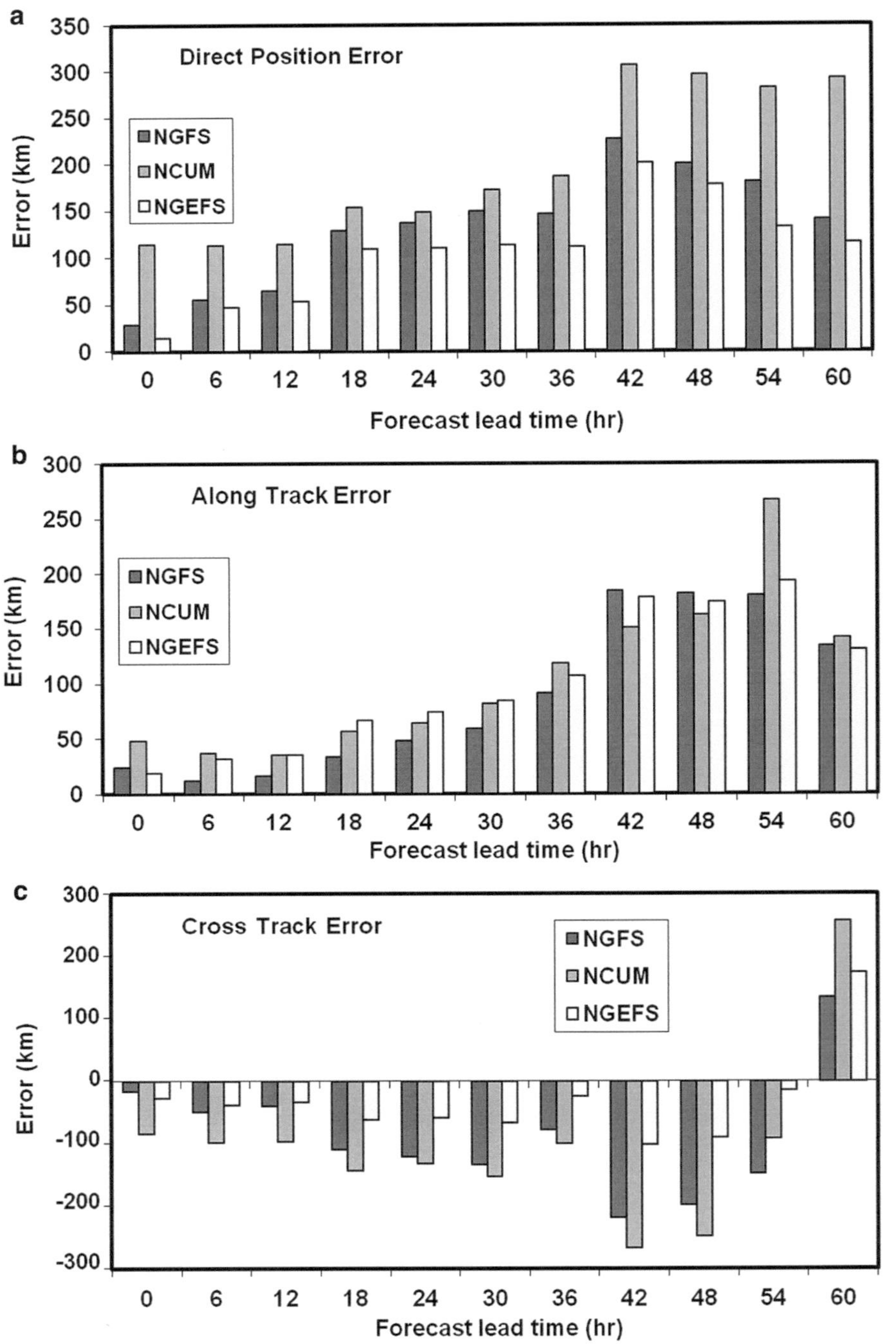

Fig. 7 Average forecast track errors for VSCS 'Lehar' expressed in terms of (**a**) Direct position error (**b**) Along track (*lag* or *lead*) error and (**c**) Cross track (*left* or *right*) error

Table 5 Forecast landfall position error for VSCS Lehar in the ESSO-NCMRWF global models (IMD reported landfall at 09 UTC of 28nd November at location 15.9° N, 81.1° E)

	NGFS		NGEFS		NCUM	
Initial conditions	Position error (km)	Time error (h)	Position error (km)	Time error (h)	Position error (km)	Time error (h)
IC = 24112013	257	+21	192	+27	499	+21
IC = 25112013	240	−3	148	+21	–	–
IC = 26112013	123	+15	154	+21	246	+15
IC = 27112013	154	+15	347	+9	35	+15
IC = 28112013	163	+9	123	+3	35	+3

('+' indicates delayed landfall, '−' indicates early landfall)

landfall time error varies from −3 to +27 h. NGFS and NGEFS show landfall position errors over 100 km in all the forecasts. On 27th and 28th NCUM forecasts show least landfall position errors of 35 km with landfall time errors of +15 h and +3 h, respectively.

3.5 'Madi' Forecast Tracks (6–12 December 2013)

The observed and forecast tracks from NGFS, NCUM and NGEFS (mean and control) are presented based on initial conditions starting from 6, 8, 9, 10 and 12 December 2013 in Fig. 8a–e. The forecast positions are shown at 6-h interval. Forecasts based on 6 December 2013 initial conditions do not show clear movement and landfall of the cyclonic system (Fig. 8a). On 8 and 9 December 2013 the forecasts generally indicated northward movement in the beginning and then southwestwards (Fig. 8b, c). NGFS on both days (and NGEFS on 9 December 2013) suggested the TC would strike Sri Lanka coast, while NCUM consistently predicted the TC crossing Tamil Nadu. Tracks based on 10th, 11th (not shown) and 12th consistently showed TC would cross the Tamil Nadu coast near Nagapattinam (Fig. 8d, e).

3.5.1 Forecast Track Errors (6–12 December 2013)

Forecast track errors are computed based on the 00UTC TC positions from of 6–12 December 2013 and average track errors are presented in Fig. 9a–c. DPE, ATE and CTE; are presented at 6-h interval up to 126 h.

The initial position errors in NGFS and NGEFS models are less than 50 km. The highest (lowest) initial position error of 86 km (13 km) is seen in NCUM (NGEFS). Up to 24 h NCUM has high DPE of about 100 km while NGFS and NGEFS mean show DPE increasing from under 50 km to over 100 km. From 24 to 78 h the NGFS (and NGEFS) DPE increase rapidly 463 km (434 km). Growth of DPE in NCUM is

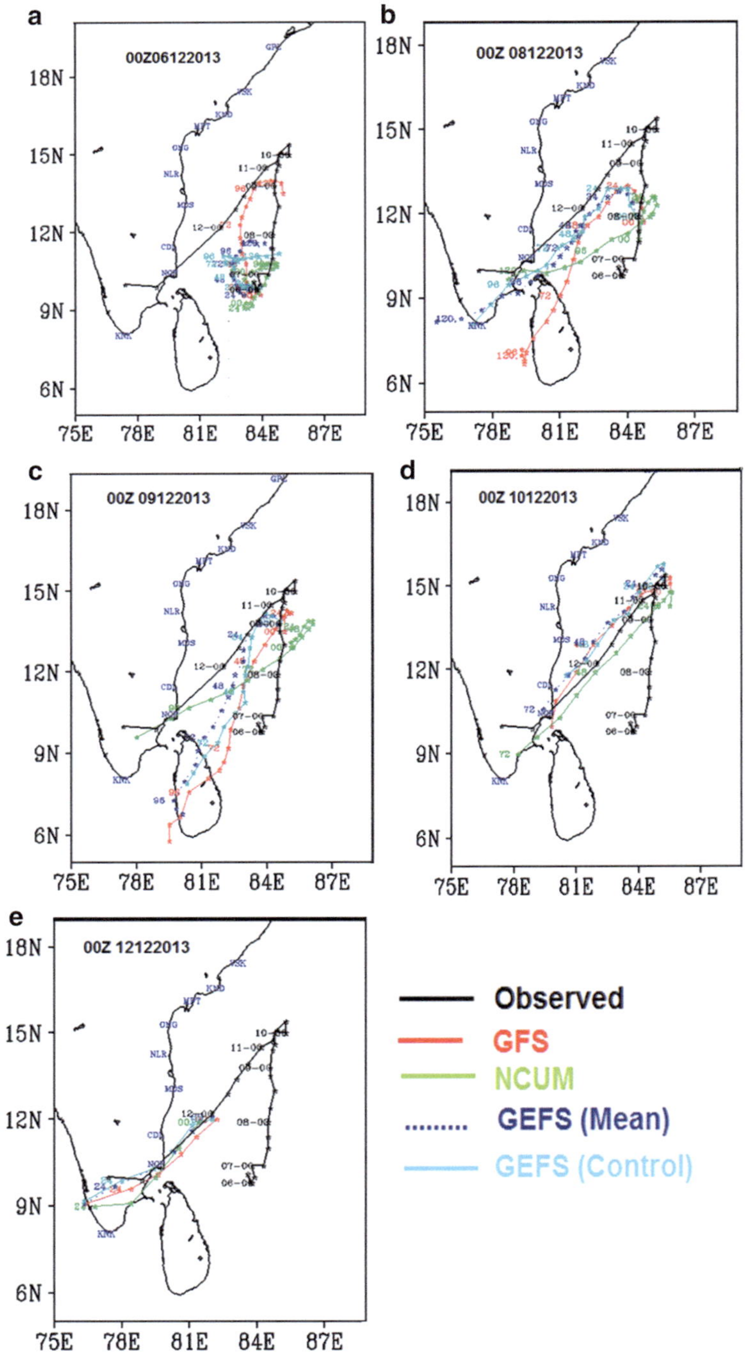

Fig. 8 Observed and forecasts tracks for VSCS 'Madi' based on 00 UTC of (**a**) 6th Dec (**b**) 8th Dec (**c**) 9th Dec (**d**) 10th Dec and (**e**) 12th Dec 2013.

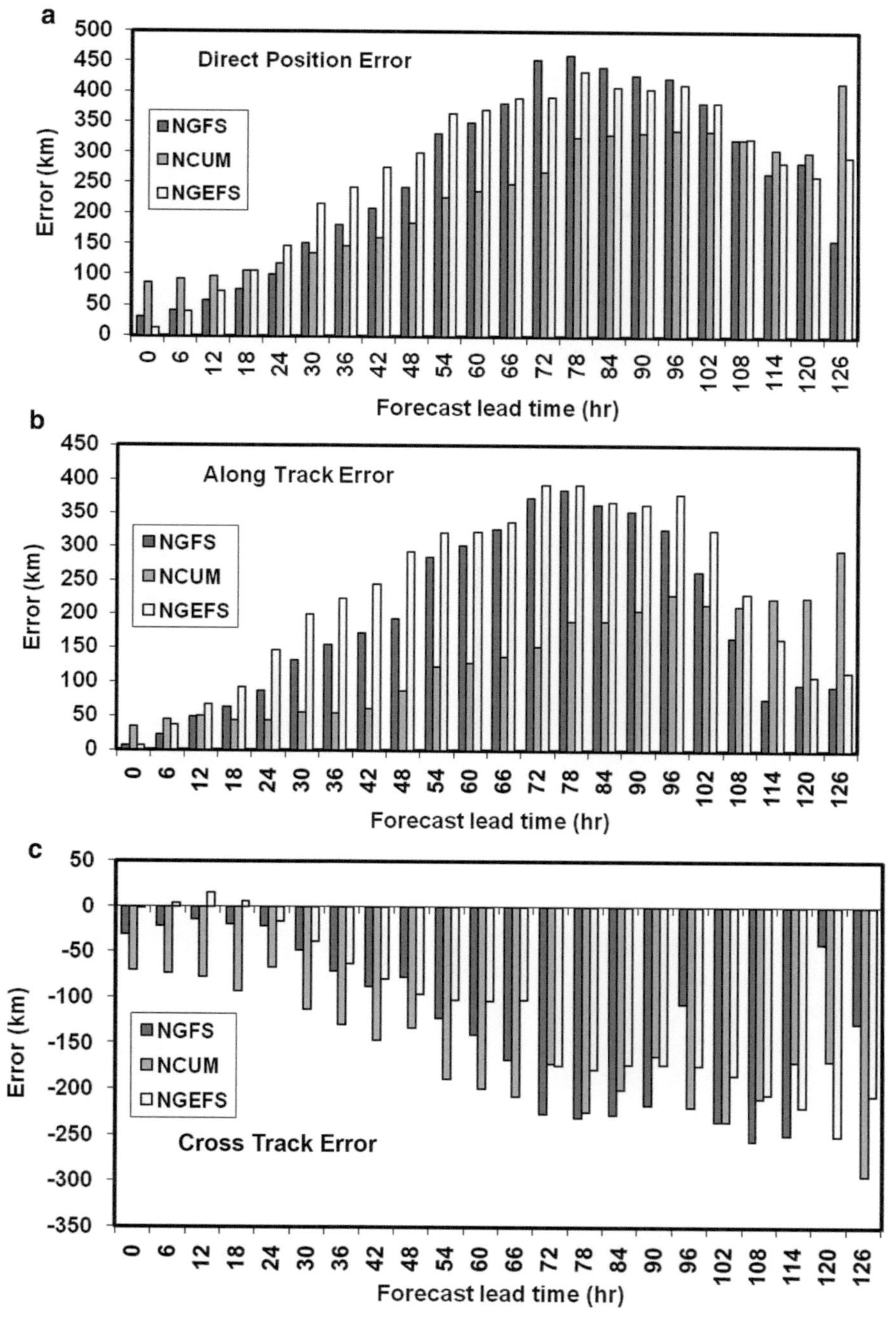

Fig. 9 Average forecast track errors for VSCS 'Madi' expressed in terms of (**a**) Direct position error (**b**) Along track (*lag* or *lead*) error and (**c**) Cross track (*left* or *right*) error

gradual in the 78 h with highest value of 325 km and 78 h. The DPE in NGFS and NGEFS gradually reduce after 78 h (after 102 h in NCUM). Similar pattern of error growth is seen for ATE (Fig. 9b). The predominantly negative values of CTE in NCUM shown in Fig. 9c indicate the forecast TCs tracks lie to the left of observed tracks. This is also evident from tracks based on 6–9 December 2013 (Fig. 8). The NGEFS mean track lies to the right of observed track on 10–12 December 2013. During the same period the NGFS and NCUM tracks show varying movement on both sides of the observed track.

3.5.2 Error in Forecast Landfall Position and Time

The IMD reported that the VSCS Madi crossed the Tamil Nadu coast near Tondi around 1700 UTC of 12 December 2013 at 10.0° N 78.5° E. Table 6 shows the landfall position and time errors based on the track forecasts from 8 to 12 December 2013 (track forecasts on 6 and 7 December 2013 did not show landfall). On 8 December NGFS shows a large time error of −41 h with a landfall over Sri Lanka. NGFS and NGEFS tracks on 9 December 2013 showed landfall over Sri Lanka while NCUM showed landfall over Tamil Nadu coast with position error of 125 km. NGFS and NGEFS have large position error on 10 December 2013 while NCUM has least error in terms of time as well as distance. Similarly, on 11 and 12 December 2013 NCUM has least distance and time error in the predicted landfall.

3.6 Average Forecast Track Errors

The forecast track errors expressed in terms of average DPE for the five TCs of 2013 are shown in Fig. 10. The Figure also provides an intercomparison of the three models used for TC track prediction. The graphics shown clearly suggests that NGEFS has the least initial position error followed by NGFS (both below 50 km). The forecast position errors in NCUM are the lowest in 24, 48 and 72 h.

The initial position errors to a large degree determine the forecast position errors. An improvement in the accuracy of initial position in the model analysis is sure to improve the accuracy of forecast positions. Figure 11 shows the improvement in the analysis position in the NCMRWF model (NGFS). The consistent and sustained reduction in the initial position errors.

The improvement in NGFS system is attributed to TC relocation implemented in 2008. Further the assimilation of satellite radiances and other observations from non-conventional platforms has contributed towards reduction of initial position errors.

Table 6 Forecast landfall position error for VSCS Madi in the ESSO-NCMRWF global models (IMD reported landfall at 1700 UTC of 12 December 2013 at location 10° N, 78.5° E near Tondi in Tamil Nadu)

	NGFS		NGEFS		NCUM	
Initial conditions	Position error (km)	Time error (h)	Position error (km)	Time error (h)	Position error (km)	Time error (h)
IC = 08122013	291	−41	110	+13	77	+7
IC = 09122013	403	−5	271	−5	125	+13
IC = 10122013	270	−5	218	+1	79	+1
IC = 11122013	171	−5	265	−11	171	−5
IC = 12122013	121	+1	104	+1	100	+1

('+' indicates delayed landfall, '−' indicates early landfall)

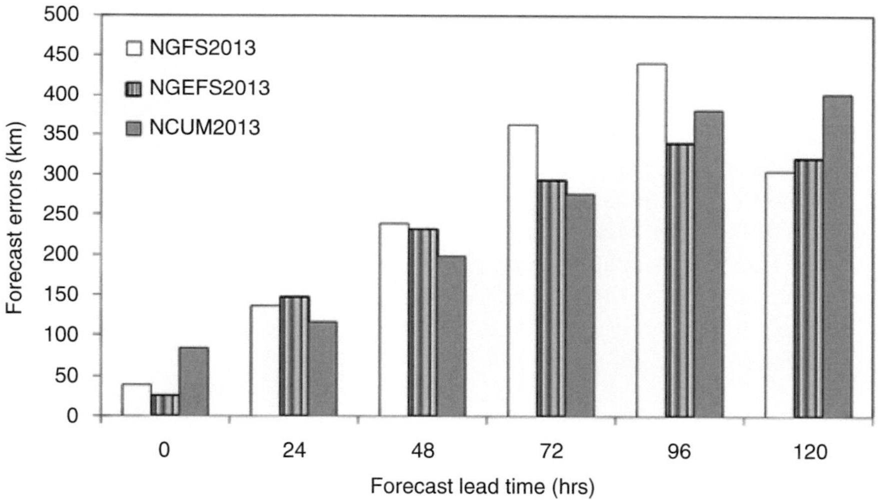

Fig. 10 Average forecast track errors in the NCMRWF models during 2013

4 Summary and Conclusions

The 2013 TC track forecast verification presented in this study to report the performance of three modelling systems used at NCMRWF. Two deterministic systems namely NGFS and NCUM and an ensemble model NGEFS are used to provide TC track forecasts to IMD. The verification-cum-intercomparison results can be summarised as below.

1. The average DPEs for the 2013 TC cases show that NGEFS has the least initial position error followed by NGFS (both below 50 km). During 2008–2013 there is nearly 48 % reduction in the initial position errors in the NGFS.
2. The forecast position errors in NCUM are the lowest in 24, 48 and 72 h except in case of 'Lehar' and 'Helen'.

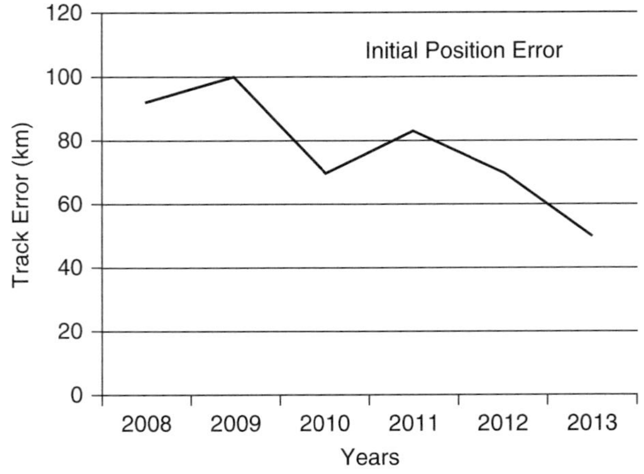

Fig. 11 Reduction in the initial position error of TCs in the NCMRWF global analysis system during recent years

3. The DPE computed for the NGFS and the NGEFS ensemble mean tracks indicate that the desired improvement in the ensemble mean track over the NGFS track prediction is evident only in case of 'Viyaru' and marginal in the two cases of 'Phailin' and 'Helen'. In the two cases of Madi' and 'Lehar' DPE in the NGFS forecast tracks are lower.
4. The CTE, ATE, landfall position errors and the errors in the predicted landfall time are all varying with TC cases, model and forecast lead time. Due to limited cases reported in this study it would be inappropriate to generalise and conclude on the performance of the models.

References

Ashrit, R.G, Iyengar, G.R, Sankar, S, Ashish, A, Dube, A, Dutta, S.K, Prasad, V.S, Rajagopal, E.N. and Basu, S. (2013). Performance of Global Ensemble Forecast System (GEFS) during Monsoon 2012. RESEARCH REPORT, NMRF/RR/1/2013.

Ashrit, R.G, Chourasia, M, Johny, C.J. and George, J.P. (2014). Improved track and intensity predictions using TC bogusing and regional assimilation. *In:* Monitoring and Prediction of TCs in the Indian Ocean and Climate Change, Mohanty, U.C, Mohapatra, M, Singh, O.P, Bandyopadhyay, B.K. and Rathore, L.S. (Eds.), Springer, Netherlands, pp.246–254.

Chourasia, M, Ashrit, R.G. and George, J.P. (2013). Impact of TC bogusing and regional assimilation on TC track and intensity predictions. *Mausam*, **64**, 135–148.

Buckingham, C, Marchok, T.P, Ginis, I, Rothstein, L. and Rowe, D. (2010). Short- and medium-range prediction of tropical and transitioning TC tracks within the NCEP Global Ensemble Forecasting System. *Weather Forecasting*, 25, 1736–1754.

Froude, L.S.R, Bengtsson, L. and Hodges, K.I. (2007). The Prediction of Extratropical Storm Tracks by the ECMWF and NCEP Ensemble Prediction Systems. *Monthly Weather Review*, **135**, 2545–2567.

Hamill, T.M, Whitaker, J.S, Kleist, D.T, Fiorino, M. and Benjamin, S.G. (2011). Predictions of 2010's TCs using the GFS and ensemble-based data assimilation methods. *Monthly Weather Review*, **139**, 3243–3247.

Heming, J.T, Chan, J.C.L. and Radford, A.M. (1995). A new scheme for the initialisation of TCs in the UK Meteorological Office global model. *Meteorological Applications*, **2**, 171–184.

Heming, J.T. and Radford, A.M. (1998). The performance of the United Kingdom Meteorological Office Global Model in Predicting the Tracks of Atlantic TCs in 1995. *Monthly Weather Review*, **126**, 1323–1331.

Marchok, T. (2002). In 25th Conference on Hurricanes and Tropical Meteorology, San Diego, CA, American Meteorological Society, 2002, p. 1.13, available at: http://ams.confex.com/ams/pdfpapers/37628.pdf

Mohapatra, M, Nayak, D.P, Sharma, R.P. and Bandopadhyay, B.K. (2013). Evaluation of official TC track forecast over north Indian Ocean issued by India Meteorological Department. *Journal of Earth System Science*, **122**, 589–601.

Mohapatra, M, Nayak, D.P, Sharma, M, Sharma, R.P. and Bandyopadhyay, B.K. (2015). Evaluation of official tropical cyclone landfall forecast issued by India Meteorological Department. *Journal of Earth System Science*, **124**, 861–874, doi:10.1007/s12040-015-0581-x

Mohanty, U.C, Osuri, K.K, Routray, A, Mohapatra, M. and Pattanayak, S. (2010). Simulation of Bay of Bengal TCs with WRF Model: Impact of Initial and Boundary Conditions. *Marine Geodesy*, **33**, 294–314.

Mohandas, S. and Ashrit, R. (2014). Sensitivity of different convective parameterization schemes on TC prediction using a mesoscale model. *Natural hazards*, **73**, 213–235.

Rajagopal, E.N, Iyengar, G.R, George, J.P, Das Gupta, M, Mohandas, S, Siddharth, R, Gupta, A, Chourasia, M, Prasad, V.S, Aditi, Sharma, K. and Ashish, A. (2012). Implementation of Unified Model based Analysis-Forecast System at NCMRWF, TECHNICAL REPORT, NMRF/TR/2/2012.

RSMC, New Delhi (2014). Report on Cyclonic Disturbances over North Indian Ocean during 2013, 150–239 pp.

Weber, H.C. (2005). Probabilistic prediction of TCs. Part I: Position. *Monthly Weather Review*, **133**, 1840–1852.

Sensitivity of WRF-ARW Model to Cumulus Parameterisation Schemes in Prediction of TC Intensity and Track Over the North Indian Ocean

S.K. Bhattacharya, S.D. Kotal, S.K. Roy Bhowmik, and P.K. Kundu

1 Introduction

Tropical cyclones (TCs) are one of the most devastating extreme weather events causing tremendous loss to human civilisation. With the growing population and economic developments, more life and property are getting exposed to the nature's fury, particularly, along the coastal areas and these vulnerable groups always look forward to efficient real time forecasting to minimise their losses. Numerical Weather Prediction (NWP) models with its reasonable strength to forecast for several days in advance especially for the data sparse regions over the oceans are extremely handy for the operational forecasters. There has been considerable improvement during the past decades in the prediction of track and intensity of TC (Mohapatra et al. 2013a, b, c) using NWP models (Kotal et al. 2014). Hence, the operational forecasters across the globe look forward to the NWP models' products.

There are several NWP models, being run at different centers. The analyses and the forecasts for any particular weather event vary from model to model because of different boundary and initial conditions as well as different model configurations. The same may also vary for a single model with same boundary and initial conditions because of difference in physics options used. It is important to investigate and determine the correct parameterisation option compatible with a model for the best performance of the model in predicting TC.

S.K. Bhattacharya (✉) • S.D. Kotal
India Meteorological Department, NWP Division, 110003 New Delhi, India
e-mail: sumit.kumar.bhattacharya@gmail.com

S.K. Roy Bhowmik
SAARC Meteorological Research Centre, 1207 Dhaka, Bangladesh

P.K. Kundu
Department of Mathematics, Jadavpur University, 700032 Kolkata, India

© Capital Publishing Company 2017

M. Mohapatra et al. (eds.), *Tropical Cyclone Activity over the North Indian Ocean*,
DOI 10.1007/978-3-319-40576-6_20

Cumulus (Cu) convection is one of the most important features of any weather system. In TC there are cluster of cumulus cells those contribute in cumulative impact in determining the character of the system. The impact of the convection within the system and with that of the surrounding is studied and parameterised by various researchers, time to time, to improve the NWP models' performance in respect of track and intensity prediction of TC. WRF mesoscale model includes various cumulus parameterisation schemes (CPS) viz Kain–Fritsch (new Eta) (KF) scheme (Kain 2004), Betts–Miller–Janjic (BMJ) scheme (Janjic 1994), Grell–Devenyi ensemble (GDE) scheme (Grell and Devenyi 2002), Old Simplied Arakawa–Schubert (SAS) scheme (Pan and Wu 1995), New Grell scheme (G3) scheme (Grell 1993), Tiedtke scheme (Tiedtke 1989), New GFS SAS (NGSAS) scheme from YSU (Han and Pan 2011), Old Kain–Fritsch (OKF) scheme (Kain and Fritsch 1990) are compatible for use in WRF-ARW Model. However, it is essential to adapt the best suitable scheme for forecasting of events like TCs over any particular basin like the North Indian Ocean (NIO).

Osuri et al. (2012) carried out experiments, with combinations of 3 selected CP schemes and 2 Planetary Boundary Layer (PBL) schemes for 5 TCs and found that the combination of KF CPS scheme with YSU PBL scheme yielded least error in both intensity and track prediction. Mukhopadhyay et al. (2011) studied the sensitivity of 3 convective parameterisation schemes with various microphysics schemes on the TC track and intensity on simulation of 2 TCs viz. Gonu and Sidr, over the Arabian Sea (AS) and found KF scheme as a better one. Raju et al. (2011) also examined the sensitivity of parameterisation for simulation of TC Nargis and found that the combination of KF CPS scheme with YSU PBL scheme yielded least error in both intensity and track prediction. Reddy et al. (2014) studied the effect of CPS on TC Jal using 4 schemes and found that KF with all microphysics schemes were best in track prediction while KF and G3 schemes made overestimation and BMJ made under estimation in intensity prediction. One interesting finding of Reddy et al. (2014) has been that there is no significant variation between 9 and 27 km resolution model run in track prediction. While Osuri et al. (2014) have shown that the WRF model performance is sensitive to model resolution.

In this study, in an attempt to find the best possible CPS option for prediction of intensity and track of TC over the NIO, 31 cases of TC that occurred over the NIO using 2001 and 2012 are considered and total 248 simulation runs are made with WRF-ARW Model using each of the 8 CP schemes with US- National Centre for Environment Prediction (NCEP) final analysis data as initial and boundary conditions. The PBL scheme used is the Yonsei University (YSU, Hong and Lim 2006) scheme for all the runs. The primary objective of this study are: (i) to find the best possible CPS for WRF-ARW model in prediction of TC intensity and track over the North Indian Ocean, both basin wise and season wise, (ii) to verify whether any particular CPS is best suited for prediction of both intensity as well as track and (iii) to investigate if Initial Bias Correction (IBC) can add any improvement to the model predictions of intensity and track of TC over the NIO.

The study is presented in four sections viz. (1) brief description of the model, the TCs under study and the Data, (2) sensitivity of CPS in prediction of mean sea level pressure (MSLP) at the centre of the TCs, (3) sensitivity of CPS in prediction of track of the TCs and (4) summary of the findings.

2 Brief Description of the Model, TCs and the Data

2.1 The Model Description and Its Configuration

For the present experiments, the WRF-ARW model is run with 27 km resolution, at 90 s integration time steps. The Regional Specialized Meteorological Centre (RSMC), New Delhi is responsible for monitoring and forecasting of TCs over the NIO. These TCs affects the coasts of Yemen, Oman, Pakistan, Bangladesh, Myanmar and the western and eastern coasts of India. Therefore, the domain has suitably been fixed to cover the whole area under the jurisdictions of the RSMC, New Delhi. Table 1 shows the description of the model options used in simulation of the TCs. The model has been run using all the eight CPS for each of the base hours keeping the other options fixed.

2.2 TCs Under Study

The genesis of TCs over the NIO takes place during the pre-monsoon (PR) season, generally between the first week of April and the first week of June, and in the post-monsoon (PO) season, between October and December. Out of the 31 TCs considered under this study, 18 were over the Bay of Bengal (BOB) (9 each in PR and PO seasons) and 13 were over the AS (8 in PR season and 5 in PO season). Among all the systems 8 had recurved (RE) tracks. The Table 2 shows the list of TCs simulated, the date and time on which the simulation is based and the total hours upto which the forecasts are generated.

2.3 The Data Used for Model Running and Subsequent Verification Model of Output

The Final (FNL) global tropospheric analyses of the NCEP have been used to provide the initial and boundary conditions to the model simulations. The RSMC, New Delhi Best Track data (www.rsmcnewdelhi.imd.gov.in) are used for the model output verifications.

3 Sensitivity of CPS

The sensitivity test of the model with respect to various CPS has been divided, broadly into two parts viz. prediction skill of MSLP at the centre of the TCs (MSLP) and the prediction skill of track of the TCs.

Table 1 The WRF-ARW model description and its configuration

Model configuration	Options
Model core	ARW
Dynamics	Non-hydrostatic
Model domain	23 °S–46° N/40–120° E
Model resolution	27 km
Eta levels	38
Map projection	Mercator
Integration time step	90s
Microphysics	WSM 3-class scheme
Radiation (shortwave/ longwave)	Dudhia/RRTM
Planetary boundary layer Physics	Yonsei University Scheme (YSU)
Cumulus parameterisation scheme	Kain–Fritsch (new Eta), Betts–Miller–Janjic, Grell–Devenyi ensemble, Old Simplied Arakawa–Schubert, New Grell scheme, Tiedtke, New GFS SAS, Old Kain–Fritsch

3.1 Prediction Skill of MSLP at the Centre of the TCs

3.1.1 Skill of Direct Model Output (Without Initial Bias Correction)

The model analyses and 12-h predictions of the MSLP at the centre of each of the TCs have been computed, using each of the eight schemes and each result has been compared with the corresponding observed MSLP as per the RSMC, New Delhi best track data using the expression,

$$(\text{Error})_{\text{mslp}} = (\text{MSLP})_{\text{model}} - (\text{MSLP})_{\text{observed}} \tag{1}$$

In Eq. (1), $(\text{Error})_{\text{mslp}} = 0$, implies a perfect simulation, $(\text{Error})_{\text{mslp}} > 0$, implies an underestimation, while $(\text{Error})_{\text{mslp}} < 0$, implies an overestimation of the intensity, by the simulation.

For comparison of prediction skills of the direct model output (DMO), using different CPS, the average absolute error (AAE) for individual TCs as calculated through Eq. (1), have been determined for 12-h predictions of the MSLP, corresponding to each scheme.

Figure 1a–f shows the scheme wise variations of 12-hourly prediction of MSLP of DMO without any bias correction. It is seen that (a) GD Ensemble, BMJ and Tiedtke schemes are overall better than average of all the schemes, (b) for recurved TCs BMJ, GD Ensemble, Tiedtke are better than average of all schemes, New GFS SAS is also better upto 60-h forecast than the same (c) G3, GD Ensemble and BMJ are better than average of all over BOB (for pre-monsoon). Previous KF is also better than average of all up to 48-h forecast for the same, (d) GD Ensemble and

Table 2 List of TCs simulated, the date and time on which the simulation is based and the total hours up to which the forecasts are generated

Location/period	SN	System	Based on yyyymmdd	hh	Total hours of f/c
Bay of Bengal /pre-monsoon	1.	LAILA	20100517	12	72
	2.	AILA (1)	20090524	12	42
	3.	AILA (2)	20090523	06	72
	4.	BIJLI (1) (RE)	20090415	12	48
	5.	BIJLI (2) (RE)	20090414	12	72
	6.	NARGIS (RE)	20080428	00	72
	7.	MALA (2) (RE)	20060426	00	72
	8.	MALA (1) (RE)	20060425	12	72
	9.	Unnamed TC, May 2003 (RE)	20030511	00	72
Bay of Bengal /post-monsoon	1.	NILAM	20121029	00	72
	2.	THANE	20111227	00	72
	3.	JAL	20101105	00	72
	4.	GIRI	20101020	12	66
	5.	WARD	20091211	00	72
	6.	SIDR	20071112	00	72
	7.	FANOOS	20051207	00	66
	8.	BAZZ	20051128	12	72
	9.	Unnamed TC, November 2002	20021124	06	72
Arabian Sea/pre-monsoon	1.	PHET (2) (RE)	20100603	00	72
	2.	PHET (1) (RE)	20100601	12	72
	3.	BANDU	20100519	12	72
	4.	GONU	20070602	12	72
	5.	Unnamed TC, May 2004 (2)	20040507	00	72
	6.	Unnamed TC, May 2004 (1)	20040505	12	72
	7.	Unnamed TC, May 2002	20020507	00	72
	8.	Unnamed TC, May 2001	20010522	00	72
Arabian Sea/post-monsoon	1.	MURJAN	20121023	00	72
	2.	PHYAN	20091109	12	48
	3.	MUKDA	20060921	12	72
	4.	ONIL	20040930	12	72
	5.	Unnamed TC, November 2003	20031112	12	72

BMJ are better than average of all over BOB (for post-monsoon) up to 60-h forecast, (e) GD Ensemble and BMJ are better than average of all over AS (for pre-monsoon) and (f) GD Ensemble is better than average of all over AS (for post-monsoon).

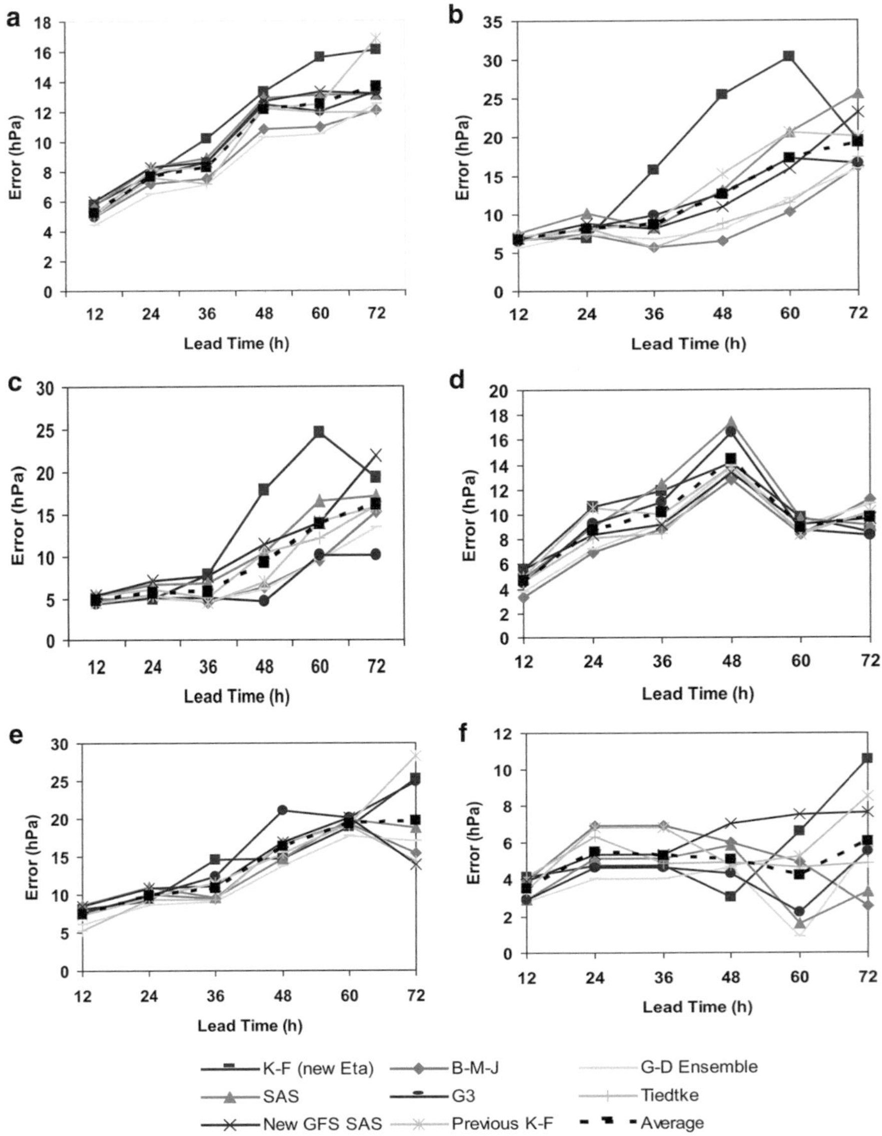

Fig. 1 Scheme wise variation of 12-h prediction of mean sea level pressure at the centre of the tropical cyclones for (**a**) all 31 cases, (**b**) all the recurved systems, (**c**) all pre-monsoon cases over the Bay of Bengal (**d**) all post monsoon cases over the Bay of Bengal, (**e**) all pre-monsoon cases over the Arabian Sea (**f**) all post monsoon cases over the Arabian sea

3.1.2 Skill of Direct Model Output (Initial Bias Corrected)

The error in analysis of the MSLP varied from TC to TC indicating different degrees of underestimation or overestimation of the initial intensity of the systems

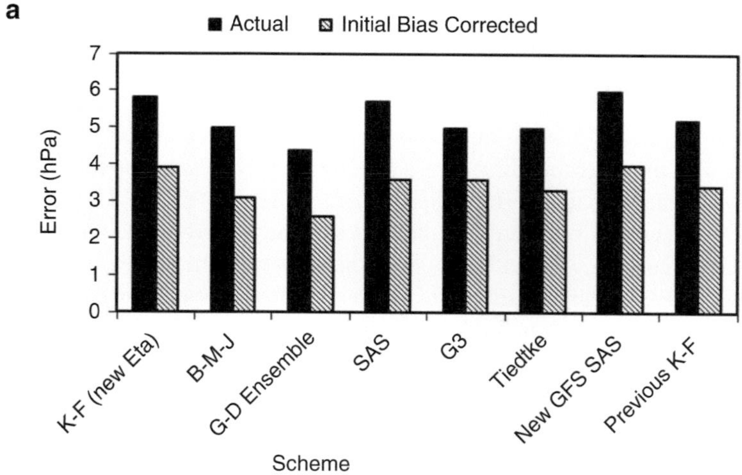

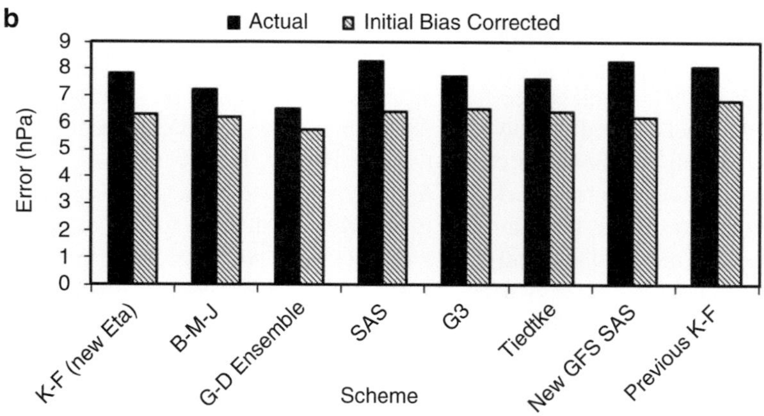

Fig. 2 Direct model output versus Initial bias corrected mean sea level pressure for (**a**) 12-h forecast (**b**) 24-h forecast

by the model simulations. By the time the model direct outputs are available the observed MSLP are determined and hence it is easy to verify the level of analysis error. Therefore, this study has also examined whether the forecasts by the simulations have been consistent with the analysis. This is carried out by subtracting the initial error (analysis error) from each of the 12-h forecast errors and thereafter, finding AAE of such errors, scheme-wise. A Standard *t*-test (Welch 1947) computation for determination of the significance level of improvement shows 82–96 % and 38–81 % improvement level respectively for 12 and 24 h forecasts. This exercise is done for each scheme after the initial correction and compared with the corresponding results of direct model output with the same scheme. A comparison is also shown in Fig. 2a, b, displaying overall improvement for the schemes after initial bias corrections for 12 and 24 h forecasts over the respective DMO.

A detailed exercise is also carried-out to investigate the impact of initial bias corrections (IBC) to improve the forecast skill for each basin and for each season. A brief methodology adopted in this process is as follows: The least AAE based on DMO at the 12-h forecast with respect to each parameterisation scheme, basin, season and cases of recurved and non-recurved are picked up and the same are plotted against the corresponding hour's bias corrected AAEs as shown in Fig. 3a–f. The Fig. 3a shows that the IBC improved the forecast for the 12 h for all the schemes in general as compared to the best scheme's performance among all the Direct Model Outputs, without bias correction. However, the same can not be taken for granted when specific cases like recurved systems, basin or season specific cases. A finer distinction among the cases show that although the IBC has reduced forecast error in 12-h forecast for the BOB pre-monsoon and AS post monsoon systems for all the schemes Fig. 3c, f, for the rest systems this method deteriorated the forecast skill.

One of the primary aims of this study is to find the best possible scheme for forecasting of MSLP. The Fig. 4a–f shows a comparison of least DMO AAE (without bias correction) and the best Initial Bias Corrected AAE for 12-h prediction of MSLP at the centre of the TCs. The figure shows that IBC is capable to improve forecast skill for all 31 cases up to 60 h of forecast, in general. GD Ensemble, BMJ, Tiedtke are, better options for central pressure (CP) prediction. When the track is predicted to recurve, the bias corrected BMJ scheme is a better option. For all pre-monsoon cases over the BOB, GD Ensemble and New Grell schemes are found to be superior for the forecasts up to 60 h. Over the BOB during post-monsoon season, BMJ and GD Ensemble are better up for the forecasts up to 48 h i.e. IBC may add skill only up to 48 h for these schemes. For all pre-monsoon cases over the AS IBC may add skill up to 48 h with Tiedtke, GD Ensemble and BMJ schemes and (f) for the Arabian Sea, during post-monsoon season, Tiedtke scheme is better and IBC with BMJ scheme may also add skill up to 36 h.

3.2 *Prediction Skill of Track of the TCs*

The Direct Position Error (DPE) has been derived as being the great-circle distance between the forecast position and the observed position, both at the forecast verification time (Neumann and Pelissier 1981). The AAE of the 12-h DPE have been shown in Fig. 5a–f. It is seen that in the case of track forecasting, the SAS and New GFS SAS schemes performed better than other schemes, over NIO. For the BOB post-monsoon season, KF (new Eta) scheme is a better option.

No Initial Bias Corrected DPE variation has been studied, as no significant correlation has been found between $(\mathrm{delLAT})_{\mathrm{analysis}}$ and $(\mathrm{delLAT})_{\mathrm{forecast}}$ as well as $(\mathrm{delLON})_{\mathrm{analysis}}$ and $(\mathrm{delLON})_{\mathrm{forecast}}$, where delLAT is the actual error in latitudinal position and delLON is the actual error in longitudinal position.

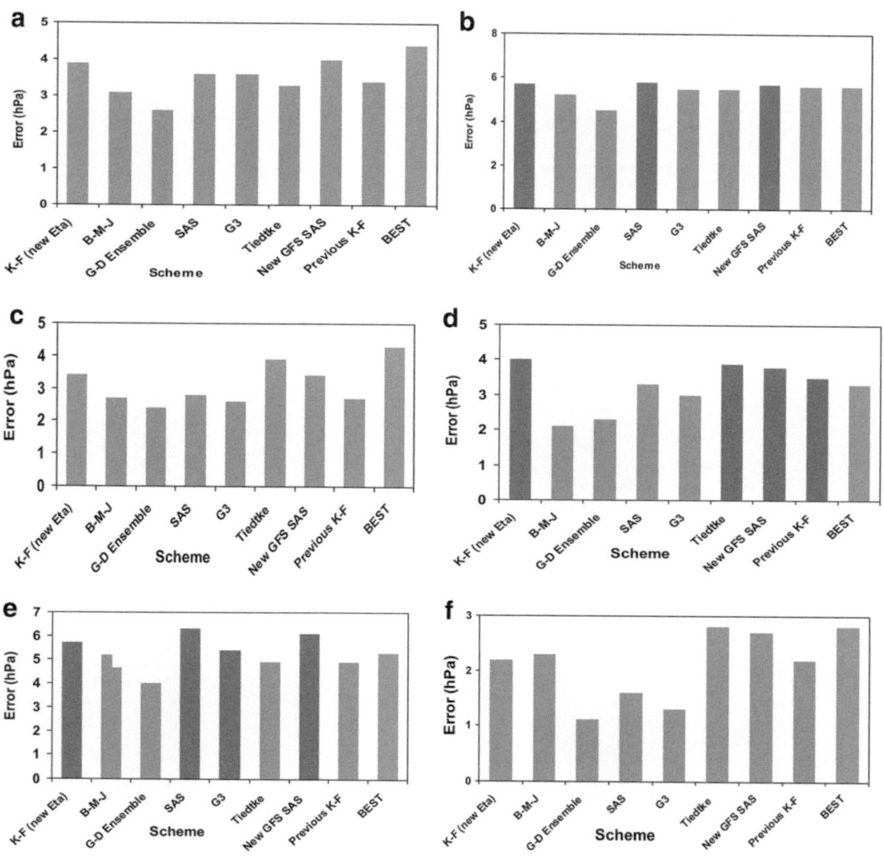

Fig. 3 Least direct model output average absolute error (without bias correction) vs Initial bias corrected average absolute error for 12-h forecast for (**a**) all 31 cases, (**b**) all the recurved systems, (**c**) all pre-monsoon cases over the Bay of Bengal (**d**) all post monsoon cases over the Bay of Bengal, (**e**) all pre-monsoon cases over the Arabian Sea and (**f**) all post monsoon cases over the Arabian sea

4 Conclusion of the Findings

1. The predictions of intensity and track of TC over Indian Ocean by WRF-ARW model are found to be sensitive to cumulus parameterisation schemes.
2. Certain schemes (BMJ, GDE and Tiedtke) are consistent with initial model bias of CP. In prediction of MSLP at the centre of a TC it is found that for BOB, during pre-monsoon season, GD Ensemble and New Grell schemes are found to be superior whereas during post-monsoon season, BMJ and GD Ensemble are better. For the Arabian Sea, during pre-monsoon Tiedtke, GD Ensemble and BMJ schemes are found to be superior whereas during post-monsoon season Tiedtke scheme is better.

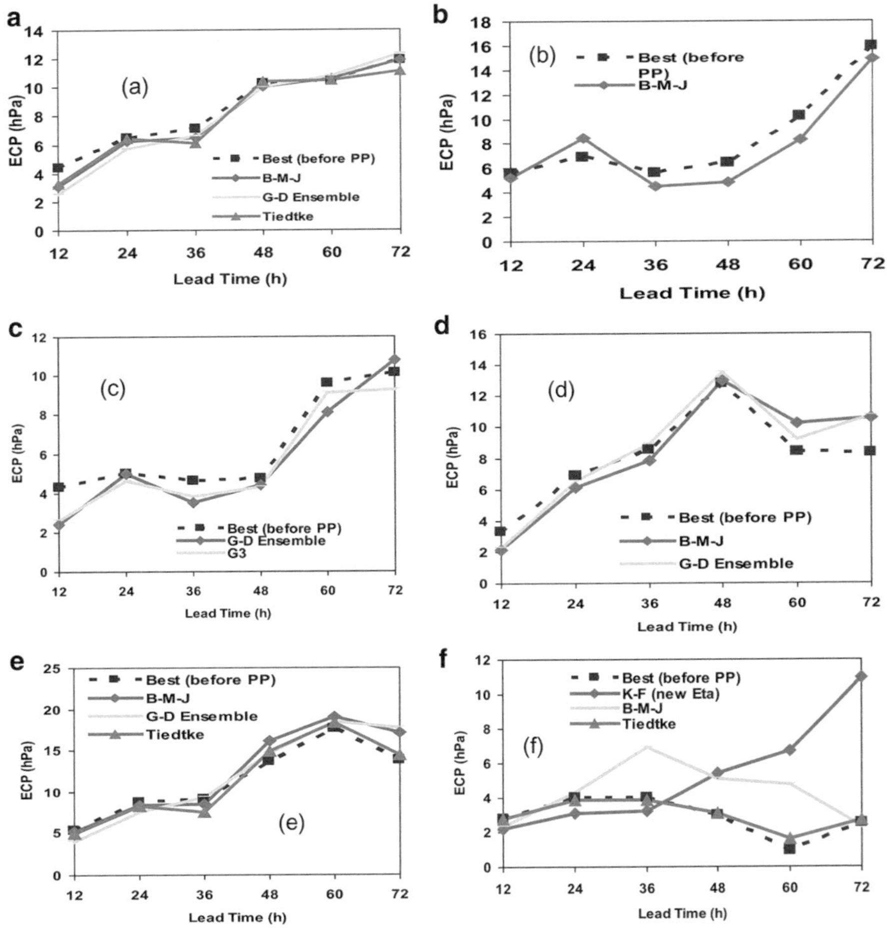

Fig. 4 Least direct model output average absolute error (without bias correction) vs Best initial bias corrected average absolute errors for 12-h prediction of mean sea level pressure at the centre of the TCs for (**a**) all 31 cases, (**b**) all the recurved systems, (**c**) all pre-monsoon cases over the Bay of Bengal, (**d**) all post monsoon cases over the Bay of Bengal, (**e**) all pre-monsoon cases over the Arabian Sea and (**f**) all post monsoon cases over the Arabian sea

3. IBC in post processing is found to have no impact on the track forecast improvement.
4. The schemes better for prediction of CP are not better for track forecasting and vice versa.
5. In the case of track forecasting, the SAS and New GFS SAS schemes are found to have performed better than other schemes in general. For BOB post-monsoon season KF (new Eta) scheme is a better option.

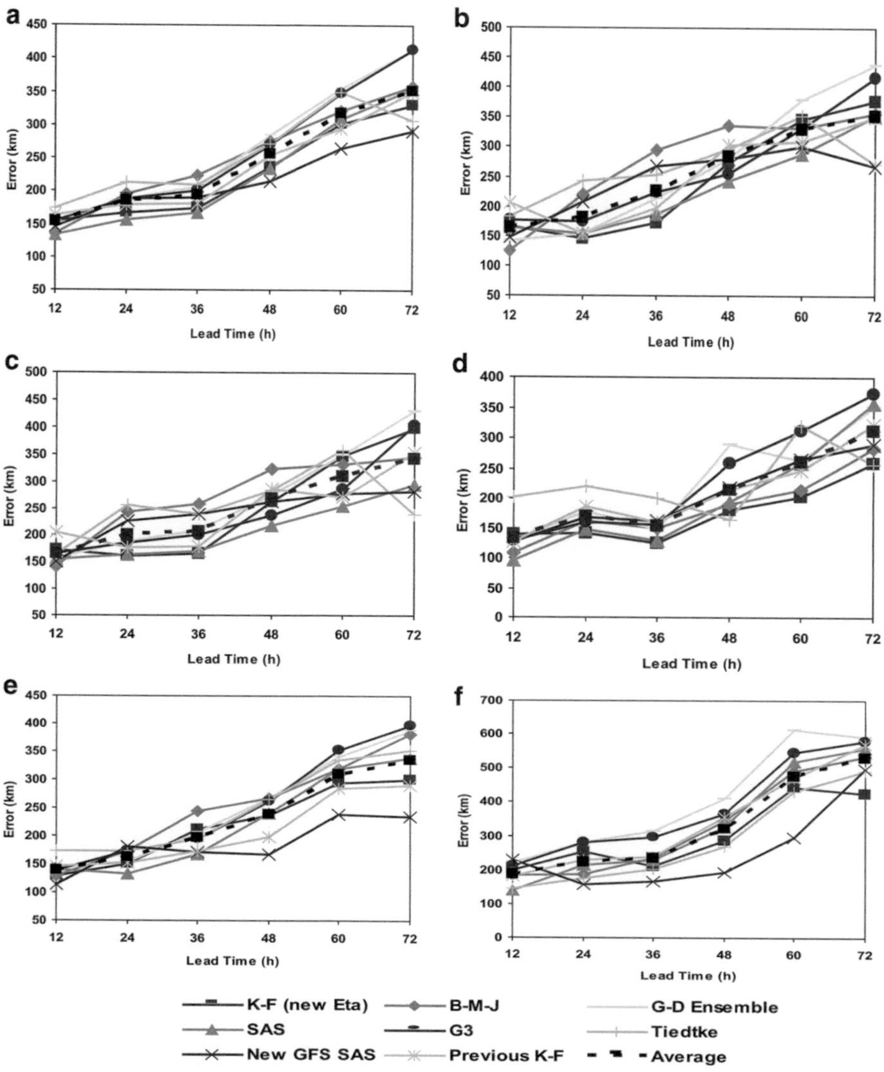

Fig. 5 Direct position average absolute errors for 12-h prediction of track of the tropical cyclones for (**a**) all 31 cases, (**b**) all the recurved systems, (**c**) all pre-monsoon cases over the Bay of Bengal, (**d**) all post-monsoon cases over the Bay of Bengal, (**e**) all pre-monsoon cases over the Arabian Sea and (**f**) all post-monsoon cases over the Arabian Sea

Acknowledgements The researchers gratefully acknowledge the support and encouragement of the Director General of Meteorology, India Meteorological Department. The researchers also acknowledge the use of NCEP Final analysis data for model initial and boundary conditions. The authors are grateful to the anonymous reviewer for improvement of the quality of the paper.

References

Grell, G.A. and Devenyi, D. (2002). A generalized approach to parameterizing convection combining ensemble and data assimilation techniques. *Geophysical Research Letters*, **29**, 1693.
Grell, G.A. (1993). Prognostic evaluation of assumptions used by cumulus parameterizations. *Monthly Weather Review*, **121**, 764–787.
Han, J. and Pan, H. (2011). Revision of convection and vertical diffusion schemes in the NCEP Global Forecast System. *Weather Forecasting*, **26**, 520–533.
Hong, S.-Y. and Lim, J.-O.J. (2006). The WRF Single-Moment 6-Class Microphysics Scheme (WSM6). *Journal of Korean Meteorological Society*, **42**, 129–151.
Janjic, Z.I. (1994). The Step-Mountain Eta Coordinate Model: Further developments of the convection, viscous sublayer, and turbulence closure schemes. *Monthly Weather Review*, **122**, 927–945.
Kain, J.S. (2004). The Kain–Fritsch convective parameterization: An update. *Journal of Applied Meteorology*, **43**, 170–181.
Kain, J.S. and Fritsch, J.M. (1990). A one-dimensional entraining/detraining plume model and its application in convective parameterization. *Journal of Atmospheric Sciences*, **47**, 2784–2802.
Kotal, S.D, Bhattacharya, S.K. and Roy Bhowmik, S.K. (2014). Development of NWP Based objective TC prediction system (CPS) for North Indian Ocean TCs – Evaluation of performance, *Tropical Cyclone Research and Review*, **3**, 162–177.
Mohapatra, M, Nayak, D.P, Sharma, R.P. and Bandyopadhyay, B.K. (2013a). Evaluation of official tropical cyclone track forecast over north Indian Ocean issued by India Meteorological Department. *Journal of Earth System Science*, **122**, 433–451, doi:10.1007/s12040-013-0291-1
Mohapatra, M, Bandyopadhyay, B.K. and Nayak, D.P. (2013b). Evaluation of operational tropical cyclone intensity forecasts over north Indian Ocean issued by India Meteorological Department. *Natural Hazards*, **68**, 589–601, doi:10.1007/s11069-013-0624-z
Mohapatra, M, Sikka, D.R, Bandyopadhyay, B.K. and Tyagi, A. (2013c). Outcomes and challenges of Forecast Demonstration Project (FDP) on landfalling cyclones over Bay of Bengal. *Mausam*, **61**, 1–15.
Mukhopadhyay, P, Taraphdar, S. and Goswami, B.N. (2011). Influence of moist processes on track and intensity forecast of TCs over the north Indian Ocean. *Journal of Geophysical Research*, **116**, 1–21.
Neumann, J.C. and Pelissier, J.M. (1981). An analysis of Atlantic TC forecast errors, 1970–1979. *Monthly Weather Review*, **109**, 1248–1266.
Osuri, K.K, Mohanty, U.C, Routray, A, Kulkarni, A.M. and Mohapatra, M. (2012). Customization of WRF-ARW model with physical parameterization schemes for the simulation of TCs over North Indian Ocean. *Natural Hazards*, **63**, 1337–1359.
Osuri, K.K, Mohanty, U.C, Routray, A, Mohapatra, M. and Niyogi, D. (2014). Real-time track prediction of tropical cyclones over the North Indian Ocean using the ARW Model. *Journal of Applied Meteorology and Climatology*, **52**, 2476–2492.
Pan, H.L. and Wu, W.S. (1995). Implementing a mass flux convective parameterization package for the NMC medium range forecast model, NMC office note, **409.40**, 20–233.
Raju, P.V.S, Potty, K.V.J. and Mohanty, U.C. (2011). Sensitivity of physical parameterizations on prediction of TC Nargis over the Bay of Bengal using WRF model. *Meteorology and Atmospheric Physics*, **113**, 125–137.
Reddy, M.V, Prasad, S.B.S, Murali Krishna, U.V. and Reddy, K.K. (2014). Effect of cumulus and microphysical parameterizations on the JalTC prediction. *Indian Journal of Radio & Space Physics*, **43**, 103–123.
Tiedtke, M. (1989). A comprehensive mass flux scheme for cumulus parameterization in large-scale models. *Monthly Weather Review*, **117**, 1779–1800.
Welch, B.L. (1947). The generalization of 'Student's' problem when several different population variances are involved. *Biometrika*, **34**, 28–35.

Simulation of Tropical Cyclone 'Phailin' Using WRF Modeling System

S. Kumar, A. Routray, G. Tiwari, R. Chauhan, and I. Jain

1 Introduction

Tropical cyclones (TCs) are one of the most devasting and deadliest meteorological phenomena worldwide. This devastation is mainly due to torrential rains, high winds and associated storm surges (Mohapatra et al. 2012, 2015). The TC genesis has been attributed to both thermodynamic and dynamical factors. Palmen (1948) showed that TCs form over regions where sea surface temperature (SST) is greater than 26 °C. In addition to SST, other important factors for genesis of TCs are: large Coriolis force (LCF), high low-level relative vorticity, weak vertical wind shear, moisture in the middle troposphere and convective instability. In the regions, moist convection dominates the process of transporting mass, energy, and momentum through the atmosphere. On longer timescales, the large-scale environment can influence and control the mesoscale organisation and activities. The long chain of multi-scale interactions of physical parameters is challenging task to handle in the numerical models of weather systems and hence, genesis of TCs is a particular example that has motivated the present study.

Synoptic and statistical methods have limitations in predicting track and intensity beyond 24-h over the North Indian Ocean (NIO; Mohanty and Gupta 1997; Gupta 2006). The Advance Research WRF (ARW) model is being widely used for

S. Kumar (✉) • G. Tiwari • R. Chauhan
Department of Applied Mathematics, School of Vocational Studies and Applied Sciences, Gautam Buddha University, 201312 Greater Noida, Uttar Pradesh, India
e-mail: sushil.kumar@gbu.ac.in

A. Routray
National Centre for Medium Range Weather Forecasting (NCMRWF) A-50, Sector 62, 201309 Noida, India

I. Jain
RMSI Pvt. Ltd., Sec-16, Noida, Uttar Pradesh, India

© Capital Publishing Company 2017
M. Mohapatra et al. (eds.), *Tropical Cyclone Activity over the North Indian Ocean*,
DOI 10.1007/978-3-319-40576-6_21

the simulation of a variety of weather events, such as heavy rainfall, TC and real time numerical weather prediction (NWP) over the Indian monsoon region, and indeed globally, (Routray et al. 2010; Osuri et al. 2012). Weather agencies of various nations give forecasts of TC operationally. The high-resolution non-hydrostatic mesoscale atmospheric models are being used for research and operational forecasting of several mesoscale atmospheric phenomena. Various studies are reported on the simulation/prediction of TCs over the NIO (Mohanty et al. 2004, 2010; Osuri et al. 2011, 2012; Bhaskar Rao and Hari Prasad 2006, Srinivas et al. 2010, Osuri et al. 2014). The Bay of Bengal (BOB) is a potentially energetic region for the development of TCs and contributes about 7 % of the global annual storms. It is well known that TCs characteristics drastically change at different stages in the life cycle. Though there have been significant improvement in track and intensity forecast (Mohapatra et al. 2013a, b, c, 2015) still there is scope for improvement in prediction of intensity and motion of TCs especially from deepening period to mature stage and from mature stage to dissipation stage. There has been significant development in the field of TC track simulation for this region using mesoscale models. The models through grid resolution, physical parameterisations, and data assimilation, etc. among all, the physical parameterisations, which includes cumulus convection, surface fluxes of heat, moisture, momentum, and vertical mixing in the planetary boundary layer (PBL) play important role in simulating the development and intensification of TCs. The present study tells about the application of non-hydrostatic nested-grid mesoscale model to precisely simulate the various physical and dynamical processes of the storm environment and its evolution. For this purpose, an attempt is made to simulate the TC Phailin that formed in October 2013 over BOB using the Weather and Research Forecast (WRF) mesoscale model.

2 Description of the Very Severe Cyclonic Storm 'Phailin'

Phailin is a very severe cylconic storm (VSCS) formed in the BOB and had a duration of 7 days from 8 October 2013 to 14 October 2013. A low pressure system from north Andaman Sea intensified into depression at 0300 UTC of 8 October 2013 near latitude 12.0° N and longitude 96.0° E (RSMC, New Delhi, 2014). It moved northwestwards and intensified into a deep depression at 0000 UTC of 9 October 2013 near latitude 13.0° N and longitude 93.5° E with the speed of 170 km north-northeast of Port Blair, 1100 km east-southeast of Paradip and 1200 km east-southeast of Visakhapatnam and further intensified into a cyclonic storm, Phailin at 1200 UTC of 9 October 2013. The cyclonic storm continued to move in northwesterly direction and intensified into severe cyclonic storm at 0300 UTC of 10 October 2013 and centred at latitude 14.5° N and longitude 91.5° E intensified into very severe cyclonic storm at 0600 UTC near latitude 15.0° N and longitude 90.5° E of same day. Moving northwestward direction the system further rapidly intensified to 1200 UTC, 1500 UTC and 2100 UTC of 10 October 2013, respectively. The Phailin crossed the area of Odisha and adjoining north Andhra Pradesh coast near Gopalpur

(Odisha) around 22:30 h according to IST of 12 October 2013 with a sustained maximum surface wind speed (MSW) of 200–210 kmph around 1700 UTC near latitude 19.2° N and longitude 84.9° E. The system maintained its intensity of very severe cyclonic storm up to 7 h after landfall and cyclonic storm intensity till 1200 UTC of 13 October 2013, latitude 22.5° N and longitude 83.8° E.

3 Description of the ARW Model and Numerical Experiments

The WRF model (version 3.4, ARW dynamical core) is configured with two interactive nested domains in the present study. A resolution of 27 km and 173 × 189 grids are for the outer domain (D01) covering a larger region and the inner domain has 9-km resolution with 296 × 308 grids (D02) covering the BOB and its environments (Fig. 1) with 39 vertical levels. The initial and lateral boundary conditions for the ARW model are obtained from the analysis fields of the US-National Center for Environment Prediction (NCEP), Final Analysis (FNL). The lateral boundary conditions are updated in 6-h intervals with a fixed SST throughout the model integration, with no regional data assimilation used in this study. The land surface boundary conditions are taken from the US. Geological Survey with a horizontal grid spacing of 10 min (D01) and 5 min (D02). The physics schemes used in the present study are Yonsei University (Hong et al. 2006), non-local diffusion scheme for PBL processes, Kain–Fritsch for cumulus convection (Kain and Fritsch 1993), Purde Lin scheme (Lin et al. 1983) for explicit moisture processes, five-layer soil thermal diffusion model for surface processes, Rapid Radiation Transfer Model (RRTM) for long-wave radiation (Mlawer et al. 1997) and Dudhia (1989) scheme for shortwave radiation. The TC Phailin is simulated with a total of five different initial conditions through 12-h cycles. The TC is initialised at 00 and 12 UTC between 00 UTCof 09–11 October 2013 (named as Case-1 to Case-5, respectively) and forecasted upto 12 UTC 13 October 2013.

4 Result and Discussion

4.1 Intensity Prediction of TC

The evolution of intensity at 12-h interval of the TC from Case-1 (IC: 00UTC 09 October 2013) is shown in Fig. 2a, b in terms of mean sea level pressure (MSLP) (hPa) and surface wind at 10 m (m/s) respectively from D01 and D02 simulations. The root mean square errors (RMSE) and standard deviation (Std. Dev.) of MSLP and 10 m surface wind from all the five ICs alongwith the mean of 5 simulations are represented in Fig. 2c, d.

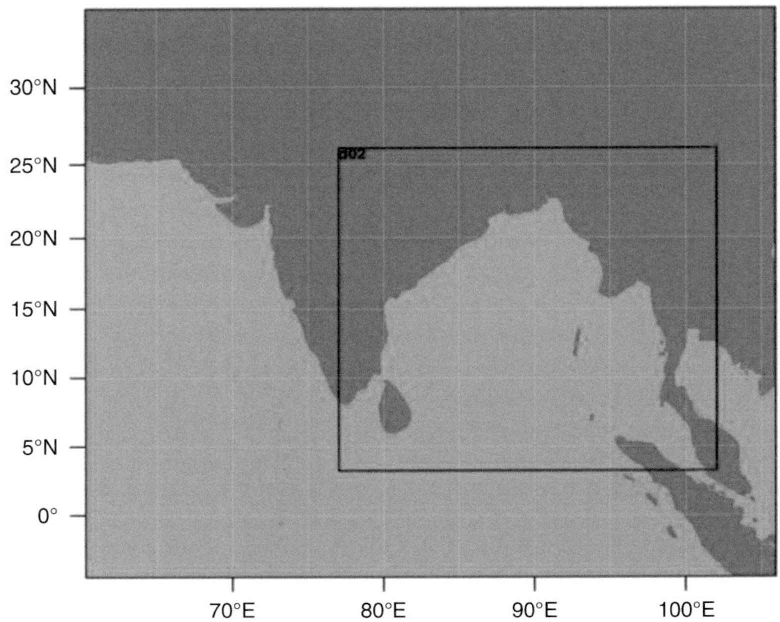

Fig. 1 Model domain considered in the study

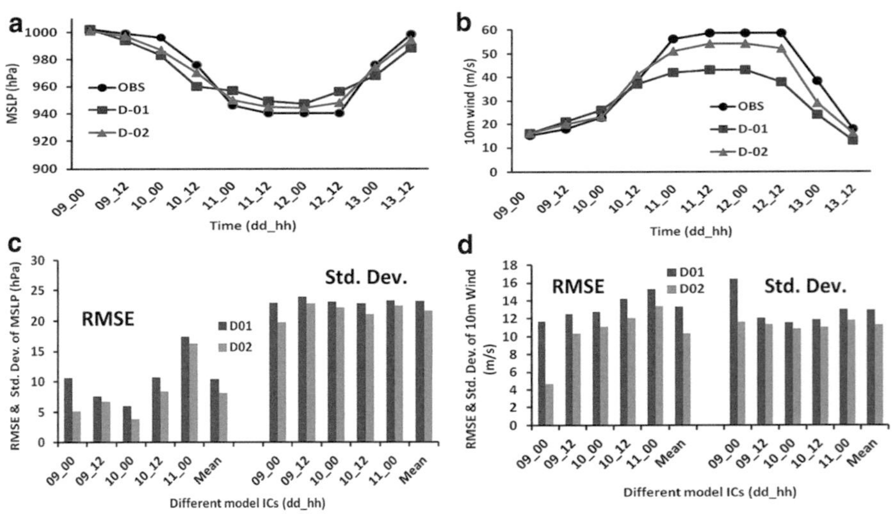

Fig. 2 Evolution of intensity with time (12-h interval) for (**a**) MSLP (hPa); (**b**) 10 m wind (m/s) from Case-1 (IC: 00UTC 09 October 2013). (**c–d**) same as (**a–b**) but for mean RMSE and standard deviation (Std. Dev.) from different model ICs

Table 1 Mean absolute errors (MAE) of MSLP (hPa) and 10 m surface wind (m/s)

	MAE of MSLP (hPa)		MAE of 10 m wind (m/s)		No. of f/c verified
Forecast length (h)	D01	D02	D01	D02	5
00	14	12.2	6.74	5.54	5
12	9	6.4	10.27	9.06	5
24	7.4	6	10.51	7.67	5
36	4.4	1.8	11.66	9.56	5
48	5.2	2.8	13.03	9.83	5
60	6.6	4.6	12.31	8.31	5
72	8	5	13.35	9.09	5
Mean	7.8	5.5	11.12	8.44	4

Both the experiments simulate strong TCs in terms of MSLP in the case of VSCS Phailin (Fig. 2a). The trend of the evolution of the intensity of the storm in the D02 simulation is closer to the IMD observed trend as compared to the D01 simulation. The 10 m surface wind (Fig. 2b) is well simulated in the D02 during the intensity period of the TC which is not depicted in the D01 simulation. Similar results are also found in all cases when the model is initialised with different ICs. It is also clearly noticed from Fig. 2c, d, the RMSE and Std. Dev. of MSLP and 10 m wind are less in the D02 simulation for all the cases. The mean RMSE and Std. Dev. of MSLP (8.06 and 21.65 hPa) and 10 m wind (10.37 and 11.32 m/s) are significantly improved in the D02 simulations as compared to the D01 simulations (MSLP: 10.46 and 23.24 hPa; 10 m wind: 13.30 and 12.97 m/s, respectively).

The mean absolute errors of MSLP and 10 m wind at different forecast time as based on five cases are provided in Table 1. It is clearly seen from the Table 1 that the high resolution model simulations (D02) capture well the intensity of the TC throughout the forecast period as compared to the D01 simulations. The mean absolute errors of MSLP and 10 m wind are reduced by 29 % and 24 % in the D02 runs over the D01 runs.

4.2 *Track Prediction*

Figure 3a, b illustrates the model simulated (D01 and D02) storm tracks (location of minimum MSLP centre) along with IMD best track from five model ICs. The mean vector displacement errors (VDEs) 12-h interval and gain skill score (% of improvement) with respect to the D01 & D02 TC is depicted in Fig. 3c. It is clearly seen from the figure that the simulated tracks in D02 simulations (Fig. 3b) are reasonably better and closer to the IMD observed track as compared to the D01 simulation (Fig. 3a) in all IC cases. The mean VDEs (Fig. 3c) are significantly reduced in the high resolution model simulations (D02) as compared to the D01 simulations. The mean VDEs are gradually increased along with the increase of the forecast hours in all simulations, however the values of the VDEs are less in the

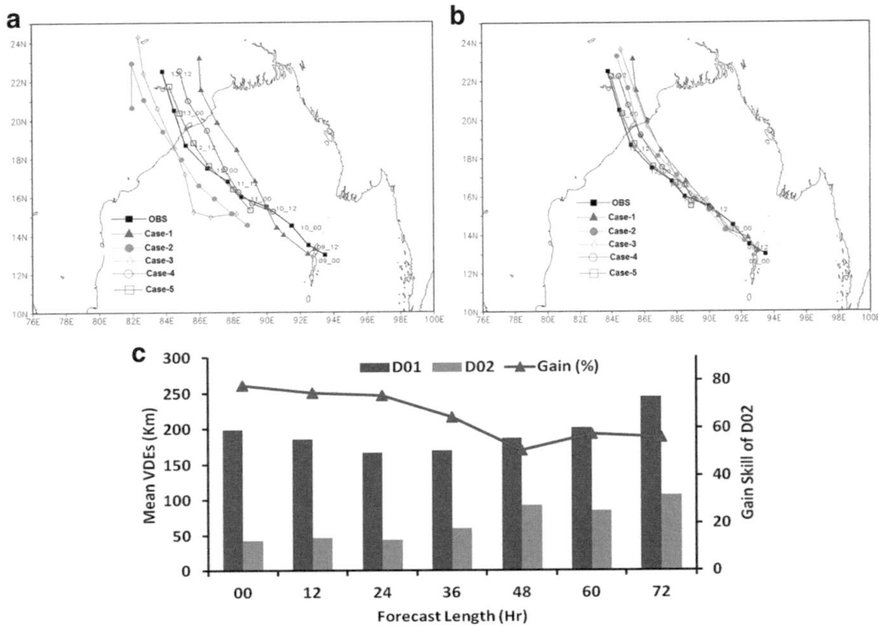

Fig. 3 Model simulated tracks from (**a**) D01 and (**b**) D02. (**c**) Mean vector displacement errors (VDEs, km) in 12 hr interval from different model initial conditions (histograms) and the gain in skill (%) of D02 over D01

D02 as compared to the D01 simulation. The mean VDEs at different forecast hours are varies in the range of 43–107 km in D02 as compared to 266–243 km in D01 runs. The ensemble mean (averaged all forecast hours) VDEs are also less in the D02 simulation in comparison to the other simulation.

The skill of the D02 is significantly higher compared to that of the D01 simulation. The gain in skill of the D02 simulation (with respect to D01) ranged from 51 to 78 % throughout forecast period. Thus, the mean gain in skill of the D02 simulation is about 65 % over the D01. The gain in skill of D02 is gradually decreased as forecast length increased. The model characterised more realistic meso-convective features embedded within the large-scale environmental field IN D02 case due to increase of horizontal resolution, which leads to improved model forecasts.

The landfall (LF) position and time errors are provided in Table 2 from D01 and D02 simulation covering the different initial conditions. The LF position errors are significantly reduced in the D02 simulations as compared to the D01 simulations. The LF position errors are reduced by 24 % in the D02 simulation with respect to D01 runs. The error at the time of landfall is not exactly simulated by any of the runs, however, the landfall time error of the storm is less in the D02 simulation (Table 2) as compared to the D01 run.

Table 2 Landfall point errors (km) and time errors (hours) from D01 and D02

Different ICs (October 2013)	Observed LF time	LF position errors (km)		LF time errors (hours) +/– sign represents ahead/delay in time	
		D01	D02	D01	D02
00 UTC 09th (Case-1)	17 UTC 12 October 2013	375.88	158.23	+02 (15 UTC 12 October 2013)	+05 (12 UTC 12 October 2013)
12 UTC 09th (Case-2)		156.38	98.88	−01 (18 UTC 12 October 2013)	+02 (15 UTC 12 October 2013)
00 UTC 10th (Case-3)		102.34	180.56	+17 (00 UTC 12 October 2013)	+05 (12 UTC 12 October 2013)
12 UTC 10th (Case-4)		302.99	240.82	+02 (15 UTC 12 October 2013)	+02 (15 UTC 12 October 2013)
00 UTC 11th (Case-5)		224.02	203.66	−01 (18 UTC 12 October 2013)	+02 (15 UTC 12 October 2013)
Mean position errors (% of Improvements)		232.32	176.43 (24 %)		

4.3 *Reflectivity*

The simulated cloud band structure of the TC from different simulations is analysed at 06 UTC 12th and landfall time corresponding to 12 UTC 12 October 2013 using simulated radar reflectivity from D01 to D02 respectively and compared with India Meteorological Department (IMD) Visakhapatnam Doppler Radar reflectivity (Fig. 4a–f). Both the simulations show a comma cloud organisation around the TC. The reflectivity pattern as well as intensity obtained from D02 is close to the IMD Visakhapatnam Doppler Radar reflectivity. However, the distribution and intensity are not well simulated in D01. During the landfall time, the D02 (Fig. 4f) well simulated the reflectivity, but the D01 simulation shows away from coast. The structure is well matched with the observed reflectivity (Fig. 4d).

5 Conclusion

This study is aimed at assessing the performance of the non-hydrostatic model WRF-ARW with nested domain for simulation of VSCS Phailin over BOB. For this purpose, the model is integrated with 27 km (D01) and 09 km (D02) horizontal resolutions with five different model ICs using FNL analyses as initial and boundary conditions. The broad conclusions drawn are given as follows.

The intensity of the TC in terms of MSLP and 10 m surface wind is well simulated in the D02 as compared to the D01 simulation. The magnitude of 10 m

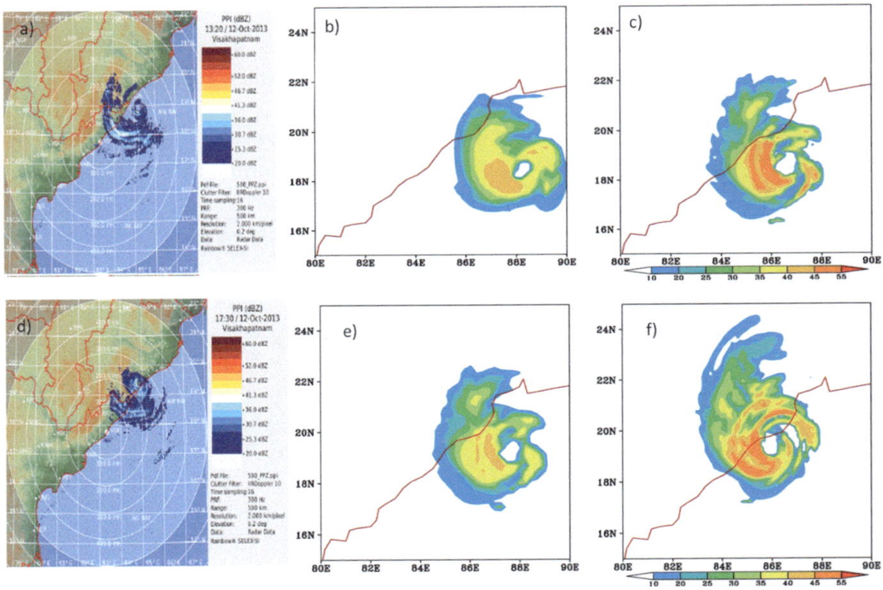

Fig. 4 Reflectivity (dBz) from observation (IMD Visakhapatnam Doppler radar) and model simulations (**a**) Observed, (**b**) D01 and (**c**) D02 valid at 06 UTC of 12th October 2013 (78 h forecast) (**d**–**f**) are same as (**a**–**c**) but at landfall time

surface wind is not well simulated in the D01 during intensification of TC as compared to observed. The mean RMSE and standard deviations of MSLP and 10 m surface wind is reasonably less in D02 simulation as compared to the D01 simulations with different ICs. Similarly, the absolute errors are less in D02 run in different forecast hours.

The track of the TC is well confined with IMD best track in the D02 simulations as compared to the D01 simulations. The mean VDEs are significantly reduced in the D02 simulation throughout the forecast hours compared to the error from D01 run. The ensemble mean (averaged all forecast hours) VDEs are also less in the D02 simulation in comparison to the other simulation. The landfall position and time errors are significantly reduced in D02 simulation in comparison with D01 simulation.

The skill of the D02 is significantly higher compared to that of the D01 simulation. The gain in skill of the D02 simulation (with respect to D01) ranged from 51 to 78 % throughout forecast period. Thus, the mean gain in skill of the D02 simulation is about 65 % over the D01. The gain in skill of D02 is gradually decreased as forecast length increased the landfall of TC is delayed around 3 h in D01 simulation.

The intensity and spatial distribution of the reflectivity (dBz) is well simulated in the D02 simulation mainly in the landfall time which is comparable with observed reflectivity obtained from IMD Visakhapatnam DWR. The features are not simulated in the D01.

This study conclude that the model WRF frame work with the increased of resolution have positive impact in the prediction of cyclonic storms. However, to further support this conclusion, more TC case studies and also assimilation of observations from different platforms are required.

Acknowledgement The authors thank NCEP/NCAR made available FNL analyses as well as WRF-3DVAR analysis system, which used in the present study. We also thank the IMD for providing observation data to validate the model results in this study. We thank anonymous reviewer for valuable suggestions and comments for improvement of the manuscript.

References

Bhaskar Rao, D.V. and Hari Prasad, D. (2006). Numerical prediction of the Odisha super-Tropical Cyclone (2006) sensitivity to the parameterization of convection, boundary layer and explicit moisture processes. *Mausam*, **57**, 61–78.

Dudhia, J. (1989). Numerical study of convection observed during winter monsoon experiment using a mesoscale two-dimensional model. Journal of Atmospheric Science, **46**, 3077–3107.

Gupta, A. (2006). Current status of Tropical Cyclone track prediction techniques and forecast errors. *Mausam*, **57**, 151–158.

Hong, S.Y, Noh, Y. and Dudhia, J. (2006). A new vertical diffusion package with explicit treatment of entrainment processes. *Monthly Weather Review*, **134**, 2318–2341.

Kain, J.S. and Fritsch, J.M. (1993). Convective parameterization for mesoscale models: The Kain-Fritcsh scheme. *In:* Emanuel, K.A, Raymond, D.J. (eds.) The Representation of Cumulus Convection in Numerical Models. American Meteorological Society, pp 246.

Lin, Y.L, Farley, R.D. and Orville, H.D. (1983). Bulk parameterization of the snow field in a cloud model. *Journal of Climate and Applied Meteorology*, **22**, 1065–1092.

Mlawer, E.J, Taubman, S.J, Brown, P.D, Iacono, M.J. and Clough, S.A. (1997). Radiative transfer for inhomogeneous atmosphere: RRTM, a validated correlated-k model for the longwave. *Journal of Geophysical Research* 102(D14), 16663–16682.

Mohanty, U.C. and Gupta, A. (1997). Deterministic methods for prediction of Tropical Cyclone tracks. *Mausam*, **48**, 257–272.

Mohanty, U.C, Osuri, K.K, Routray, A, Mohapatra, M. and Pattanayak, S. (2010). Simulation of Bay of Bengal tropical cyclones with WRF model: Impact of initial and boundary conditions. *Marine Geodesy*, **33**, 294–314.

Mohanty, U.C, Mandal, M. and Raman S. (2004). Simulation of Odisha super-Tropical Cyclone (1999) using PSU/NCAR mesoscale model. *Natural Hazards*, **31**, 373–390.

Mohapatra, M, Mandal, G.S, Bandyopadhyay, B.K, Tyagi, A. and Mohanty, U.C. (2012). Classification of cyclone hazard prone districts of India. *Natural Hazards*, **63**(3), doi:10.1007/s11069-011-9891-8

Mohapatra, M, Nayak, D.P, Sharma, R.P. and Bandyopadhyay, B.K. (2013a). Evaluation of official tropical cyclone track forecast over north Indian Ocean issued by India Meteorological Department. *Journal of Earth System Science*, **122**, 433–451, doi:10.1007/s12040-013-0291-1

Mohapatra, M, Bandyopadhyay, B.K. and Nayak, D.P. (2013b). Evaluation of operational tropical cyclone intensity forecasts over north Indian Ocean issued by India Meteorological Department. *Natural Hazards*, **68**, 589–601, doi:10.1007/s11069-013-0624-z

Mohapatra, M, Sikka, D.R, Bandyopadhyay, B.K. and Tyagi, A. (2013c). Outcomes and challenges of Forecast Demonstration Project (FDP) on landfalling cyclones over Bay of Bengal. *Mausam*, **61**, 1–15.

Mohapatra, M, Nayak, D.P, Sharma, M, Sharma, R.P. and Bandyopadhyay, B.K. (2015). Evaluation of official tropical cyclone landfall forecast issued by India Meteorological Department. *Journal of Earth System Science*, **124**, 861–874, doi:10.1007/s12040-015-0581-x

Osuri, K.K, Mohanty, U.C, Routray, A, Kulkarni, M.A. and Mohapatra, M. (2012). Customization of WRF-ARW Model with physical parametrization schemes for the simulation of Tropical Cyclones over the North Indian Ocean. *Natural Hazards*, **63**, 1337–1359, doi:10.1007/s11069-011-9862-0

Osuri, K.K, Mohanty, U.C, Routray, A. and Mohapatra, M. (2011). The impact of satellite-derived wind data assimilation on track, intensity and structure of Tropical Cyclones over the North Indian Ocean. *International Journal of Remote Sensing*, **2011**, 1–26.

Osuri, K.K, Mohanty, U.C, Routray, A, Mohapatra, M. and Dev, N. (2014). Real-Time track prediction of tropical cyclones over the North Indian Ocean in the ARW model, *Journal Of Applied Meteorology and Climatology*, **52**, doi:10.1175/JAMC-D-12-0313.1

Palmen, E.N. (1948). On the formation and structure of the tropical hurricane. *Geophysica*, **3**, 26–38.

Routray, A, Mohanty, U.C, Rizvi, S.R.H, Niyogi, D, Osuri, K.K. and Pradhan, D. (2010). Impact of Doppler weather radar data on simulation of Indian monsoon depressions. *Quarterly Journal of the Royal Meteorological Society*, **136**, 1836–1850.

Srinivas, C.V, Venkatesan, R, Vesubabu, V. and Nagaraju, C. (2010). Impact of assimilation of conventional and satellite meteorological observations on the numerical simulation of a Bay ofBengal Tropical Cyclone of Nov 2008 near Tamilnadu using WRF model. *Meteorology and Atmospheric Physics*, **110**, 19–44.

Data Assimilation Experiments with ARW–3DVAR for Tropical Cyclone Extreme Weather Predictions Over Bay of Bengal

C.V. Srinivas, G.M. Mohan, V. Yesubabu, K.B.R.R. Hariprasad, R. Baskaran, and B. Venkatraman

1 Introduction

Tropical cyclones (TCs) are highly disastrous weather phenomena characterised with extreme winds, heavy precipitation and storm surges at landfall along coastal lands. Accurate prediction of the TC formation, movement and intensity is vital for early warning and disaster management. The favourable environmental conditions have been identified as presence of an initial disturbance in the form of incipient lows or tropical easterly waves, high sea surface temperature (SST) ($\geq$26.5 °C) for transport of energy through air–sea fluxes, weak vertical wind shear in the 850–200 hPa layer conducive to large-scale cloud development and divergence in the upper atmosphere to facilitate further surface-level convergence and for overall sustenance of the system (Anthes 1982; Gray 2000). The annual average frequency of TCs in the North Indian Ocean (NIO) is about five (Asnani 1993). Numerical prediction of TCs requires accurate specification of initial conditions that define the characteristics of an incipient storm in terms of its location, radius, central pressure, and tangential and vertical winds.

Weather prediction using numerical models depends on a number of factors such as model dynamics, physics and initial conditions. Improvements in numerical weather prediction in the last decade are attributed to the development of various sources of observations and the data assimilation methods to effectively use them. Various methods of data assimilation such as successive correction, optimum interpolation, and variational methods (3-DVAR, 4-DVAR) have been developed for assimilating atmospheric observations in numerical atmospheric models

C.V. Srinivas (✉) • G.M. Mohan • V. Yesubabu • K.B.R.R. Hariprasad • R. Baskaran
B. Venkatraman
Radiological Safety and Environment Group, Indira Gandhi Centre for Atomic Research, 603102 Kalpakkam, India
e-mail: cvsri@igcar.gov.in

© Capital Publishing Company 2017

M. Mohapatra et al. (eds.), *Tropical Cyclone Activity over the North Indian Ocean*,
DOI 10.1007/978-3-319-40576-6_22

(Kalnay 2003; Park and Zupanski 2003). Variational techniques like 3DVAR and 4DVAR have the additional advantage of assimilating observations by satisfying model dynamic and thermodynamic constraints through a set of independent balance equations (Courtier et al. 1998). As TCs form and develop over oceanic regions, observations over oceanic areas become important in their prediction. It is also important to effectively use the conventional surface, upper air data and satellite radiances in initial conditions to reduce the forecast errors.

A number of impact studies due to data assimilation for TCs over NIO region are available (e.g., Singh et al. 2008, 2011, 2012; Xiao et al. 2009; Osuri et al. 2010, 2012; Srinivas et al. 2010; Prashant et al. 2012; Yesubabu et al. 2013; among others). They are focussed on using various sources of data sets (satellite, radar, conventional etc) for assimilation. Firstly, it is required to use commonly available observational data and conduct numerical experiments for a number of cyclonic storms to assess the model performance quantitatively. Secondly, direct assimilation of satellite-observed radiance data for atmospheric parameters (temperature and humidity) is also important to study its contribution in TC predictions.

The Advanced Research Weather Research and Forecast (ARW) model is implemented in operational mode along with 3DVAR data assimilation system for early warning of extreme weather including TCs for the Department of Atomic Energy (DAE) site Kalpakkam. The performance of ARW model without data assimilation over Bay of Bengal (BOB) was assessed for tropical TC predictions in recent studies (Srinivas et al. 2012). The objective of our study is to assess the skill of ARW model with observation assimilation for operational TC track and intensity predictions over the BOB region. Towards this objective, data assimilation experiments are conducted for a set of eight TCs using the operationally available global surface and upper air observations and satellite radiances over the NIO.

2 Numerical Experiments

The ARW model developed and sourced by the National Centre for Atmospheric Research (Skamarock et al. 2008) is used to study the impact of data assimilation on the evolution of TCs. A two-way interactive double nested (27-, 9-km resolutions) domain configuration with 35 vertical levels is used (Fig. 1). The outer domain covers a larger region with 27-km resolution and 173×162 grids. The inner domain has 9-km resolution with 292×229 grids covering the BOB and adjoining coastal land areas. The data for the terrain elevation, land-use and soil types for the 27- and 9-km resolution domains are taken from the U.S. Geological Survey (USGS) topography data at $10'$ and $5'$ resolutions. In the present study, eight tropical TCs (Phailin, Thane, Jal, Laila, Aila, Nilam, Khai-Muk and Nisha) originated in the BOB in the period from 2008 to 2013 are considered for simulation. Among the eight cases, Phailin and Thane are very severe cyclonic storms, Jal, Laila, Aila are severe cyclonic storms, Nilam, Khaimuk and Nisha are cyclonic storms (India Meteorological Department 2009–2014). In all the cases the model is initialised at 00 UTC on different days

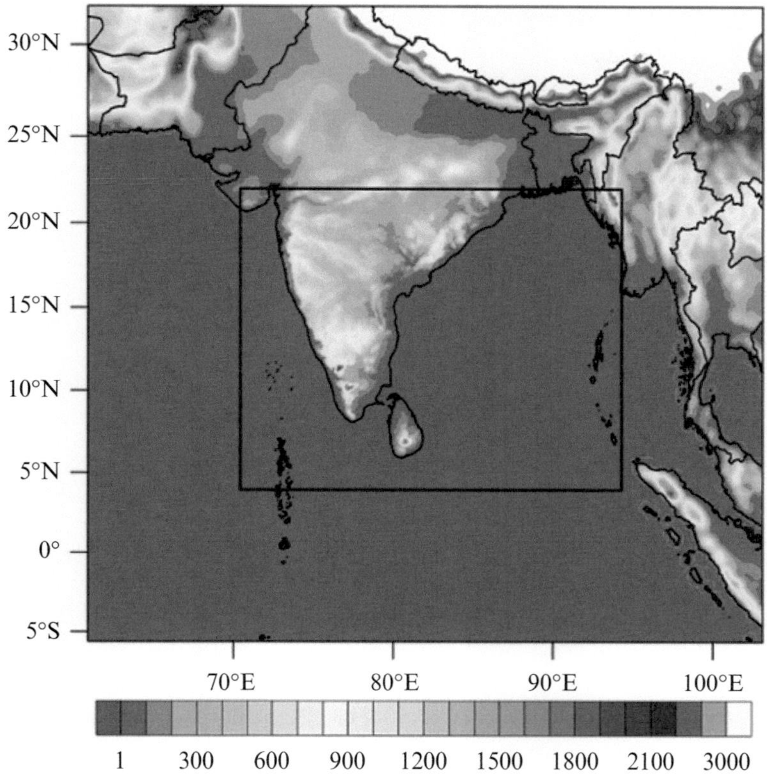

Fig. 1 Modelling domains used in ARW for TC simulations. The shading gives the terrain height

corresponding to deep depression stage and integrated up to 96 h (4 days) to cover up to the landfall stage. Details of model initialisation time and data assimilated for the model run are presented in Table 1.

3 Data and Model Initialisation

For each TC two sets of simulations are conducted i.e. one without assimilation (CTL) and the other with assimilation using 3DVAR. The CTL experiments are initialised with National Centers for Environmental Prediction (NCEP) Global Forecast System (GFS) analysis at $0.5 \times 0.5°$ resolution and boundary conditions except for SST are updated from the same source. The GFS data is interpolated to the model horizontal and vertical grids. The initial and boundary conditions for SST are taken from the $0.08 \times 0.08°$ NCEP real-time global sea surface temperature analysis.

The 3DVAR assimilation experiments are performed in a cold start mode ingesting the NCEP PrepBufr observations with +/– 3 h assimilation window at

Table 1 Names of the TCs along with their period and landfall time, model forecast length, model initialisation time and data density assimilated in the initial conditions

Name of the TC	Model initialisation time for 96 h forecast YYMMDD	GFS Resolution (°)	Landfall	Density of Data used in assimilation in domains (D1 and D2)													
				SYNOP		BUOY		METAR		SHIP		SCAT		Geoamv		SOUND	
				D1	D2	D1	D2	D1	D2	D1	D2	D1	D2	D1	D2	D1	D2
Phailin	20131009–13	0.5 × 0.5	Gopalpur, Odisha	655	183	29	10	100	17	20	11	0	0	1901	341	48	11
Nilam	20121029–20121101	0.5 × 0.5	Mahabalipuram, TN	891	292	32	20	111	16	21	5	0	0	4634	1336	25	2
Thane	20111227–31	0.5 × 0.5	Puducherry, TN	389	91	53	25	96	15	36	13	495	134	2903	1065	55	91
Jal	20101104–08	0.5 × 0.5	Chennai, TN	377	84	25	13	89	12	25	12	177	61	2657	771	61	19
Laila	20100517–21	0.5 × 0.5	Machilipatnam, AP	347	73	28	2	87	11	18	8	475	143	3712	1027	51	14
Aila	20090523–26	1 × 1	Sagar Island, West Bengal	360	71	13	6	86	12	27	15	2832	819	3188	911	57	17
Khai-muk	20081113–16	1 × 1	Ongole, AP	361	62	18	13	86	12	25	13	1557	543	3865	1166	61	20
Nisha	20081124–28	1 × 1	Karaikal, TN	381	79	22	13	90	11	24	13	1654	559	3635	1018	56	21

initialisation corresponding to first guess data. The PrepBufr data comprise land surface, marine surface, radiosonde, pilot-balloon and aircraft reports from the Global Telecommunications System (GTS), Atmospheric Motion vectors (AMV) from Geostationary satellites, profiler and radar derived winds wherever available, Special Sensor Microwave Imager (SSM/I) oceanic winds and National Environmental Satellite Data and Information Service (NESDIS) satellite wind data which are operationally collected by the NCEP.

Presently, radar data over Indian region is not included in GTS. Similarly, wind profiler data are not available over the Indian region for operational predictions. The details of the data can be obtained from http://rda.ucar.edu/datasets/ds337.0. The distribution of different observations are shown in Fig. 2 for Laila TC as an example. For this case, a good coverage of Advanced Microwave Sounding Unit (AMSU) radiance data and atmospheric motion vectors can be seen, though the SSM/I winds did not cover the inner core of the TC. A few experiments are conducted with AMSU radiances (Channels A, B) of atmospheric temperature and humidity profiles for the cases of Phailin, Thane and Nilam TCs to study their relative contributions to predictions with respect to operationally available observations. The community radiative transfer model (CRTM) available in 3DVAR (Liu et al. 2006) is used for direct ingestion of radiance data in the model. The background error covariance data for 3DVAR are adopted from NCEP. In both the experiments (CTL, 3DVAR) same physics options as identified in our previous model physics sensitivity study (Srinivas et al. 2012) are used. These include Lin et al. scheme (Lin et al. 1983) for cloud microphysics, RRTM scheme for longwave radiation (Mlawer et al. 1997), Dudhia scheme (Dudhia 1989) for short-wave radiation, Monin–Obukhov with Carlson–Boland viscous sub-layer similarity for surface layer physics, the Yonsei University non-local diffusion (Hong et al. 2006) for boundary layer turbulence, and Kain–Fritisch (Kain 2004) mass flux scheme for convection and NOAH model (Chen and Dudhia 2001) for land surface physics. Results from CTL and 3DVAR runs for TC track and intensity parameters are evaluated using the India Meteorological Department (IMD) best track data and the IMD, TRMM gridded rainfall data sets. The azimuthally averaged temperature anomaly and the radius of maximum wind (RMW) from the model are compared with the data from Cooperative Institute for Research in the Atmosphere (CIRA). The CIRA data also provides maximum sustained wind (MSW) and central sea level pressure (CSLP) and gives a quantitative structure of TCs from Multi-Platform Tropical TC Analysis products (Knaff et al. 2011). The details of the period of each TC, initial and model integration time, density of observations of each type used in assimilation experiments are provided in Table 1.

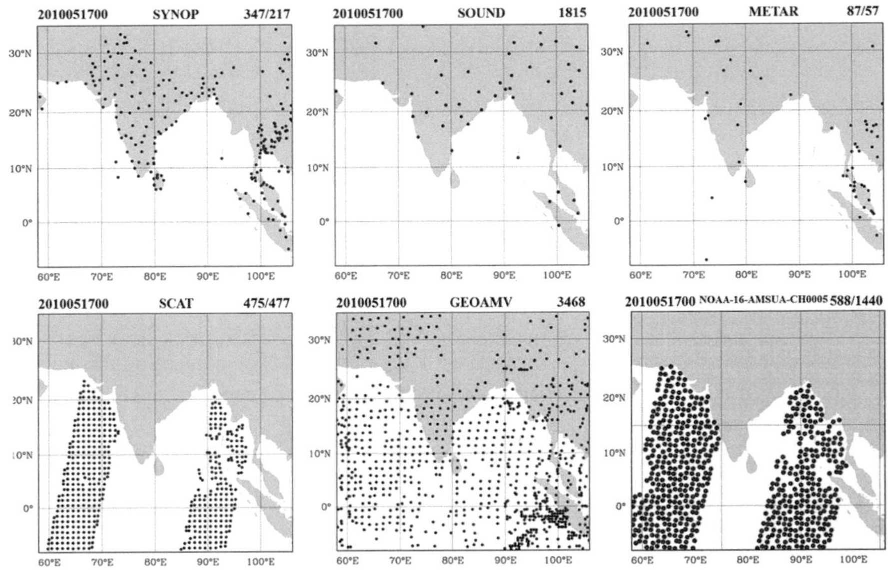

Fig. 2 Distribution of NCEP PrepBufr observations used in assimilation experiments for Laila TC

4 Results and Discussion

The qualitative results of simulations for intensity and track parameters are presented for the case of Laila TC followed by the quantitative analysis of errors for all the eight cases. The analysis increments obtained using data assimilation in the wind field, temperature and relative humidity both horizontally and vertically is discussed below. The improvement in the 3DVAR from the CTL is described in terms of error in CSLP, MSW and track position.

4.1 Qualitative Results of Storm Prediction

4.1.1 Differences in Initial Conditions

The changes in the initial condition (pressure, winds and temperature distribution) for the CTL and 3DVAR experiments are analysed. The Laila TC formed in the central BOB near Andaman. The analysis increments in winds, temperature and sea-level pressure between control and 3DVAR experiments at the initial time 00 UTC 17 May 2010 for Laila TC are presented in Fig. 3a in horizontal and vertical planes. The differences indicate an anti-cyclonic wind distribution in the region of TC with positive pressure differences (D*p* ~0.4–0.7 hPa) in the 3DVAR experiment. This indicates a less intense initial TC vortex in the 3DVAR run. The vertical initial differences in winds, temperature and humidity around the TC from

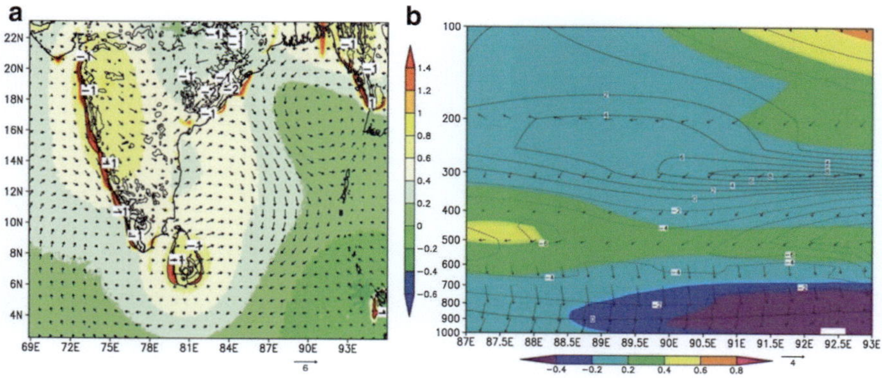

Fig. 3 (**a**) Initial difference in the sea-level pressure, 10 m winds, and temperature (*black dashed contours*) distribution pattern between CTL and 3DVAR experiments and (**b**) difference in the vertical distribution of temperature (*shaded*), humidity (contours) and wind (vectors) distribution at 10.4174° N for the case of TC Laila at 00 UTC 17 May 2010

1000 to 100 hPa, centered at 10.42°° N and 90.66°° E are shown in Fig. 3b. It indicates negative temperature difference (DT ~ −0.35 to −0.05 °C), positive humidity increments (DRH ~2 %) and stronger winds in the control run relative to the 3DVAR all indicating a less intensive vortex in 3DVAR experiment. The vertical distribution of temperature, horizontal winds and relative humidity reveal that nearer to the low pressure area (centred at 90.75°° E) the temperature below 800 hpa is reduced by ~0.3 K in 3DVAR experiment. Also convergence of surface wind over the low pressure is noted, which indicates conditions that may lead to higher intensity of the TC in the CTL.

4.1.2 Intensity Predictions

The intensity of the simulated Laila TC is analysed from central pressure and maximum winds. The time series of central pressure, maximum winds and radius of maximum winds is presented in Fig. 4a–c respectively from control and 3DVAR experiments for TC Laila. Both the experiments overestimated the intensity. The control run produces a more intense TC with slightly higher pressure drop relative to the 3DVAR simulation (Fig. 4a). The least pressure observed in 3DVAR and control experiments are 961 and 957 hPa, respectively. The observed central pressure from IMD and CIRA reports are 988 and 970 hPa which indicate overestimation of intensity in ARW simulations and more prominently in the CTL run. Generally, models have a tendency of overestimation of weaker storms and underestimation of stronger storms. The reasons for over prediction of intensification could be deficiencies in model physics such as boundary layer turbulence

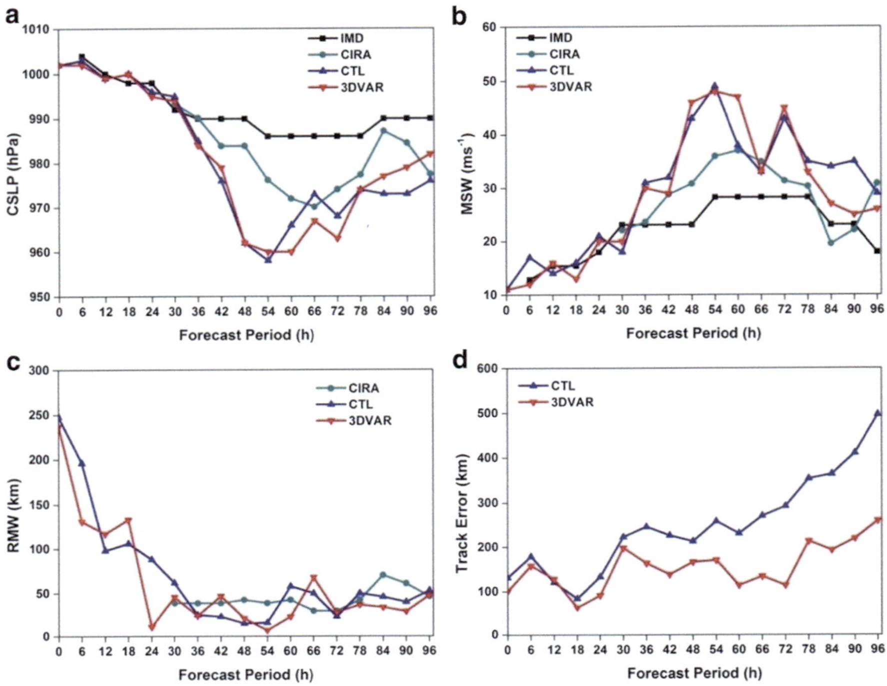

Fig. 4 Time variation in (**a**) Central pressure (hPa) (**b**) Maximum sustained winds (m/s) (**c**) Radius of maximum winds (km) and (**d**) vector track errors (km) of TC Laila in Control and 3DVAR experiments

and convection leading to stronger inflow, low level convergence and enhanced convection etc., which need to be carefully examined. The maximum wind speed found in simulation is in the range of 47–50 m/s (Fig. 4b). But the IMD observations indicate a maximum wind speed of 30 m/s and CIRA data indicates ~40 m/s. Although large differences are noted between IMD observations and model results, the 3DVAR gives less errors in winds and pressure compared to the control run. Simulated TCs in control run and 3DVAR are intensified faster (by 12 h) than the actual storm. The maximum wind speed obtained in 3DVAR and control runs are 47 and 50 m/s, respectively. The time series of radius of maximum winds (Fig. 4c), which gives a measure of the size of storm is presented in Fig. 4c. Generally, the intense storms have a smaller RMW compared to weak storms. The time series of RMW shows that initially is less for the 3DVAR during the developing stages. Subsequently, during the peak intensity stage the RMW is smaller for the 3DVAR compared to control run. This also suggested that the control run produced a more intensive storm relative to the 3DVAR. The time series of RMW derived from 3DVAR experiment is in better agreement with the CIRA data.

4.1.3 Track Predictions

The time series (3 h) of errors in vector track positions for TC Laila from control and 3DVAR experiments are presented in Fig. 4d. The vector track positions of all eight TCs from both CTL and 3DVAR experiments are presented in Fig. 5. The TC Laila moved in a north-westward direction towards Andhra Pradesh coast (Fig. 5e). The simulated track throughout the lifecycle of the storm is found north of the observed track with higher track errors in the CTL relative to the 3DVAR (Fig. 5e). It may be noted that the tracks of the simulated storms from both CTL and 3DVAR in some cases appear to be same. The actual track errors are due to both cross-track as well as along track vector position differences. The improvement in track forecast with the 3DVAR is due to improvements in cross track and along track position errors in the simulations. The initial track position error in 3DVAR run is 101 km and that in control run is 131.55 km.

This initial position error in control run is further propagated in the simulation as seen in the final landfall positions that are about 220 and 500 km away in 3DVAR and control runs respectively from actual landfall point. The 3DVAR run is accomplished with assimilation of Prepbufr data which consists of multiple sources of data as explained in Sect. 3. Of the various observations used in assimilation the atmospheric motion vectors cover larger regions in more density (Fig. 2).

This can be useful to define the initial position of the TC more precisely over the ocean region where conventional observations are lacking. It has been found that the 3DVAR experiment gives closer landfall position relative to control run with an error of about 100 km. The track errors are of the order of 133.4, 213, 292 and 496.5 km in the control run; and 91.5, 166.46, 114.07 and 259.55 km in the 3DVAR run at 24, 48, 72 and 96 h forecasts, respectively. An improvement of about 75 % in the track errors is noted in the 3DVAR experiment. From the tracks of all eight TCs (Fig. 5), large improvements in track predictions are found for the cases of Laila, Thane; meagre improvements are found for the cases of Phailin, Nisha, Aila, Jal and no improvements for Nilam and Khaimuk. The improvements in tracks for Laila, Phailin and Thane are due to reduction of along track errors.

4.1.4 Convective Warming

Cyclones are associated with warming in the upper atmosphere due to large-scale convection and release of latent heat by condensation. At the lower levels cooling takes places due to evaporation of water from ocean surface. Warming is measured by temperature anomaly which is the difference between ambient temperature and normal atmospheric temperature before formation of TC. The azimuthally averaged temperature anomaly for TC Laila from control and 3DVAR simulations along with corresponding data from CIRA observations is presented in Fig. 6. Warming is noted in the layers 4–16 km in the control and 5–11 km in the 3DVAR runs. A wider lateral spread of warming is seen in the 3DVAR run. CIRA observations indicate

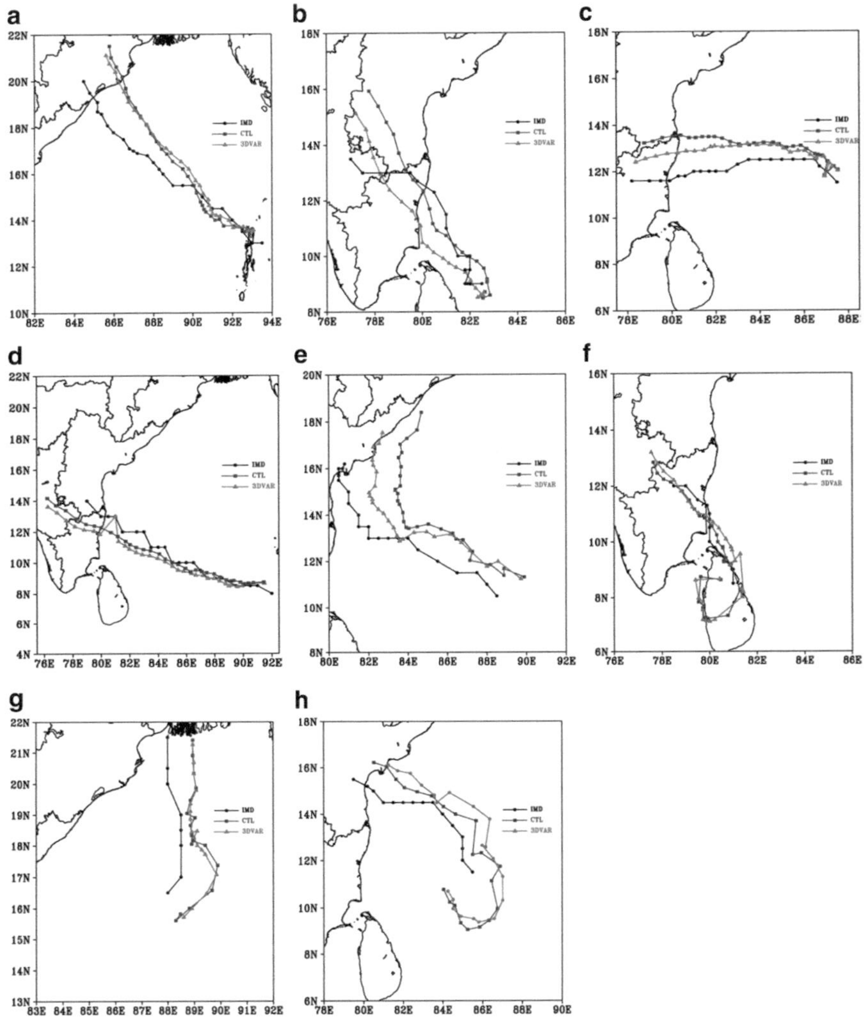

Fig. 5 Simulated vector track positions along with IMD best track from control and 3DVAR experiments for eight tropical TCs over BOB (**a**, Phailin; **b**, Nilam; **c**, Thane; **d**, Jal; **e**, Laila; **f**, Nisha; **g**, Aila and **h**, Khai-Muk)

the convective warming is confined in the 9–15 km with a maximum warming of 4° K at about 12 km. Also the cooling of the atmosphere in the lower layers is simulated well. The maximum warming in CTL is confined at ~10-km height, whereas it is confined to ~6–9 km in 3DVAR. Comparison with CIRA data shows that the distribution of vertical warming is better simulated in CTL and the horizontal warming distribution in 3DVAR. Overall, the CTL is close to CIRA and indicates better simulation of convection.

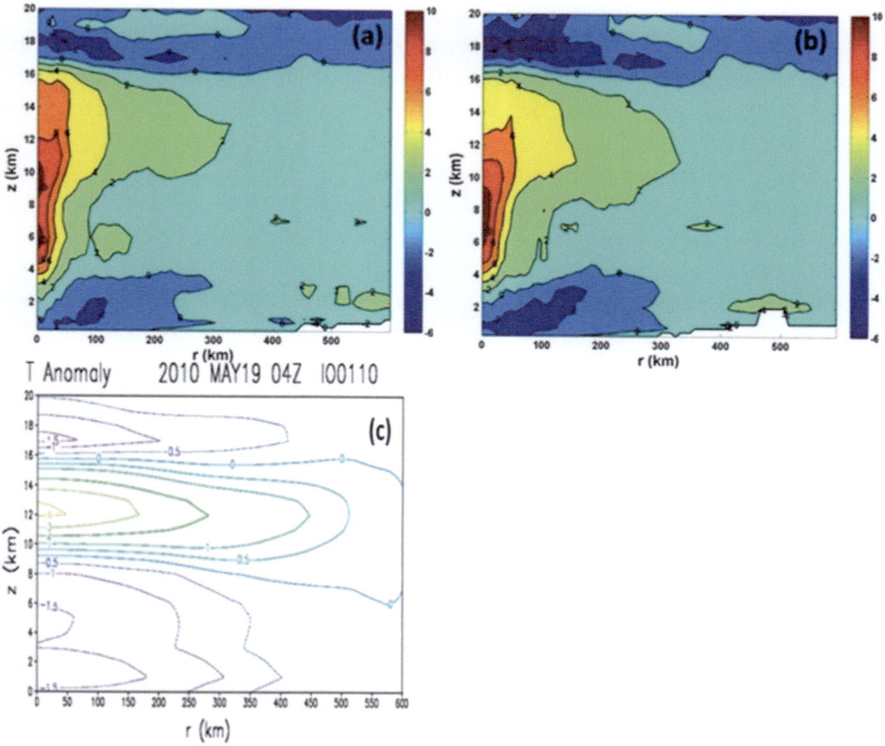

Fig. 6 Azimuthal variation of temperature anomaly for TC Laila in (**a**) Control (**b**) 3DVAR experiments (**c**) CIRA multi-satellite data at mature stage of the TC corresponding to 04 UTC 19May 2010

4.2 Quantitative Error Analysis in Control and 3DVAR Experiments

In this study the errors in the model simulation with and without data assimilation are calculated using IMD best track data as reference for the predicted parameters of track positions, maximum sustained winds and central sea level pressure. The time series of errors (Model-IMD) are analysed. The errors are shown as mean error (ME) and standard deviation (SD). The ME is the average of the track errors of all eight TCs at a given forecast time. The SD gives an idea about how much the error values are away from the average error. The model performance is assessed from the error scatter with respect to time using ME + SD and ME − SD as thresholds. The time series of mean errors in various parameters are given in Tables 2, 3, and 4. The errors in track positions (Fig. 7) indicate a wide scatter, maximum errors are found with Khaimuk (>400 km) and minimum with Nisha (30 km). In all cases track errors increased continuously. From Fig. 7a for control run it is seen that in the first 12 h of the integration period the errors are in the range of the error threshold.

From 18 to 48 h error build-up can be noticed. It has been found that during the first 2 days of simulation the model gives better track prediction in control runs for the TCs Nilam, Phailin, Thane and Nisha. The mean track error of all control runs ranges as 110, 122, 184 and 402 km at 24, 48, 72 and 96 h forecasts, respectively. A drastic reduction in track errors can be found in 3DVAR simulations (Fig. 7b) as seen from a small scatter of errors about the mean. The mean track errors in 3DVAR experiments ranged as 97.5, 151, 170 and 282 km at 24, 48, 72 and 96 h forecasts, respectively. Except Khaimuk and Nilam track errors reduced with data assimilation in all TCs. An improvement of 11.6 %, −24.39 %, 7.73 %, 29.76 % for 24, 48, 72 and 96 h, respectively are noticed in 3DVAR simulations over control experiments (Table 4). The deterioration of track forecast with 3DVAR at 48 h (−24 %) is because of relatively large errors in track for Khaimuk and Nilam simulated in 3DVAR at 48 h compared to CTL. The deterioration of 3DVAR for these two cases could be because of poor coverage of sounding data (confined to $> 20°$ N) and less density of motion vectors (confined to $<12°$ N).

The error in the CSLP (Fig. 8) shows that for majority of the TCs the CSLP error is negative, which means that for most TCs the CSLP is underestimated implying simulation of stronger storms. The ME for CSLP in control runs ranges as −12.3, −15.0, −8.3 and −10.0 hPa at 24, 48, 72 and 96 h forecasts, respectively. The error in CSLP increases gradually from 24 to 48 h. In all the control runs Nisha is the best simulated TC and Jal is simulated poor. In the case of 3DVAR runs, the mean CSLP error ranges as −11.3, −15.6, −8.2 and −7.0 hPa at 24, 48, 72 and 96 h forecasts, respectively. Among the 3DVAR experiments Nisha, Khaimuk and Nilam are the better predicted cases for central pressure. An improvement of 8.14 %, −4.17 %, 2 %, 30 % are noted at 24, 48, 72 and 96 h forecasts, respectively with 3DVAR.

The error in the MSW distribution (Fig. 9) shows that the ME for tangential winds of the selected TCs is generally positive, which indicates the model tendency to simulate stronger winds and hence intensive TCs. Based on error in MSW, it is seen that the intensity is better simulated in the cases of Nilam, Nisha, Khaimuk and Laila. The TC Jal is poorly simulated upto a forecast period of 42 h, afterwards the simulated winds are found more realistic. The mean error in winds ranges as 10.3, 9.7, 2.3 and 8.1 m/s in control runs; 9.9, 9.7, 2.3 and 6.6 m/s in 3DVAR runs at 24, 48, 72 and 96 h, respectively. An improvement of 4.15 %, 0 %, 0 % and 18.56 % are noted at 24, 48, 72 and 96 h, respectively in the intensity predictions in 3DVAR simulations over control runs.

The time series of average errors for radius of maximum winds of all eight TCs are shown in Fig. 9 for control and 3DVAR experiments. It is seen that improvements of 8.1 %, 29.34 %, 48.78 % and 79.83 % are noted in radius of maximum winds in the 3DVAR runs over the control runs. The above results of all parameters (track positions, winds, pressure and radius of maximum winds) indicate assimilation of most commonly available observations in the operational runs in the ARW produces significant improvements in TC track and intensity predictions.

Table 2 Mean error metrics for wind speed, pressure, track and radius of the maximum in the control experiments for the eight TCs

	Error in max. winds (m/s)		Error in central pressure (hPa)		Vector track position error (km)		Error in radius of max. winds (km)	
Forecast period (h)	ME	SD	ME	SD	ME	SD	ME	SD
12	8.6	7.4	−5.3	5.8	96.37	68.22	25.1	49.8
24	10.3	7.9	−12.3	11.9	110.34	93.13	6.2	27.8
36	10.1	8.9	−13.1	14.9	159.59	146.42	−14.4	73.5
48	9.7	13.0	−15.0	17.9	122.09	83.89	43.8	51.6
60	6.2	12.3	−11.9	16.4	144.03	78.27	57.7	67.1
72	2.3	10.6	−8.3	15.6	184.26	95.73	34.5	47.4
84	1.1	11.1	−10.3	7.0	228.76	84.62	100.7	103.9
96	8.1	4.1	−10.0	5.7	402.81	132.54	−14.06	69.31

Table 3 Mean Error metrics for wind speed, pressure, track and radius of the maximum in 3DVAR experiments for the eight TCs

	Error in max. winds (m/s)		Error in central pressure (hPa)		Error in vector track position error (km)		Error in radius of max. winds (km)	
Forecast period (h)	ME	SD	ME	SD	ME	SD	ME	SD
12	8.5	7.8	−5.7	6.6	82.61	66.02	20.1	52.8
24	9.9	6.5	−11.3	9.9	97.51	88.87	5.7	29.3
36	8.3	9.8	−12.8	15.6	150.97	127.15	−21.6	92.9
48	9.7	14.5	−15.6	20.5	151.86	101.85	31.0	46.5
60	7.4	13.3	−11.9	16.8	130.86	62.72	46.7	80.7
72	2.3	13.0	−8.2	16.5	170.01	44.27	17.7	46.8
84	−1.1	9.5	−7.0	7.9	199.80	35.96	46.7	86.4
96	6.6	2.0	−7.0	1.4	282.94	33.08	25.29	57.49

Table 4 Improvement in predictions in 3DVAR over control runs

Forecast period	Track error (%)	CSLP error (%)	MSW error (%)	RMW error (%)
24	11.63	8.14	4.15	8.08
48	−24.39	−4.17	0.00	29.34
72	7.73	2.00	0.00	48.78
96	29.76	30.00	18.56	79.83

4.3 Results of Experiments with Radiance Data Assimilation

Assimilation of satellite measured radiances in numerical weather prediction models is given importance in recent times to improve atmospheric thermodynamical structure in initial conditions. The satellite measured radiances can be directly

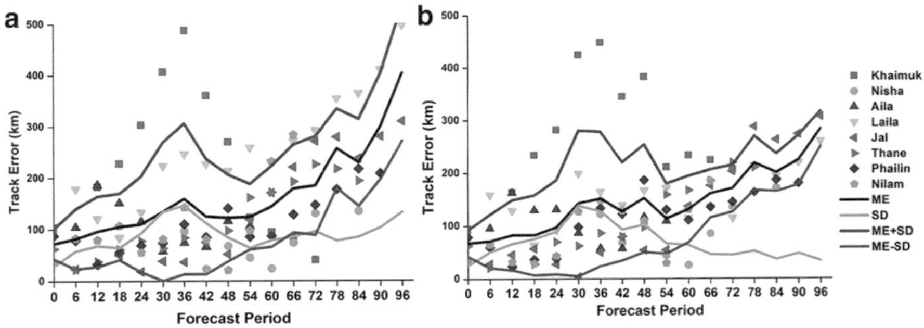

Fig. 7 Time variation of errors in track position from (**a**) control run (**b**) 3DVAR experiments for all eight TCs

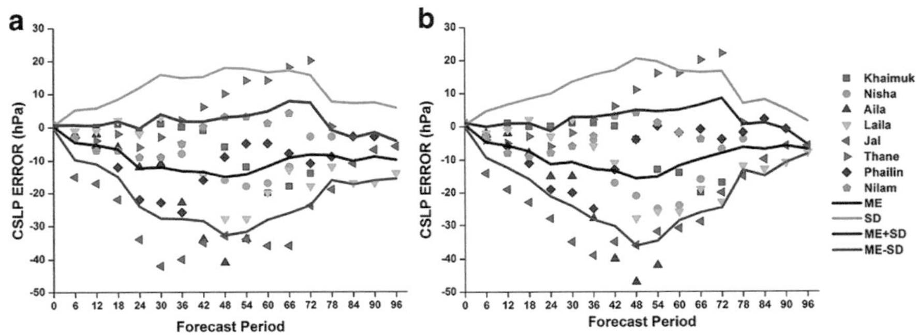

Fig. 8 Time variation of errors in central pressure from (**a**) control run and (**b**) 3DVAR experiments for all 8 TCs

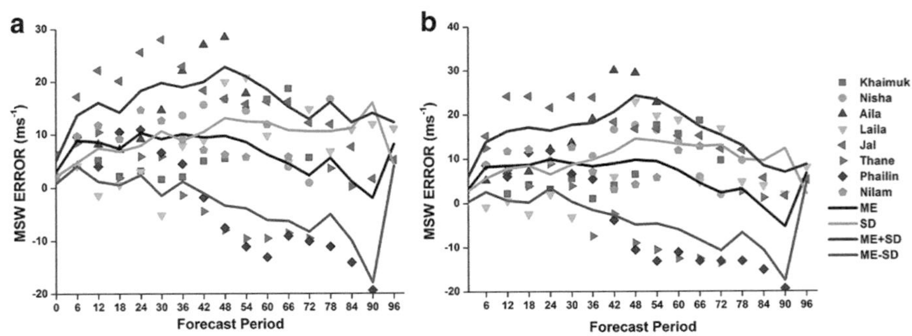

Fig. 9 Time variation of errors in maximum sustained wind from (**a**) control run and (**b**) 3DVAR experiments for all eight TCs

assimilated with 3DVAR using radiation transfer models. Presently radiance observations like AMSU-A/B, AMT are not used in operational weather forecasting. Here results of simulations using radiance data assimilation for three TCs Phailin, Thane, Nilam which had landfall along the east coast are analysed. Assimilation experiments include the use of PrepBufr (conventional surface, upper air), both radiance and Atmospheric Motion Vectors data (RAD + AMV), Atmospheric Motion Vectors (AMV), and AMSU-A, B radiance (RAD) data. While TC Thane has sufficient coverage of AMV, scatterometer winds and RAD Phailin has a poor coverage of these data especially of *radiances* (not shown). Assimilation experiments for these two events are examined to understand the differences in simulations. Simulated tracks are shown in Fig. 10a, b, vector track position are presented in Fig. 11a, b and intensity predictions are depicted in Fig. 12a–d. Results for Phailin indicate deviation of the simulated track to the north of the observed track (Fig. 10a) and the vector track position errors from RAD and RAD + AMV are more than the errors from the control run. The track position errors are 70.6, 183.7, 191.3 and 254.8 km in CTL; 47.7, 222.2, 243.4 and 313.5 km in AMV; 87.9, 253.6, 223.3 and 295.3 in RAD; 54.0, 182.4, 117.8 and 143.5 km in Prepbufr and 49.5, 260.8, 252.3 and 392.4 km at 24, 48, 72 and 96 h forecasts, respectively. It is seen that the simulation PrepBufr produced least track errors of all runs. The track errors with RAD, RAD + AMV, AMV progressively increased with forecast time (Fig. 11a). As regards, intensity of Phailin all the assimilation experiments produced nearly similar order of errors (Fig. 12a, b) in CSLP and MSW. The CSLP errors are −16.0, −3.0, −6.0 and −4.0 hPa in CTL; −17.0, −2.0, −6.0 and −4.0 hPa in AMV; −19.0, −4.0, −1.0 and −4.0 hPa in RAD; −17.0, −4.0, −4.0 and −10.0 hPa in Prepbufr; −18.0, −3.0, −6.0 and −7.0 hPa in RAD + AMV at 24, 48, 72 and 96 h forecasts, respectively. The errors in MSW are 7.9, −11.5, −10.1 and −17.4 m/s in CTL; 9.9, −10.5, −6.1 and −15.4 m/s in AMV; 7.9, −9.5, −14.1 and −17.4 in RAD; 8.9, −11.5, −12.1 and −12.4 in Prepbufr; and 10.9, −10.5, −10.1 and −15.4 m/s in RAD + AMV at 24, 48, 72 and 96 h forecasts, respectively. These results for CSLP and MSW indicate the cases RAD and Prepbufr showed slightly better intensity predictions during the peak intensity stage (60–72 h) of Phailin. The Prepbufr almost reproduced the observed intensity. The less impact of RAD and AMV observations could be due to the poor area coverage of these observations for Phailin. For the case of TC Thane (Fig. 10b) the AMV and Prepbufr conventional data both produced considerable improvements in the track predictions. For this case, while most of the simulations showed northward deviation of track from the initial time, the simulated track with AMV closely followed the observations up to 48 h and thereafter deviated to the north. The simulations AMV followed by Prepbufr produced better track predictions (Fig. 11b). The error range of track positions is 40, 50 and 240 km for AMV and 50, 100 and 220 km for Prepbufr at 24, 48 and 72 h forecasts, respectively. It is seen that RAD data is not influencing the motion of the TC as track errors obtained with RAD are of the same order of magnitude as of CTL. Next to AMV and Prebufr, the simulation RAD + AMV produced smaller errors in track which is again due to use of AMVs. The errors in MSW are 7.9, −1.0 and −3.6 m/s in RAD + AMV; 4.9, −1.0 and −2.6 m/s in RAD; 9.9, −1.0 and −5.6 m/s in AMV; 8.9, −6.0 and −11.6 m/s in Prepbufr; and 6.9, −8.0 and −14.6 m/s at 24, 48 and 72 h, respectively. The errors in CSLP are −7.0,

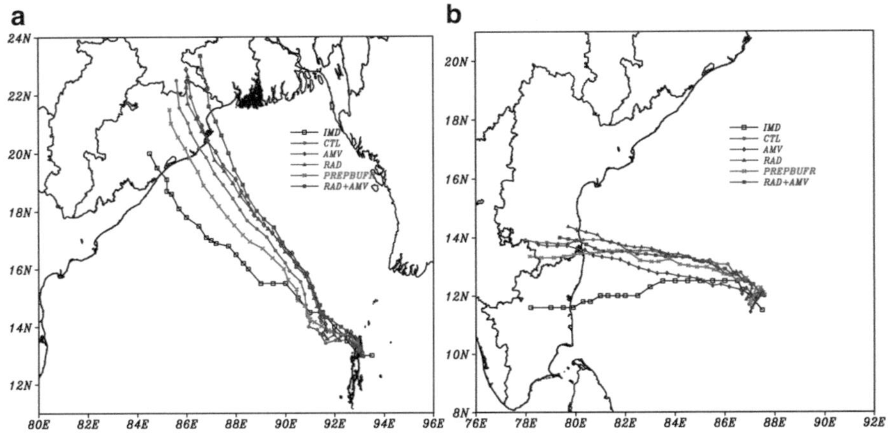

Fig. 10 Simulated tracks of TCs (**a**) Phailin (**b**) Thane in assimilation experiments with different types of observations

0.0 and 11.0 hPa in RAD + AMV; −6.0, 1.0 and 8.0 hPa in RAD; −11.0, −2.0 and 14.0 hPa in AMV; −5.0, 11.0 and 24.0 hPa in CTL at 24, 48 and 72 h, respectively. These results suggest that the intensity of TC Thane is better simulated by the experiments RAD + AMV followed by RAD and AMV.

The AMV data improves the windfield representation while the RAD data improves the model initial thermodynamical structure. The results for Thane with good coverage of AMV and RAD show that track predictions are mainly influenced by AMV while the intensity predictions are influenced by both AMV and RAD. TCs are characterised by strong coupling of thermodynamical and dynamical quantities at various levels in the atmosphere. In this regard the present results of better intensity prediction with AMV and RAD are pertinent. The influence of the wind data on the dynamics and RAD data on the thermodynamics would lead to better representation of coupling of thermodynamics and dynamics leading to better intensity predictions. The influence of wind data in Thane simulation can be justified from availability and density of the SCAT and GEOAMV observations. Also the less impact of Ocean observations (BUOY and SHIPS) is found similar in both Thane and Phailin cases.

In the case of Nilam TC the experiments with Prepbufr showed better track predictions and with error reduction of 20 %, 30 % and 2 % at 24, 48 and 72 h, respectively. As in the case of Thane TC, intensity predictions for Nilam are improved with assimilation of AMV and RAD.

Results of all the experiments clearly show that the Prepbufr observations which also contain the AMV always show better predictions relative to the RAD data in both intensity and track estimates and the latter shows impact only on intensity. An average error reduction of 20–50 % in vector tracks, 10–25 % in maximum winds,

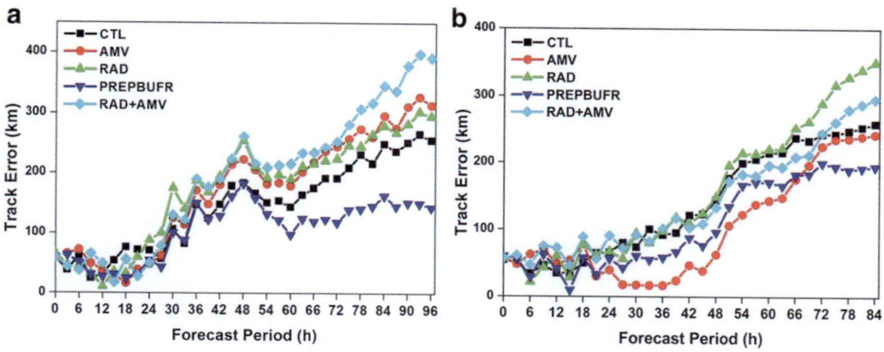

Fig. 11 Time series of vector track errors of TCs (**a**) Phailin and (**b**) Thane in assimilation experiments with different types of observations

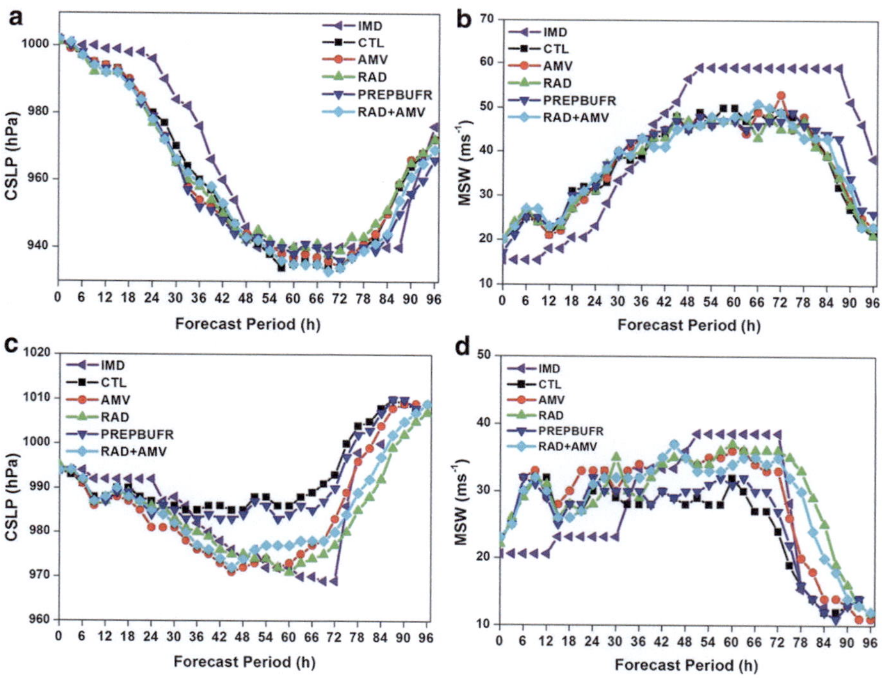

Fig. 12 Time series of central pressure (**a** and **c**) and maximum surface winds (**b** and **d**) from experiments with assimilation of different types of observations. Top panels (**a** and **b**) are for Phailin and bottom panels (**c** and **d**) are for Thane

5–15 % in central pressure and about 5 % in RMW parameters was estimated from the three simulated cases with Prepbufr data. These results indicate the advantages of using operationally available Prepbfur observations in TC forecasts in cold start mode using GFS analysis as model background.

5 Conclusions

In this work we employed most commonly available surface, upper air and few satellite based observations to study their impact on TC forecasts in our operational weather forecast model WRF using 3DVAR. Large improvements are found in the assimilation runs using NCEP PrepBufr data with 3DVAR over control runs. Of the eight recent TC cases tested, six have shown improvements in both track and intensity. An improvement of about 11–30 % in track, 8–30 % in central pressure, 4–19 % in maximum tangential winds and 8–80 % in radius of maximum winds in 24–96 h lead time forecasts are noted with observation assimilation. Experiments with assimilation of AMSU-A/B satellite radiance data pertaining to temperature and moisture profiles have shown impact on the intensity estimates alone with least contribution to track error minimisation. The better intensity prediction obtained using radiance data may be because of the changes in model thermodynamical structure. Use of AMV showed improvements both in intensity and track predictions which may be due to the changes in dynamics. Relative results of all assimilation experiments using various types of observations clearly shows the advantages of operationally available Prepbufr data sets in TC predictions which reduced average error by 20–50 % in vector tracks, 10–25 % in maximum winds, 5–15 % in central pressure and 5 % in RMW parameters for the three cases (Phailin, Thane and Nilam).

Acknowledgements The authors wish to thank Sri S.A.V. Satya Murthy, Director EIRSG for encouragement and support. The WRF ARW model was obtained from NCAR. The GFS analysis and forecasts used in operational predictions are available from NCEP. The India Meteorological Department is acknowledged for the use of best track and intensity estimates. Authors acknowledge anonymous reviewers for their technical comments that helped to improve the manuscript.

References

Anthes, R.A. (1982). TCs: Their Evolution, Structure and Effects. American Meteorological Society, Science press, Ephrata, 208 pp.

Asnani, G.C. (1993). Tropical Meteorology, vols. 1 and 2, published by Prof. G.C. Asnani, c/o Indian Institute of Tropical Meteorology, Dr. HomiBhabha Road, Pashan, Pune 411008, India.

Chen, F. and Dudhia, J. (2001). Coupling an advanced land-surface/hydrology model with the Penn State/NCAR MM5 modeling system. Part I: Model description and implementation. *Monthly Weather Review*, **129**, 569–585.

Courtier, P, Andersson, E, Heckley, W, Pailleux, J, Vasiljevic, D, Hamrud, M, Hollingsworth, A, Rabier, F. and Fisher, M. (1998). The ECMWF implementation of three dimensional variational assimilation (3D-Var). I: Formulation. *Quarterly Journal of the Royal Meteorological Society*, **124**, 1783–1807.

Dudhia, J. (1989). Numerical study of convection observed during winter monsoon experiment using a mesoscale two-dimensional model. *Journal of Atmospheric Science*, **46**, 3077–3107.

Gray, W.M. (2000). General characteristics of TCs. In: Roger, P.Jr, Roger, P. Sr (eds.) Storms vol 1, Routledge, 11 New Fetter Lane, London EC4P4EE. pp. 145–163.

Hong, S.Y, Noh, Y. and Dudhia, J. (2006). A new vertical diffusion package with explicit treatment of entrainment processes. *Monthly Weather Review*, **134**, 2318–2341.

India Meteorological Department (2009). Report on cyclonic disturbances over North Indian Ocean during 2008, RSMC-TC Report No. 1/2009, IMD, New Delhi, India, 108 pp.

India Meteorological Department (2010). Report on cyclonic disturbances over North Indian Ocean during 2009, RSMC-TC Report No. 1/2010, IMD, New Delhi, India, 122 pp.

India Meteorological Department (2011). Report on cyclonic disturbances over North Indian Ocean during 2010, RSMC-TC Report No. 1/2011, IMD, New Delhi, India, 162 pp.

India Meteorological Department (2012). Report on Cyclonic Disturbances over North Indian Ocean during 2011. RSMC-TCs Report No. 02 (2012) IMD, New Delhi, India, 186 pp.

India Meteorological Department (2013). Report on Cyclonic Disturbances over North Indian Ocean during 2012. Met. Monograph No. ESSO/IMD/RSMC-TCs Report No. 01 (2013) IMD, New Delhi, India, 214 pp.

India Meteorological Department (2014). Report on Cyclonic Disturbances over North Indian Ocean during 2013. Met. Monograph No. ESSO/IMD/RSMC-TCs Report No. 01 (2014) IMD, New Delhi, India, 274 pp.

Kain, J.S. (2004). The Kain–Fritsch convective parameterization: An update. *Journal of Applied Meteorology*, **43**, 170–181.

Kalnay, E. (2003). Atmospheric Modeling, Data Assimilation and Predictability. Cambridge University Press, pp. 364.

Knaff, J.A, DeMaria, M, Molenar, D.A, Sampson, C.R. and Seybold, M.G. (2011). An automated objective, multiple-satellite-platform TC surface wind analysis. *Journal of Applied Meteorology and Climatology*, **50**, 2149–2166.

Lin, Y.-L, Farley, R.D. and Orville, H.D. (1983). Bulk parameterization of the snow field in a cloud model. *Journal of Climate and Applied Meteorology*, **22**, 1065–1092.

Liu, H, Crawford, J.H, Pierce, R.B, Norris, P, Platnick, S.E, Chen, G, Longan, J.A, Yantosca, R.M, Evans, M.J, Kittaka, C, Feng, Y. and Tie, X. (2006). Radiative effect of clouds on tropospheric chemistry in a global three-dimensional chemical transport model. *Journal of Geophysical Research,* **111**, D20303, 1–18.

Mlawer, E.J, Taubman, S.J, Brown, P.D, Iacono, M.J. and Clough, S.A. (1997). Radiative transfer for inhomogeneous atmosphere: RRTM, a validated correlated-k model for the longwave. *Journal of Geophysical Research*, **102**, 16663–16682.

Osuri, K.K, Routray, A, Mohanty, U.C. and Kulkarni, M.A. (2010). Simulation of TCs Over Indian Seas: Data Impact Study Using WRF-Var Assimilation System. In: Charabi, Y. (ed.) Indian Ocean TCs and Climate Change, Springer, Netherlands (Book Chapter), 115–124 pp, doi:10.1007/978-90-481-3109-9_15.

Osuri, K.K, Mohanty, U.C, Routray, A. and Mohapatra, M. (2012). The impact of satellite-derived wind data assimilation on track, intensity and structure of TCs over the North Indian Ocean. *International Journal of Remote Sensing*, **33**, 2012.

Park, S.K. and Zupanski, D. (2003). Four-dimensional variational data assimilation for mesoscale and storm-scale applications. *Meteorology and Atmospheric Physics*, **82**, 173–208.

Prashant, K, Harishkumar, K.P. and Pal, P.K. (2012). Impact of Oceansat-2 Scatterometer Winds and TMI Observations on Phet TC Simulation, doi:10.1109/TGRS.2012.2221720.

Singh, R, Pal, P.K, Kishtawal, C.M. and Joshi, P.C. (2008). The impact of variational assimilation of SSM/I and QuikSCAT Satellite observations on the Numerical Simulation of Indian Ocean TCs. *Weather Forecasing*, **23**, 460–476.

Singh, R, Kishtawal, C.M, Pal, P.K. and Joshi, P.C. (2011). Assimilation of the multisatellite data into the WRF model for track and intensity simulation of the Indian Ocean TCs. *Meteorology and Atmospheric Physics*, **111**, 103–119.

Singh, R, Prashant, K. and Pal, P.K. (2012). Assimilation of Oceansat-2-scatterometer-derived surface winds in the weather research and forecasting model. *IEEE Transactions on Geoscience and Remote Sensing*, **50**, 1015–1021.

Skamarock, W.C, Klemp, J.B, Dudhia, J, Gill, D.O, Barker, D.M, Dudha, M.G, Huang, X, Wang, W. and Powers, Y. (2008). A description of the Advanced Research WRF Ver.30.NCAR Technical Note. NCAR/TN-475 + STR. Mesocale and Microscale Meteorology Davison, National Centre for Atmospheric Research, Boulder Colorado, USA, pp. 113.

Srinivas, C.V, Venkatesan, R, Vesubabu, V. and Nagaraju, C. (2010). Impact of assimilation of conventional and satellite meteorological observations on the numerical simulation of a Bay of Bengal TC of Nov 2008 near Tamilnadu using WRF model. *Meteorology and Atmospheric Physics*, **110**, 19–44.

Srinivas, C.V, BhaskarRao, D.V, Yesubabu, V, Baskaran, R. and Venkatraman, B. (2012). TC predictions over the Bay of Bengal using the high-resolution advanced research weather research and forecasting (ARW) model. *Quarterly Journal of the Royal Meteorological Society*, doi:10.1002/qj.2064

Xiao, Q, Zhang, X, Davis, C, Tuttle, J, Holland, G. et al. (2009). Experiments of hurricaneinitialization with airborne doppler radar data for the advanced-research hurricane WRF (AHW) Model. *Monthly Weather Review*, doi: 10.1175/2009MWR2828.1.

Yesubabu, V, Srinivas, C.V, Hariprasad, K.B.R.R. and Baskaran, R. (2013). A study on the impact of observation assimilation on the numerical simulation of TCs Jal and Thane using 3DVAR. *Pure and Applied Geophysics*, doi:10.1007/s00024-013-0741-3.

Sensitivity Study on 2013: Tropical Cyclones Using Different Cloud Microphysical and Planetary Boundary Layer Parameterisation Schemes in WRF Model

M. Venkatarami Reddy, S.B. Surendra Prasad, U.V. Murali Krishna, and K. Krishna Reddy

1 Introduction

The Indian subcontinent is the worst affected part of the world due to tropical cyclones (TCs). This region account for ~7 % 0.of the total number of global TCs (Gray 1968). The formation of TCs is more pronounced over Bay of Bengal (BOB) compared to Arabian Sea. A large number of TCs form over the BOB region generally move in the north and north-west directions and make landfall along the coastal regions of India, Bangladesh, and Myanmar (Mohapatra et al. 2012, 2015). These TCs have been responsible for the damage of property, loss of agriculture crops, and thousands of human lives (Paul 2010). In the BOB, TC genesis is highly seasonal with primary maximum in the post-monsoon season (October to December) and secondary maximum during pre-monsoon season (April and May). Hence, there is a need to improve the understanding and the forecast of TC over the Indian Ocean region. Several dynamic models have been used for the forecasting of the track and intensity of TC over specific regions. There has been significant improvement in recent years in terms of track, intensity and landfall forecasts (Mohapatra et al. 2013a, b; 2015). However, the accurate track and intensity predictions of TCs remain a challenging task for atmospheric scientists and the research community.

Initially, forecasting of tropical TCs was done by using a computer oriented half persistence and half climatology technique (Sikka and Suryanarayana 1972) in India. Mohanty and Gupta (1997) have used different physical parameterisation in multi-level primitive equation models for the TC track prediction. In the last few years, several studies were carried out to simulate the structure and intensity of TCs

M.V. Reddy • S.B.S. Prasad • U.V.M. Krishna • K.K. Reddy (✉)
Semi-arid-zonal Atmospheric Reseach Centre (SARC), Yogi Vemana University, Kadapa 516 003, Andhra Pradesh, India
e-mail: krishna.kkreddy@gmail.com

© Capital Publishing Company 2017

M. Mohapatra et al. (eds.), *Tropical Cyclone Activity over the North Indian Ocean*,
DOI 10.1007/978-3-319-40576-6_23

using high-resolution atmospheric models (Trivedi et al. 2002; Mohanty et al. 2004; Sathi Devi et al. 2006; Bhaskar Rao et al. 2006; Pattanik and Mohanty 2008; Pattanaik and Rama Rao 2009; Deshpande et al. 2010; Raju et al. 2011a; Venkat et al. 2014, Osuri et al. 2012, 2014). Prater and Evans (2002) studied the tropical TC Irene (1999) with various CP schemes in MM5 model and concluded that the Kain–Fritsch (KF) scheme produces relatively accurate storm. Several researchers studied the sensitivity of the model for planetary boundary layer (PBL) and cumulus parameterisation (CP) schemes. Bhaskar Rao and Hari Prasad (2007) studied the sensitivity of Odisha super TC and Srinivas et al. (2007) studied the Andhra severe TC (2003) with MM5 model and showed that the MYJ PBL scheme gives the best results. However, Osuri et al. (2012) showed that YSU-PBL scheme simulated track and intensity with less errors for the north Indian Ocean TCs. Deshpande et al. (2012) simulated the Odisha super TC (1999) with different CP and micro physics (MP) schemes in MM5 model. But, all these studies simulated better with KF CP scheme in combination with the above PBL and MP schemes. However, Loh et al. (2010) showed that the PBL parameterisation schemes did not affect much the track and intensity prediction of near equatorial TCs.

The simulation of track and intensity of TC NARGIS was carried out by Raju et al. (2011b) by using WRF-ARW model and Pattanayak et al. (2012) utilised WRF-NMM with different CP, MP and PBL schemes. Raju et al. (2011b) showed that CP schemes controls the intensity of the TC and MP schemes influences the track of the TC. Although Pattanayak et al. (2012) showed that the CP schemes controls the track of the TC and PBL schemes influences the intensity of the TC. However, variations in the MP schemes produce similar type of results. Chandrasekar and Balaji (2012) simulated the track and intensity of tropical TC JAL using different CP, MP, PBL, land surface, long wave and short wave parameterisation schemes. They concluded that the CP, PBL and MP schemes play a major role in both the track and intensity predictions. Similarly, Srinivas et al. (2012) studied the performance of WRF-ARW model with different PBL, CP, MP and SF schemes. This study revealed that the CP and PBL schemes have larger impact whereas MP and land surface physics schemes have minimum impact on the intensity and track predictions for five severe TCs (Sidr, Khaimuk, Nargis, Jal and Thane). Recently, Osuri et al. (2013) studied the performance of ARW model for the track and intensity of TC's over, North Indian Ocean (NIO), BOB and Arabian Sea (AS) basins. The ARW model has a good overall capability to predict TCs over the NIO Basin irrespective of the region of formation, nature of movement, intensity, and season of formation. The model performed better for track prediction if the model is initialised at the SCS stage rather than at CS or depression stage.

Numerous studies were carried-out to simulate the structure, intensity, track of TCs, Hurricanes and Typhoons with different physical processes using different numerical models (Zhu et al.2006; Pattanayak and Mohanty 2008). For some time, we have known that assumptions within microphysical parameterisations, which account for the creation, evolution and destruction of hydrometeors, can dramatically influence TC intensity (e.g. Lord et al. 1984; Braun and Tao 2000; Wang 2002; Zhu and Zhang 2006). The sensitivity of simulated tropical TC structure and intensity to the cloud microphysics parameterisation in a two-dimensional non-hydrostatic model was discussed by Lord et al. (1984). Their results indicate that the warm-rain-only cloud microphysics scheme produced a rapid intensification while the mixed-ice phase cloud microphysics scheme resulted in a slowly developing storm. Brown and Swann (1997) found that the simulated surface precipitation was sensitive to the mean size and fall speed of graupel and the efficiency for the collection of snow and cloud ice by graupel in the evaluation of key MPs in three-dimensional cloud model simulations. Zhu and Zhang (2006) studied the sensitivity of simulated intensity and inner core structure of Hurricane Bonnie (1998) to various CM processes in MM5 model. They indicated that the weakest storm can be produced by removing all ice particles from the CM processes and the most rapid development of the storm was produced when evaporation processes are removed. Fovell et al. (2010a) presented the interaction of hydrometeors with radiation in different MP schemes and concluded that the motion and structural variation in their MP ensemble disappeared when clouds were made transparent to radiation in their semi idealised simulations. Fovell et al. (2010b) demonstrated that the different MP schemes accounts for different amounts and relative distribution of hydrometeors, such as cloud ice, snow, cloud droplets, etc. Mukhopadhyay et al. (2011) studied the influence of moist processes on track and intensity forecast of TCs *Gonu* and *Sidr* over the north Indian Ocean and found that hybrid moist convection treatment (which included CPS and MPS both) produces better simulation. The main reason is the heating within the inner core of the TC, which is influenced by production of graupel hydrometeors in the inner core region. The latent heat released in the formation of graupel mixing ratio is responsible for net middle level heating rate in the TC core. Higher net heating in the middle level enhances the divergence in the upper level and convergence in the lower level, which in turn helps in the intensification of the system.

The primary energy source for the TC is the air–sea surface flux exchanges of moist enthalpy via surface latent heat and sensible heat fluxes (Kubota et al. 2002). Braun and Tao (2000) presented the intensity of Hurricane Bob (1991) in different PBL schemes. They concluded that the ratio of exchange coefficients of enthalpy and momentum has a large impact on the simulated storm intensity. Specifically, deeper intensity corresponded to larger exchange ratios. However, the deepening rate solely depends on the exchange ratio. According to Emanuel (1999) the intensification rate of hurricane depends on the thermodynamic properties of large scale environment and the air–sea exchange under the core of the storm. The accurate representation of momentum, heat fluxes and moisture across the air–sea interface are important in climate research community for modeling and understanding the coupled variability of the ocean atmosphere system. The simulated track and intensity depends on the evolution and distribution of heating rates (Wu and Wang 2001; Wang 2009). The heating rates are dependent on the cloud microphysical processes taking place within the cloud system and also on the feedback with the environment (Liu and Moncrieff 2007). Pattnaik and Krishnamurti (2007a,b) also showed that the intensification process depends on the conversion of hydrometeors in the case of Hurricane Charlie. Their result suggests that whenever the sources of heat sinks are reduced, latent heating produces more instability which increases buoyancy gradient and produces higher tropical TC intensity. Recently, Benjamin and Zhang examined the impacts of model parameters on simulation of Hurricane Katrina using WRF-ARW model and found that the surface flux had a significant effect on structure and intensity of the Hurricane. But Rozoff et al. (2012) demonstrated that the size of TC is decided by the latent heating outside the primary eyewall which promotes the secondary eyewall formation leading to expansion of the outer wind field. The observational study by Rahaman and Ravichandran (2013) on near-surface air temperature and specific humidity revealed that the latent heat flux decreases with increasing specific humidity during winter and summer monsoons and the air–sea humidity difference is linearly related to sea surface temperature for values greater than 28 °C in the southern BOB.

From these, it is clear that MP and PBL schemes play a major role in the prediction of track and intensity of TC compared to other physical parameterisation schemes. Hence we simulated the track and intensity of five TCs (VIYARU, PHAILIN, HELEN, LEHAR and MADI) originated in the BOB during 2013 using WRF model with different MP and PBL schemes. Further the intensity of TC is discussed in terms of specific humidity, temperature, surface energy fluxes (SEFs) and hydrometeor mixing ratio.

2 Data and Model Description

The non-hydrostatic compressible Advanced Research WRF (ARW) v3.4 mesoscale model developed by the National Center for Atmospheric Research (NCAR) (Skamarock et al. 2008) is used for this study. WRF modeling system user's guide is available online at http://www2.mmm.ucar.edu/wrf/users/docs/user_guide_V3.4/ with the Advanced Research WRF core. It features a fully compressible, Eulerian non-hydrostatic control equation set, a terrain-following, hydrostatic pressure vertical coordinate system with the top of the model being a constant pressure surface, a horizontal Arakawa-C grid, and a third-order Runge-Kutta time integration scheme. The model has several options for diffusion, spatial discretisation, nesting and lateral boundary conditions besides number of physics. It supports idealised and real data applications with various lateral boundary condition options, as well as analysis nudging techniques. The details of the model domain and resolution are given in Table 1. In the present study, four MP schemes FERR (Ferrier et al. 2002), LIN (Lin et al. 1983), WSM6 (Hong et al. 2004) and THOMP (Thompson et al. 2004), two PBL schemes (YSU and MYJ) and KF CP scheme are used at 27-km resolution.

The YSU PBL scheme (Hong et al. 2006) is using the counter-gradient terms to represent fluxes due to nonlocal gradients. It is a first-order closure scheme that is similar in concept to the scheme of Hong and Pan (1996), but appears less biased toward excessive vertical mixing as reported by Braun and Tao (2000). The MYJ

Table 1 Model configuration and Parameterisation schemes used in the WRF model

Model	NCAR Mesoscale model WRF
Dynamics	Non-hydrostatic with 3-D Coriolis force
Map projection	Mercator
Model Domain	04.67° S–30.72° N and 57.10° E–101.59° E
Center of the Domain	13.68° N and 79.34° E
Horizontal grid resolution	27 km
Number of vertical levels	28
Data	NCEP Global Forecast System (GFS) data
Horizontal grid system	Arakawa-C grid
Vertical coordinates	Terrain-following hydrostatic pressure vertical co-ordinate with 51 vertical levels
Micro physics scheme	Lin et al.
	Ferrier (new Eta)
	WSM 6-class Scheme graupel scheme
	Thompson
Planetary Boundary Layer scheme	Yonsei University scheme (YSU)
	Mellor-Yamada-Janjic (Eta) TKE (MYJ)
Surface layer scheme	MM5 similarity scheme in combination with YSU PBL
	Eta similarity scheme in combination with MYJ PBL
Land surface scheme	Noah land surface model
Cumulus parameterisation scheme	Kain–Fritsch (KF)
Long wave radiation scheme	Rapid Radiative Transport Model (RRTM) scheme
Short wave radiation scheme	Dudhia scheme
Diffusion option	Simple diffusion

PBL Scheme (Mellor and Yamada 1982; Janjic 1994; 2003a, b), which predicts TKE and has local vertical mixing. The top of the layer depends on the TKE as well as the buoyancy and shear of the driving flow. FERR (Ferrier et al. 2002) MP scheme has four hydrometeor species (water vapor, rain water, cloud water and snow) whereas the LIN (Lin et al. 1983), WSM6 (Hong et al. 2004; Hong and Lim 2006) and THOMP (Thompson et al. 2004, 2008) MP schemes have six hydrometeor species (water vapor, rain water, cloud water, graupel, snow and cloud ice). The Kain–Fritsch convective parameterisation is based on a simple cloud model which includes the effects of detrainment, entrainment, simple microphysics and moist updrafts and downdrafts Kain 2004). Dudhia (1989) scheme is used for the short-wave radiation, Rapid Radiative Transfer Model (RRTM) scheme (Malwer et al. 1997) is used for the long-wave radiation.

For the present study, operational analyses/forecasts products from NCEP GFS global model and Final analysis data from Global Data Assimilation System (GDAS) are employed for simulation of TCs. The GFS analyses/forecast products are available in 0.5°latitude × 0.5° longitude horizontal resolution with 64 vertical levels. In the case of GFS, the analysis fields serve as initial condition, and the lateral boundary conditions in 3-h intervals are provided from their forecast field. The National Center for Environmental Prediction Final Analyses (FNL) data from Global Data Assimilation System (GDAS), analysed and available at 6-h intervals with a global resolution of 1° latitude × longitude is used in WRF model for prediction of meteorological parameters.

3 Brief Description of TCs

For the first time after 1987, five TCs developed over the BOB during the year 2013. Hence, we performed the simulation study on these five TCs namely VIYARU (10–16 May), PHAILIN (08–14 October), HELEN (19–23 November), LEHAR (23–28 November) and MADI (06–13 December) using different MP and PBL schemes in the WRF model.

TC VIYARU formed as a depression over southeast BOB at 00 UTC of 10 May near 5.0° N, 92.0° E (RSMC, New Delhi 2014). It intensified into a deep depression in the evening of the same day. It further intensified into a cyclonic storm (CS), VIYARU in the morning of 11 May 2013. The CS crossed Bangladesh coast near 22.8° N, 91.4° E, about 30 km south of Feni around 08 UTC of 16 May. It weakened into a deep depression over Mizoram in the evening and into a depression over Manipur around mid-night of 16 May. It further weakened into a well-marked low pressure area over Nagaland in the early morning and moved away towards Myanmar as a low pressure area in the morning of 17 May.

A VSCS (VSCS) PHAILIN originated from a remnant cyclonic circulation from the South China Sea. The cyclonic circulation lay as a low pressure are over Tenasserim coast on 6 October. The well-marked low pressure area was upgraded as a depression over the north Andaman Sea at 00 UTC of 8 with its center near 12.0° N, 96.0° E. The depression then intensified into a deep depression at 00 UTC of 9 October near 13.0° N and 93.5° E. Moving west-northwestwards, it crossed Andaman Islands near Mayabandar at 00 UTC of 10 October. It moved slowly over east central BOB and intensified into a CS, PHAILIN at 12 UTC of 9 October.

A trough of low pressure developed over BOB near the Andaman Islands on 16 November from the remnant of the tropical storm PODUL. It became well marked on 18th over the central BOB and concentrated into a depression over the west central BOB in the early morning of 19 November with center near 14.5° N, 86.5° E, about 600 km east-southeast of Machillipatnam. The depression then intensified into a deep depression in the night of 19 November and further into a CS, HELEN in the morning of 20 November at about 330 km east-southeast of Machilipatnam. It then moved north-northwestwards till 12 UTC of 21st and intensified into a SCS in the early morning of 21st November at a distance of 260 km east-southeast of Machilipatnam. On 22 November, it crossed Andhra Pradesh coast close to south of Machilipatnam (near 16.1° N, 81.2° E) between 08 and 09 UTC of 22 November as a CS.

A remnant of tropical depression over South China Sea moved across Malay Peninsula and lay as a low pressure area over south Andaman Sea on 21 November. It became well marked over the same area on 22nd and concentrated into a depression over south Andaman Sea near 8.5° N, 96.5° E about 550 km south-southeast of Port Blair at 12 UTC of 23 November. This intensified into a deep depression at 18 UTC of 23 November and further into a CS, LEHAR at 03 UTC of 24 November and lay centered near 10.0° N, 95.0° E. It further intensified into a SCS and crossed Andaman & Nicobar Island near Port Blair around 00 UTC of 25 November. It then emerged into southeast BOB, and intensified into a VSCS at 21 UTC of 26 November over southeast BOB near 12.5° N, 91.0° E.

A low pressure area from south China Sea moved across Malay peninsula and emerged into south Andaman Sea on 01 December morning. It became a depression in the morning of 06 December over southwest BOB and lay centered near 10.0° N, 84.0° E, about 350 km northeast of Tricomalee (Sri Lanka). It gradually intensified into a deep depression at 18 UTC of 06 December. The deep depression further intensified into a CS MADI with center near 10.5° N, 84.0° E at 00 UTC of 07 December. It intensified into an SCS over the same region at 09 UTC of 07 December. It further intensified into a VSCS at 06 UTC of 08 December near 12.3° N, 84.7° E. This VSCS weakened into a SCS at 12 UTC of 09 December and lay centered near 14.6° and 84.7° E. As a result, the SCS recurved westwards initially and then southwestwards commencing from 09 UTC of 10 December. It gradually weakened into a CS near 14.6° N, 84.6° E at 21 UTC of 10 December, further into a deep depression near 14.0° N, 83.8° E at 03 UTC of 11 December and into a depression near 12.9° N, 82.7° E at 18 UTC of 11 December. The depression crossed Tamil Nadu coast close to Vedaranyam around 1330 UTC of 12 December.

4 Results

The numerical simulations are performed with 120 h of prediction for all the TCs except TC HELEN (96 h). The TC VIYARU is simulated with the initial conditions at 12 UTC 11 May 2013. Similarly. the TC PHAILIN is simulated at 00 UTC 08 October 2013, HELEN at 00 UTC 19 November 2013, LEHAR at 00 UTC 24 November 2013 and MADI at 00 UTC 07 December 2013.

4.1 MP Schemes

Track and intensity of these five TCs are simulated with different MP schemes namely LIN, FERR, WSM6 and THOMP in combination with KF CP and YSU PBL schemes.

4.1.1 Track

Figures 1 and 2 shows the simulated track and track error for the five TCs. For the TCs VIYARU, PHAILIN, LEHAR and MADI all the MP schemes simulated the same TC track (Fig. 1). However, for TC VIYARU, THOMP MP Scheme simulated with least track error (Fig. 2a). FERR MP Scheme simulated the TCs PHAILIN (Fig. 2b) and LEHAR (Fig. 2d) with less track error whereas WSM6 simulated for TC MADI (Fig. 2e). However, for the TC HELEN, THOMP MP Scheme simulated the observed track well compared to other schemes (Fig. 2c).

4.1.2 Intensity

The intensity of the tropical TCs is discussed in terms of Minimum Mean Sea level Pressure (MSLP) and Maximum Surface Wind (MSW). All the MP combinations overestimated the intensity for TC VIYARU. Among all the MP schemes, THOMP MP Scheme simulated weaker intensity compared to all other MP schemes (Fig. 3a, b). For the TC PHAILIN, all the combinations followed the observed intensity. Among the entire MP schemes FERR scheme followed the observed intensity well compared to other schemes. But the simulated intensities are stronger upto 60 h of prediction compared to observed intensity and weaker thereafter (Fig. 3c, d). For the TC HELEN, all the MP schemes underestimated the intensity of the TC (Fig. 3e, f). For the TC LEHAR, all the MP schemes followed the observed intensity and produced stronger intensity compared to observed intensity (Fig. 3g, h). Among all the combinations, FERR scheme simulated weaker intensity compared to other MP schemes. For the TC MADI, all the schemes followed the observed intensity well upto 84 h of prediction and the observed intensity decreased thereafter, whereas the simulated intensities increased (Fig. 3i, j).

4.2 PBL Schemes

We simulated these five TCs with different PBL schemes namely YSU and MYJ in combination with KF CP scheme. The MP schemes which simulated less track error for given TC are selected for the respective TCs.

4.2.1 Track

In the simulation of TC VIYARU, YSU and MYJ schemes predicted almost the same track. However, the MYJ scheme simulated the observed track well compared to YSU scheme (Figs. 1 and 2a). Even for the TC PHAILIN, YSU and MYJ

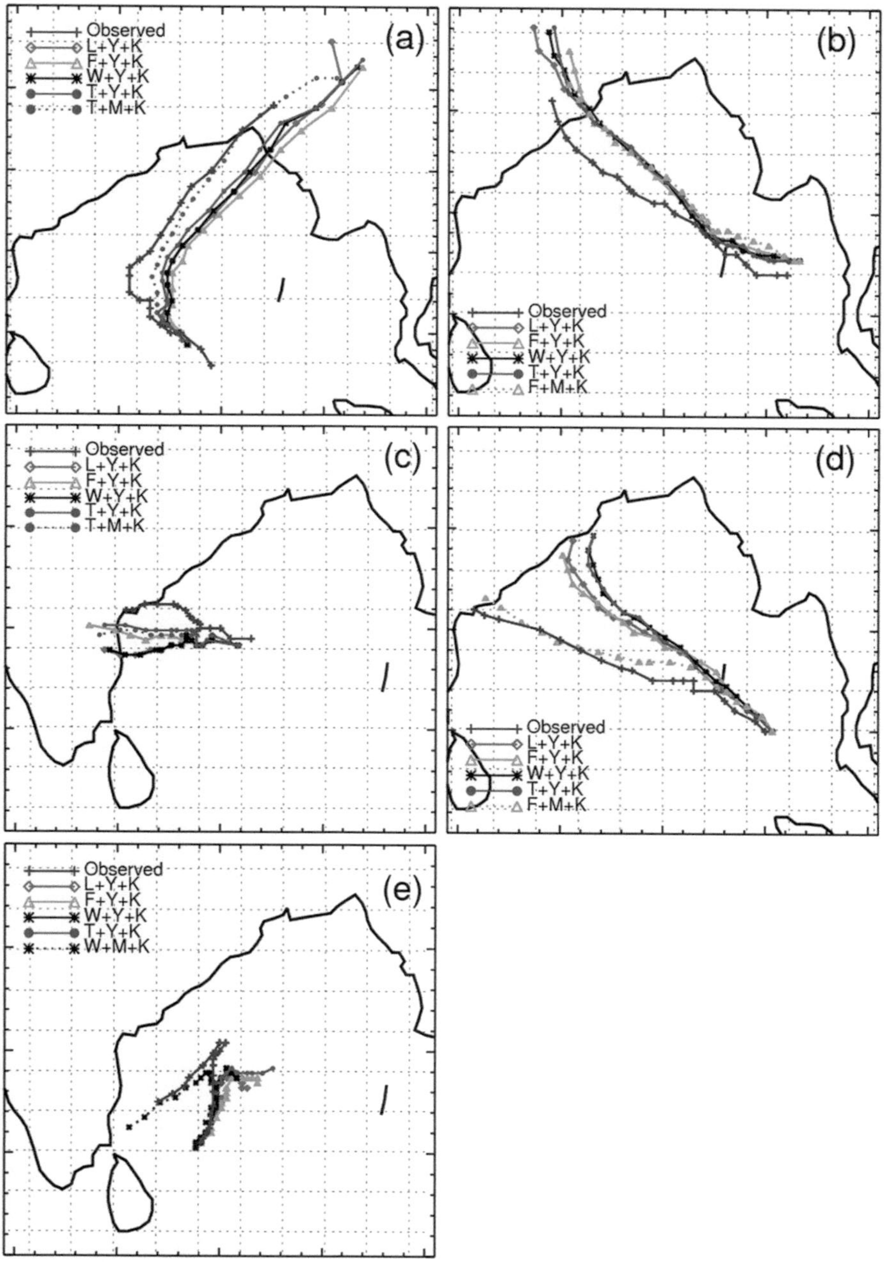

Fig. 1 The observed and simulation track for (**a**) VIYARU, (**b**) PHAILIN, (**c**) HELEN, (**d**) LEHAR and (**e**) MADI Using different MP and PBL schemes with KF CP scheme

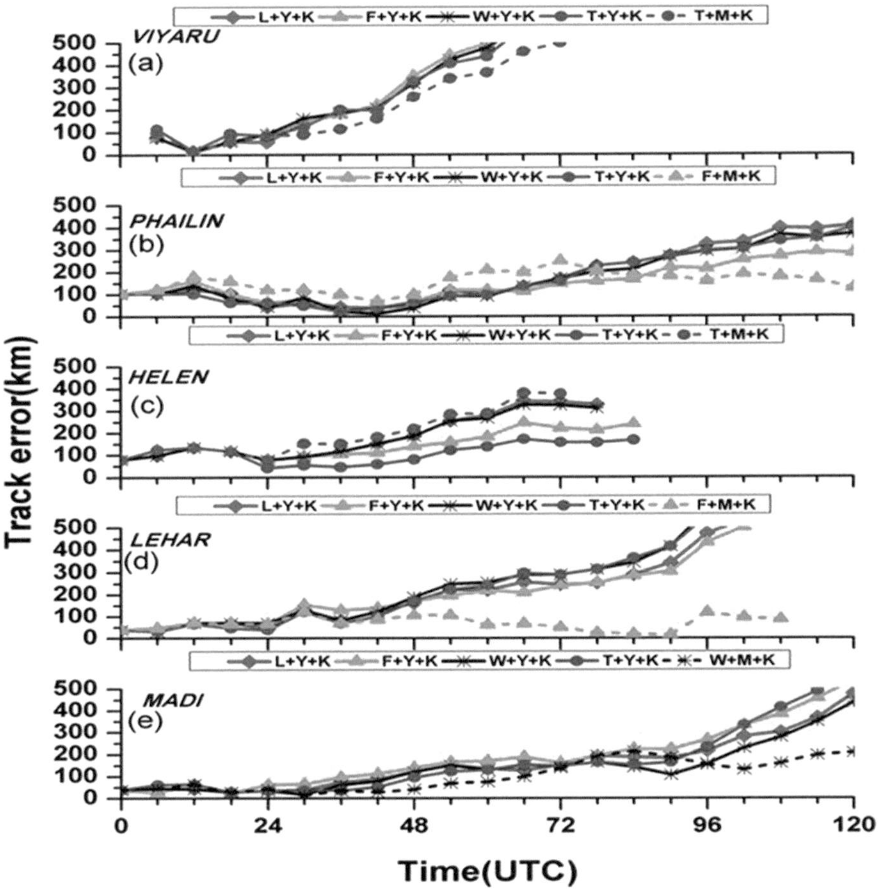

Fig. 2 The track error for (**a**) VIYARU, (**b**) PHAILIN, (**c**) HELEN, (**d**) LEHAR and (**e**) MADI with different MP and PBL schemes with KF CP scheme

schemes simulated the same track. However, YSU scheme simulated with less track error up to 72 h of prediction (Figs. 1 and 2b). In the case of TC HELEN, both the schemes deviated from the observed track. Among these two schemes, YSU scheme simulated with less track error compared to MYJ scheme (Figs. 1 and 2c). For the TC LEHAR, the variation in the tracks simulated by YSU and MYJ schemes are high. However, MYJ scheme simulated the observed track well compared to YSU (Figs. 1 and 2d). In case of MADI, both the schemes simulated the same track upto 96 h of prediction. Among these schemes, MYJ scheme simulated with less track error compared to YSU scheme (Figs. 1 and 2e).

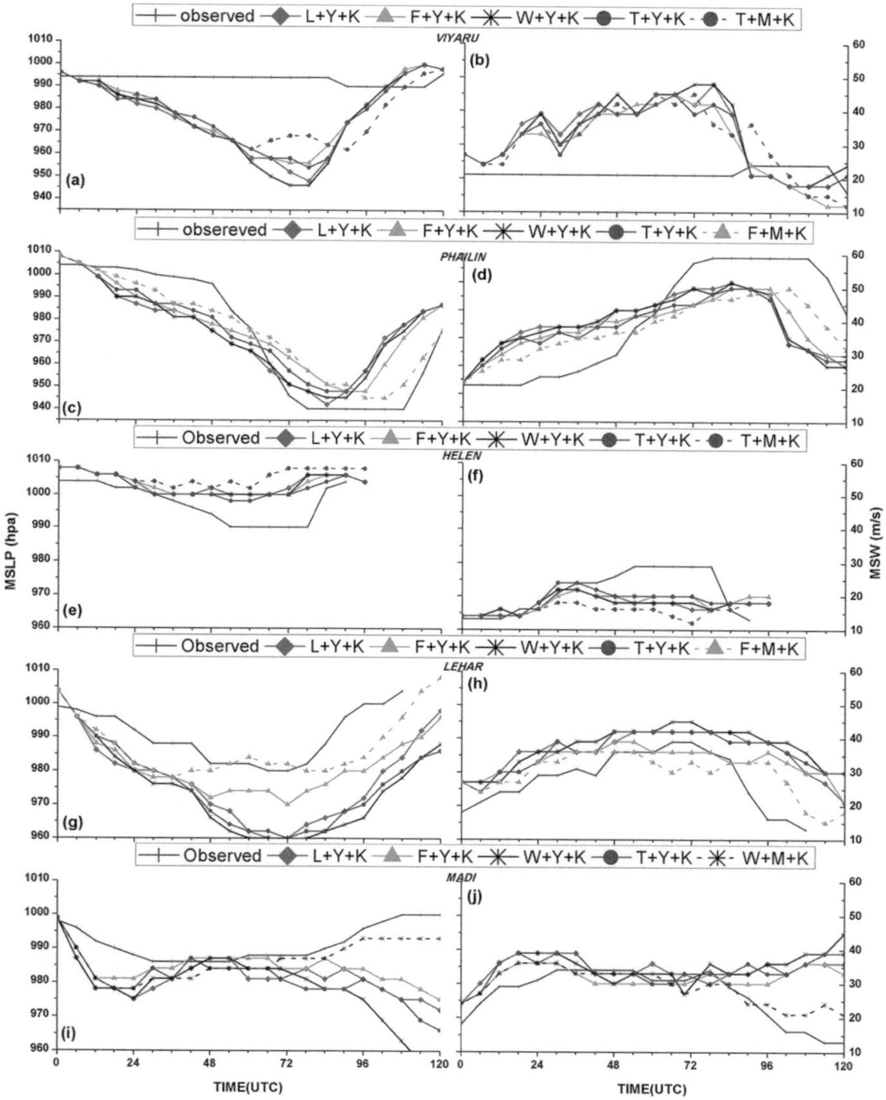

Fig. 3 The simulated MSLP and MSW for (**a**)–(**b**) VIYARU (**c**)–(**d**) PHAILIN (**e**)–(**f**) HELEN (**g**)–(**h**) LEHAR and (**i**)–(**j**) MADI with different MP and PBL schemes with KF CP scheme

4.2.2 Intensity

The intensity simulated by MYJ and YSU schemes is same upto 60 h of prediction and then varied thereafter for the TC VIYARU (Fig. 3a, b). For the TC PHAILIN, YSU scheme simulated strong intensity compared to MYJ scheme upto 90 h of prediction (Fig. 3c, d). In case of TC HELEN, YSU scheme simulated stronger intensity compared to MYJ scheme (Fig. 3e, f). For the TCs LEHAR and MADI,

YSU scheme shows stronger intensity compared to MYJ scheme. However, MYJ scheme followed the observed intensity compared to YSU scheme (Fig. 3g–j).

From these observations, it is clear that little variations are observed in the prediction of track and intensity of TCs with different MP schemes. However, THOMP MP scheme simulated well for TC VIYARU, LEHAR and WSM6 MP scheme simulated the TC MADI in combination with MYJ PBL scheme. But the TC HELEN was best simulated by THOMP MP scheme and PHAILIN was best simulated by FERR MP scheme in combination with YSU PBL scheme.

4.3 *Simulation of TC Structure*

Different schemes simulated different intensities for the same TC. For this, we studied the dependency of intensity of TC over latent heat flux, sensible heat flux, SEF, air and sea surface temperature difference, specific humidity and divergence.

4.3.1 Surface Energy Flux

To explain the changes in intensity simulated by different schemes, we studied the changes in the SEF for all the TCs.

We simulated all the TCs with different PBL schemes using KF CP scheme and the best set of MP schemes (Fig. 4). For the TC VIYARU (18 UTC, 14 May 2013), the simulated intensity and SEF are stronger in YSU scheme (Fig. 4a, b). For PHAILIN (12 UTC, 11 Oct 2013) and LEHAR (00 UTC, 27 Nov 2013), both scheme simulated almost same intensity whereas MYJ PBL schemes simulated stronger SEF compared to YSU PBL scheme (Fig. 4e, f, i, j). However, for the TC MADI (00 UTC, 9 Dec 2013), both the schemes simulated almost same intensity and SEF (Fig. 4m, n). All the PBL schemes produce stronger SEF where the wind speed is maximum and vice-versa in all the TCs. The MSW and MSLP predicted at the eye-wall are weaker in all the schemes. Hence, the predicted SEF also shows its minimum at the eye-wall region in all the schemes. However for TC PHAILIN and LEHAR, MYJ scheme predict maximum SEF whereas YSU scheme simulated slightly higher intensity compared to MYJ scheme. Hence, we observed the dependency of SEF on the difference between SST and air temperature for these TCs. The SEF increases in those schemes which have higher difference between SST and air temperature and vice-versa. However, SEF simulated for TCs PHAILIN (Fig. 4g, h) and VIYARU (Fig. 4c, d) are higher than predicted for TC MADI (Fig. 4o, p). Whereas, the temperature difference is maximum for MADI compared to PHAILIN. For this, we observed the dependency of SEF on specific humidity for these TCs. From this we can observe that the specific humidity is higher at the eye of the TC PHAILIN and LEHAR (Fig. 4k, l) compared to MADI. Also, TCs LEHAR and MADI have same specific humidity whereas SEF is maximum for

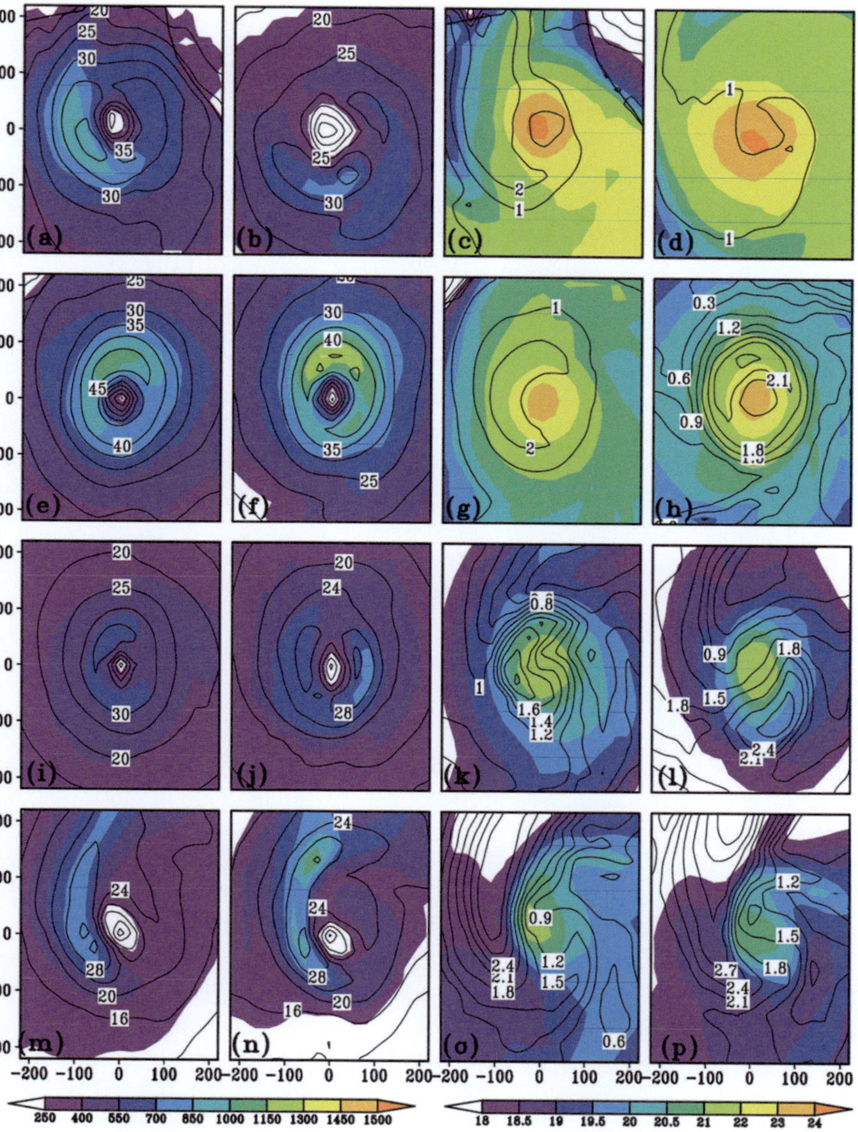

Fig. 4 The simulated SEF (*shaded*) and wind speed (*contour*) (**a**) YSU (**b**) MYJ; (**c**) and (**d**) same as (**a**) and (**b**) but specific humidity (*shaded*) and air and sea surface temperature difference (*contour*) for TC VIYARU using KF CP and THOMP MP scheme. (**e**)–(**h**) same as (**a**)–(**d**) for TC PHAILIN with FERR MP scheme, (**i**)–(**l**) same as (**a**)–(**d**) for TC LEHAR with FERR, (**m**)–(**p**) same as (**a**)–(**d**) for TC MADI with WSM6 MP scheme

TC MADI compared to TC LEHAR. This may be due to higher temperature difference and less specific humidity at the eye wall region for TC MADI. From these observations, we can conclude that SEF depends on the temperature difference and specific humidity.

4.3.2 Convergence, Divergence and Relative Vorticity

We considered the divergence field and relative vorticity integrated between 950 and 850 hPa levels for these five TCs.

Figure 5 shows the divergence and relative vorticity field integrated between 950 and 850 hPa levels for the TC. The negative values in the figure represent convergent flows and positive values represent divergent flows. A strong convergent inflow between 950 and 850 hPa levels explains the more intense storm produced by the YSU scheme in VIYARU (Fig. 5a, b) and LEHAR (Fig. 5i, j). Meanwhile, weaker convergent flows between 950 and 850 hPa correspond to weaker storm intensity predicted by MYJ scheme. However, the low-level convergent inflow is same in both YSU and MYJ scheme in PHAILIN (Fig. 5e, f) and MADI (Fig. 5m, n). A strong relative vorticity is shown at the eye-wall region in YSU scheme for the TCs VIYARU (Fig. 5c, d) and LEHAR (Fig. 5k, l). This explains the strong intense storm produced by YSU scheme compared to MYJ scheme. Whereas for TCs PHAILIN (Fig. 5g, h) and MADI (Fig. 5o, p), both the schemes simulated almost same relative vorticity at the eye-wall region. Hence the intensity simulated by both the schemes varies little. So the strong low-level relative vorticity corresponds to stronger MSLP of the TC.

4.3.3 Vertical Structure

The intensity of the TC is further studied vertically in terms of wind speed, temperature, specific humidity and hydrometeors (mixing ratio of cloud, ice, rain, snow and graupel).

Figure 6 shows the vertical cross sections of meridional wind speed, temperature, total mixing ratio (sum of cloud water, rain water, cloud ice, snow and graupel) and specific humidity for all the TCs. The temperature field has a warm core in the lower levels for all the TCs. Temperature is warmer at the eye of the TC compared to eye-wall regions for TCs VIYARU (upto ~400 hPa), PHAILIN (upto ~250 hPa), LEHAR (upto ~350 hPa) and MADI (upto ~450 hPa) in both the schemes. The distribution of the wind is symmetric and maximum wind speed in low levels is observed around the eye wall region in YSU scheme for the TCs PHAILIN and LEHAR. The maximum wind is observed in the low levels in all the schemes for entire TCs. A minimum wind is observed at the eye of the TC (width of the eye). Minimum wind is observed in MYJ scheme for the TCs LEHAR and is minimum in YSU for TC MADI.

The specific humidity is maximum near the surface in the TC center. The maximum moisture occurs in between eye and eye-wall region from lower to middle troposphere. However for the TCs PHAILIN, LEHAR and MADI the specific humidity is maximum at the eye-wall region and it decreases towards the eye of the TC. The TC center area was not as moist as the eye-wall regions from 500 to 950 hPa. The total mixing ratio is maximum in YSU PBL scheme for TCs

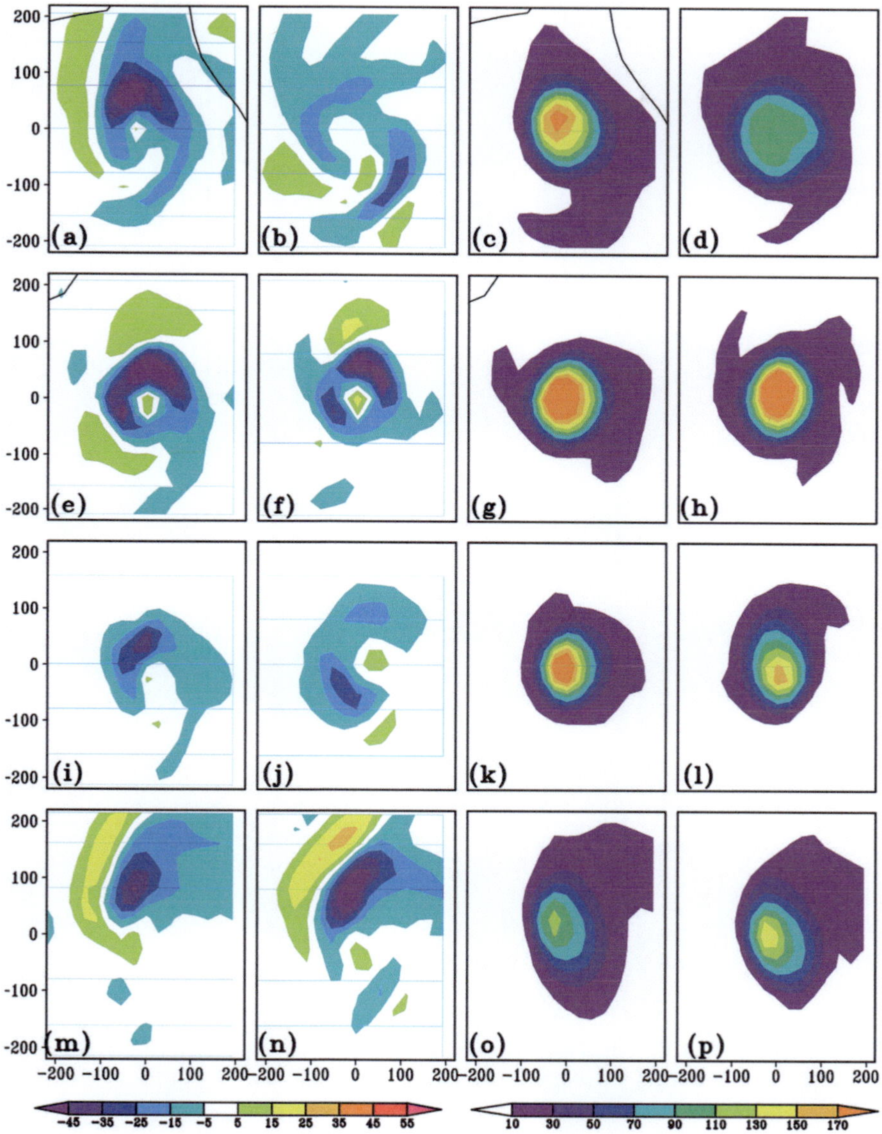

Fig. 5 Same as in Fig. 4 for divergence (*right*) (10^{-5} per s) and relative vorticity (*left*) field integrated between lower level (950–850 hPa)

VIYARU, PHILIN and LEHAR whereas for TC MADI the total mixing ratio is maximum in MYJ PBL scheme. Total mixing ratio and wind speed is maximum in lower latitude for TC VIYARU and is maximum in upper latitudes for the remaining TCs. However for TC LEHAR the total mixing ratio is maximum in lower latitude. This is due to higher specific humidity in lower latitudes for TC LEHAR. Hence, the total mixing ratio is higher where specific humidity and wind

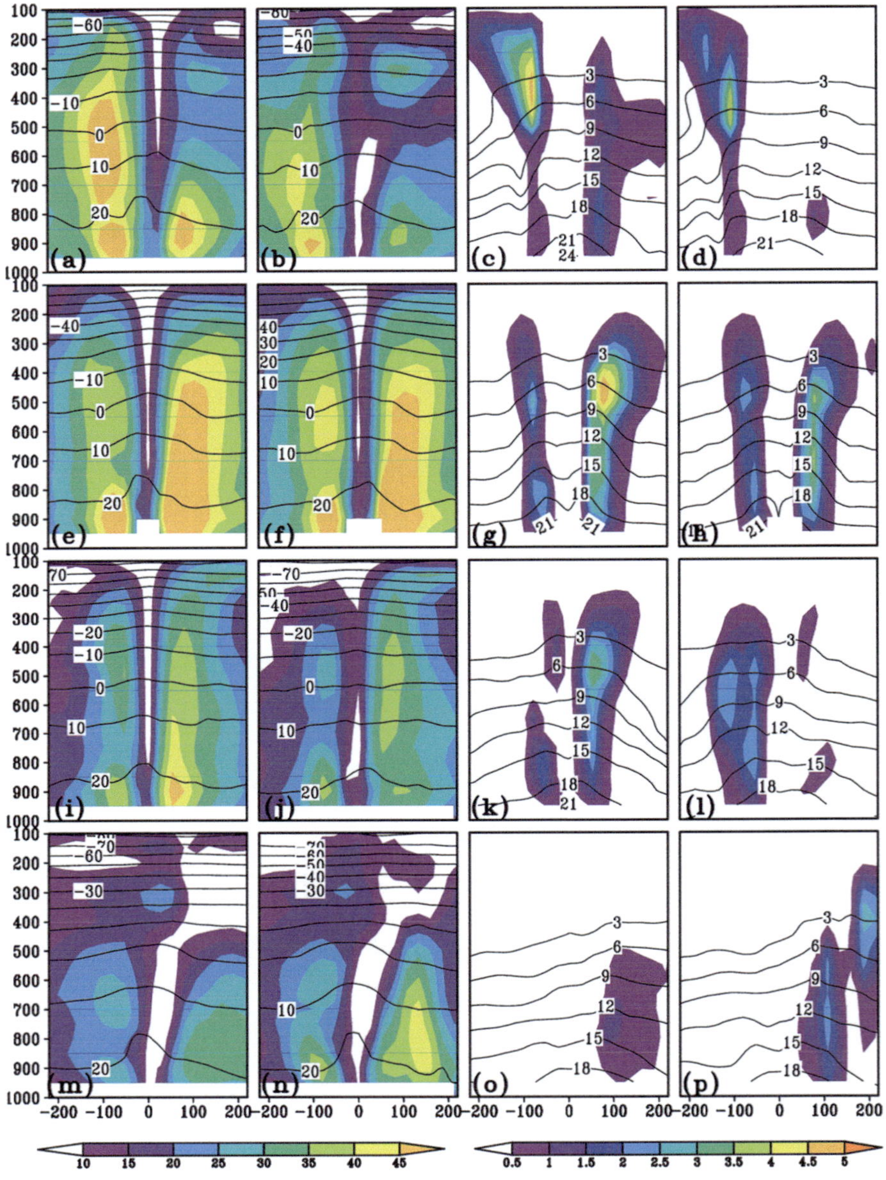

Fig. 6 Same as in Fig. 4 with vertical cross sections of meridional wind speed (m/s, *shaded*), temperature (°C, *contour*) (*left*) and total mixing ratio (*shaded*) and specific humidity (g/kg, *contour*) (right)

speed is maximum. We observed the total mixing ratio is maximum above 600 hPa levels for TC VIYARU and is maximum below 300 hPa level for TCs PHAILIN and LEHAR. To explain the variation in mixing ratio, we simulated all the TCs with different MP schemes.

4.3.4 Hydrometeors

Generally, cloud ice forms through the nucleation and freezing of super cooled cloud water which grows by the deposition of water vapour in the eye-wall. Cloud ice when reaches a critical mass forms as snow and grows by both vapour deposition and collection of cloud ice and super cooled cloud water. Snow converts into graupel when it exceeds a critical mass. Graupel also forms through freezing of rainwater which grows very quickly by both vapour deposition and collecting liquid and solid particles as it falls. Saturated water vapour condenses into cloud water. A high concentration of cloud water is associated with the activity of convective rainbands outside the eye-wall. Cloud water converts into rain water by collecting both melting snow and graupel. Hence the rain water concentrates below the melting level. Figure 7 shows the vertical cross-section of mixing ratio of cloud-ice, snow, graupel, cloud water and rain water using YSU PBL scheme, KF CP scheme with four MP schemes (LIN, FERR, WSM6 and THOMP) for the TC PHAILIN. The rain water mixing ratio is higher for LIN and THOMP MP schemes and minimum for FERR MP schemes at the eye-wall region. The cloud water mixing ratio is higher for FERR MP scheme and lower for WSM6 MP scheme at the eye-wall region. The graupel mixing ratio is higher for WSM6 MP scheme. The snow mixing ratio is higher for THOMP MP scheme and minimum for LIN MP scheme. The cloud ice mixing ratio is higher for WSM6 MP scheme. However, the intensity of TC is maximum for LIN MP scheme. Further, we considered the vertical distribution of area-averaged (within the $1.5° \times 1.5°$ from the storm center) mixing ratio (g/kg) for VIYARU (18 UTC, 14 May 2013), PHAILIN (12 UTC, 11 Oct 2013), LEHAR (00 UTC, 27 Nov 2013) and MADI (00 UTC, 12 Dec 2013) (Fig. 8). The rain water and cloud ice mixing ratio is higher for LIN MP scheme between 950 & 550 and 350 & 100 hPa, respectively. Cloud water mixing ratio is higher for FERR between 950 and 350 hPa, graupel mixing ratio is higher for WSM6 between 550 and 350 hPa and snow mixing ratio is higher for THOMP MP schemes between 500 and 150 hPa. Similarly, we estimated the mixing ratio for the TC LEHAR. The TC LEHAR also shows similar behaviour with lower mixing ratio's as in PHAILIN. So the intensity predicted in TC LEHAR shows small variations in different MP schemes. However, for the TC MADI, the rain water and cloud ice mixing ratio is higher for WSM6 MP scheme. Also the intensity of the TC MADI is maximum in WSM6 MP scheme. Although for TC VIYARU, the rain water and cloud ice mixing ratio is almost same in LIN and WSM6 scheme, the graupel mixing ratio is higher in WSM6 scheme. So the intensity is stronger in WSM6 for TC VIYARU.

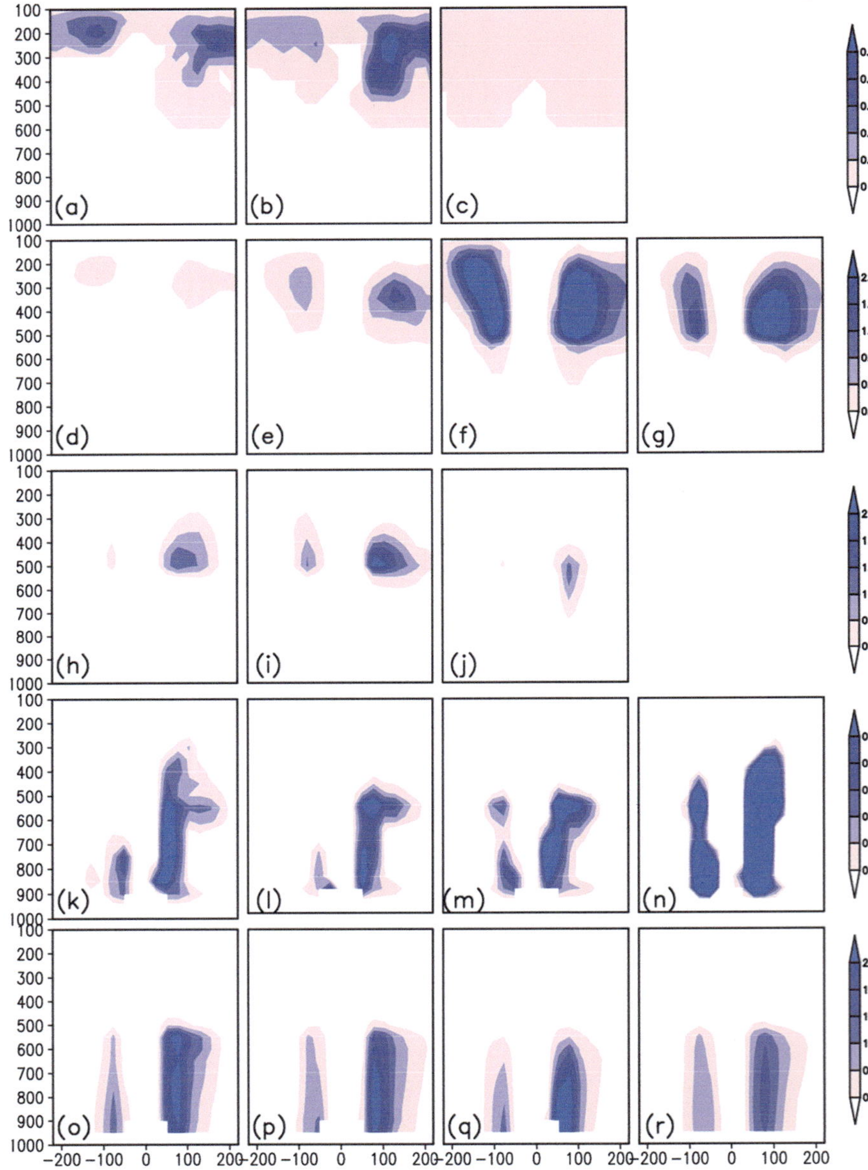

Fig. 7 Meridional vertical cross sections of mixing ratio (g/kg) of (**a**)–(**c**) cloud ice, (**d**)–(**g**) snow, (**h**)–(**j**) graupel, (**k**)–(**n**) cloud water, (**o**)–(**r**) rain water using LIN, FERR, WSM6 and THOMP MP schemes and YSU PBL scheme with KF CP Schemes for the TC PHAILIN

From these observations, it is clear that LIN MP scheme produces higher rain water, graupel and cloud–ice mixing ratio in TC PHAILIN compared to other TCs. However, the rain water and cloud–ice mixing ratio decreased in the TC MADI. In a similar way, FERR MP scheme simulated higher cloud water mixing ratio in TC

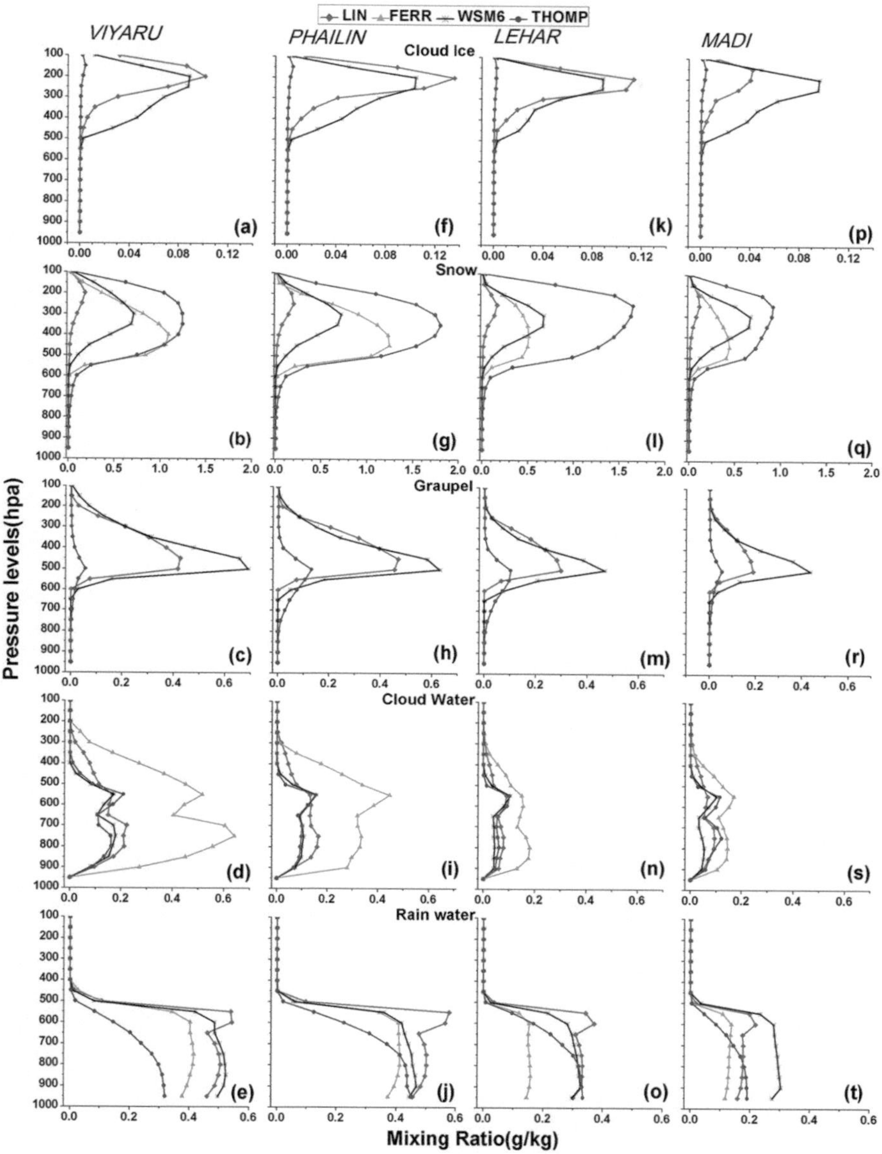

Fig. 8 Vertical distribution of area-averaged (within the 1.5° × 1.5° from the storm center) mixing ratio (g/kg) (**a**) cloud ice, (**b**) snow, (**c**) graupel, (**d**) cloud water, (**e**) rain water using LIN, FERR, WSM6 and THOMP MP schemes and YSU PBL scheme with KF CP Schemes for the TC VIYARU. (**f**)–(**j**) same as (**a**)–(**e**) for TC PHAILIN, (**k**)–(**o**) same as (**a**)–(**e**) for TC LEHAR, (**p**)–(**t**) same as (**a**)–(**e**) for TC MADI

VIYARU and minimum in MADI. The rain water, cloud–ice and graupel mixing ratio's are higher in WSM6 scheme for TC VIYARU. The snow mixing ratio is maximum for TC PHAILIN and minimum for MADI in THOMP MP scheme. Hence we can conclude that FERR scheme shows higher cloud ice, THOMP scheme shows higher snow, LIN scheme shows higher rain water and WSM6 shows higher graupel mixing ratio's when all the MP schemes shows same intensity and increases in all schemes when the intensity is maximum. As we have used THOMP MP scheme in the simulation of total mixing ratio (Fig. 6) and THOMP MP scheme has higher snow mixing ratio so VIYARU simulated higher total mixing ratio above 600 hPa level. Similarly, we used FERR MP scheme in the simulation of PHAILIN and LEHAR and the cloud ice mixing ratio is higher in FERR MP scheme, these two TCs produce higher total mixing ratio below 300 hPa level.

4.4 Initial and Boundary Conditions

Box and Whisker plot of track error for all the TCs from model simulations with different initial conditions is shown in Fig. 9. In this figure, the horizontal lines

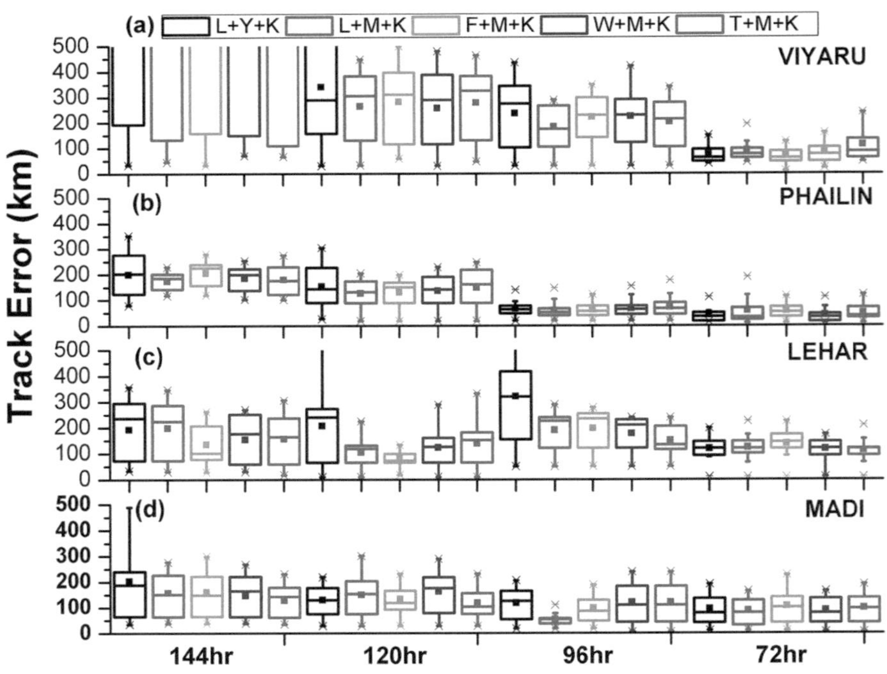

Fig. 9 Box and whisker plot of track errors for (**a**) VIYARU, (**b**) PHAILIN, (**c**) LEHAR and (**d**) MADI at 144, 120, 96 and 72 h simulation

within the box represent the median track error and the bubble represents the mean value. The boxes represent data between the first and third quartiles and the whiskers show data from 12.5 to 87.5 percentiles. The outliers are plotted individually using '*' sign. The track error for TCs VIYARU, PHAILIN, LEHAR and MADI are simulated with initial conditions at 144, 120, 96 and 72 h using FNL data. For the TC VIYARU, MYJ PBL scheme simulated less track error at 144, 120 and 96 h of simulation compared to YSU PBL scheme in combination with LIN MP scheme. Hence we simulated the track error with MYJ PBL scheme and with different MP schemes. The simulated track error is almost same in all MP schemes. However, at 144 h of simulations, THOMP MP scheme simulated less track error whereas at 120 and 96 h, LIN MP scheme simulated less track error. Also, we simulated the TC with YSU and MYJ PBL schemes in combination with LIN MP scheme for 72 h of simulation. Here YSU scheme simulated less track error compared to MYJ scheme. At 72 h of simulation, FERR and LIN MP schemes simulated the less track error. Likewise, we simulated the TCs PHAILIN, LEHAR and MADI with MYJ and YSU PBL schemes in combination with LIN MP scheme. For all these TCs, MYJ PBL scheme simulated less track error at 144, 120 and 96 h of simulation. Further we simulated these TCs with different MP schemes in combination with MYJ PBL scheme. Here all the MP schemes simulated same track error as shown in Fig. 11a. However at 72 h of simulation, YSU PBL scheme in combination with LIN MP scheme simulated less track error.

The track error decreases from 144 to 72 h of simulations in TCs VIYARU and PHAILIN. In case of TC LEHAR, the track error remains same in 144 and 120 h of simulations in LIN + YSU combination. However, the track error increases for 96 h of simulations and then decreases for 72 h of simulations. Whereas in MYJ scheme, the track error decreases for 144–120 h of simulations, then increases for 96 h of simulation and then decreases again for 72 h of simulation. For TC MADI, the track error decreases from 144 to 72 h of simulation for LIN + YSU combination. Whereas in MYJ combinations, the track error decreases from 144 to 96 h of simulation for LIN, FERR and WSM6 MP schemes. Whereas in THOMP MP scheme, the track error decreases from 144 to 120 h of simulation, increased for 96 h of simulation.

The intensity of the TC is discussed in terms of MSLP and MSW. Generally, if the simulated MSLP (MSW) is low (high) compared to observed, the simulated intensity is considered to be strong and vice-versa. Here, the difference between observed and simulated intensities (MSLP and MSW) is considered as error in the simulated intensity. If the difference between observed and simulated MSLP (and MSW) is positive (and negative), the simulated intensity is stronger and vice-versa. Fig. 10a–i show the box plot of error in MSLP and MSW for these TCs. If the upper quartile in MSLP (MSW) is negative (positive), the simulated intensity is weak and vice-versa. If the mean value of the error in simulated intensity is close to zero and the width of the box is small, then the simulated intensity followed the observed intensity well. For TC VIYARU, the simulated intensity is stronger in all combinations of YSU and MYJ at 144, 120, 96 and 72 h of simulations. Among the two PBL combinations, YSU simulated stronger intensity compared to MYJ scheme.

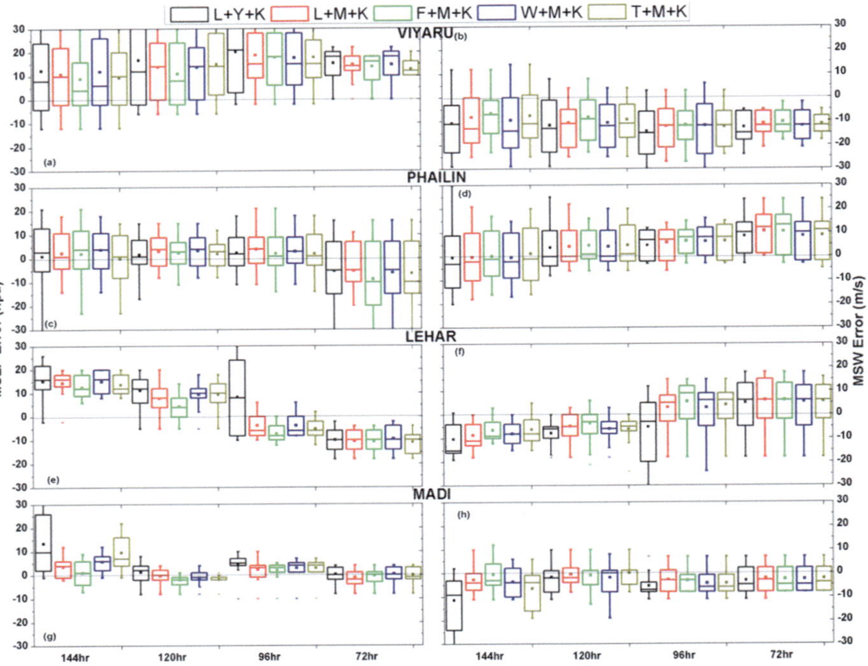

Fig. 10 Box and whisker plot of Intensity errors for (**a**)–(**b**) VIYARU, (**c**)–(**d**) PHAILIN, (**e**)–(**f**) LEHAR and (**g**)–(**h**) MADI at 144, 120, 96 and 72 h simulation

However, in FERR + MYJ scheme simulated less mean intensity error at 144 and 120 h of simulation and THOMP + MYJ simulated at 96 h of simulation. Also THOMP + YSU scheme simulated less mean intensity error at 72 h of simulation. For TC PHAILIN, the simulated intensity followed the observed intensity in all combinations at 144, 120 and 96 h of simulations. Here simulated MSLP followed the observed MSLP whereas the simulated MSW under-predicted the observed MSW. However, at 72 h of simulations, the simulated intensity followed the observed intensity and shows less intensity compared to observed intensity. In case of TC LEHAR, the simulated intensity followed the path of the observed intensity and shows higher intensity compared to observed intensity at 144 and 120 h of simulations. Here MYJ scheme simulated less intensity error compared to YSU scheme. At 144 h of simulation, LIN + MYJ and THOMP + MYJ simulated less intensity error whereas LIN + MYJ and THOMP + MYJ simulated less intensity error at 120 h of simulations. At 96 h of simulation, YSU scheme simulated stronger intensity compared to observed intensity whereas, MYJ scheme simulated weak intensity and followed the path of the observed intensity. At 72 h of simulation, all the schemes simulated weaker intensity compared to observed intensity. For TC MADI, YSU scheme simulated stronger intensity compared to observed intensity whereas, MYJ scheme simulated weak intensity and followed the path of the observed intensity at 144, 120 and 96 h of simulations. However, at 72 h of simulation all the combinations simulated stronger intensity compared to observed intensity. The intensity error shows different trends for different TCs at different simulation times. However at 72 h of simulation, all the schemes simulated less track error.

The track and intensity of TCs VIYARU, PHAILIN, LEHAR and MADI are simulated with FNL and GFS boundary conditions using different PBL schemes with LIN MP scheme and KF CP scheme at 120 h of simulation. The schemes which simulated the track and intensity well in GFS simulated the track and intensity well for the respective TC in FNL also. However, a small variation is observed when GFS and FNL are compared. For the TC VIYARU, GFS shows lower track error compared to FNL (Fig. 11a). The intensity of the TC is all most same and intensity some time stronger in GFS compared to FNL, otherwise FNL strong intensity (Fig. 11b, c). Even for the TC PHAILIN also, the track error is minimum in GFS upto 84 h of prediction and after that the track error increases compared to FNL (Fig. 11d). However, the intensity is same in both GFS and FNL upto 84 h of prediction after that FNL shows stronger Intensity and followed the observed path well (Fig. 11e, f). For the TC LEHAR, the track error is almost same upto 84 h after that FNL shows lower track error compared to GFS in YSU scheme.

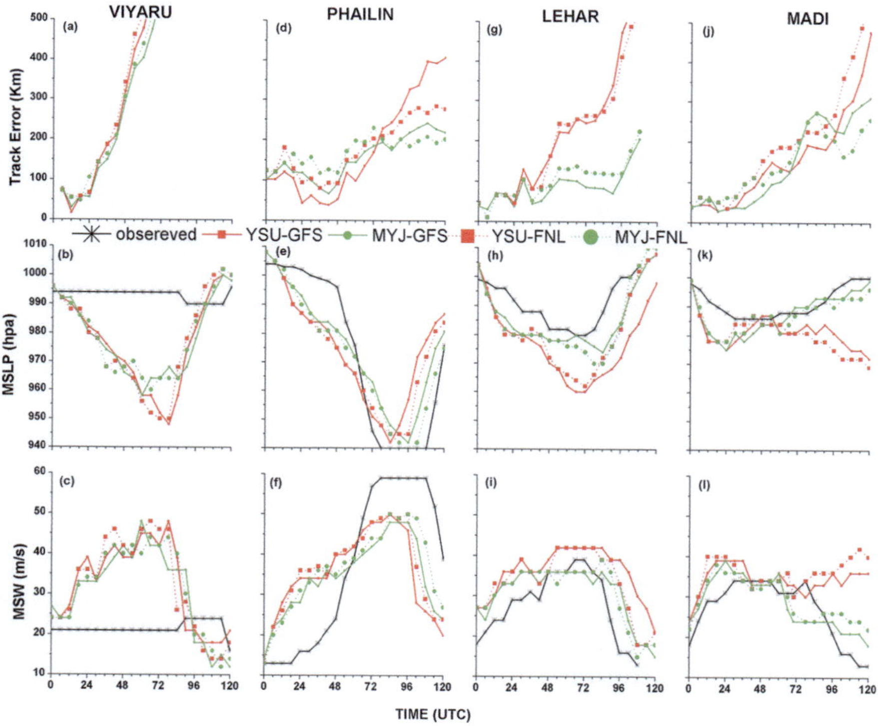

Fig. 11 Model simulated (**a**) Track error, (**b**) MSLP and (**c**) MSW for TC VIYARU, (**d**)–(**f**) same as (**a**)–(**c**) for TC PHAILIN, (**g**)–(**i**) same as (**a**)–(**c**) for TC LEHAR and (**j**)–(**l**) same as (**a**)–(**c**) for TC MADI with FNL and GFS boundary conditions

However, GFS shows minimum track error compared to FNL in MYJ scheme (Fig. 11g). In YSU scheme, the intensity predicted using FNL agreed well with the observed intensity after 84 h of prediction. However, in MYJ schemes FNL predict stronger intensity for some time and GFS predict stronger intensity at other times (Fig. 11h, i). In case of TC MADI, the track error is minimum for GFS compared to FNL in YSU scheme. In MYJ, the track error is minimum in GFS upto 78 h of prediction and after that the track error increases compared to FNL (Fig. 11j). After 84 h of simulation FNL shows strong intensity (Fig. 11k, l).

4.5 Accumulated Rainfall

Figure 12 shows the 48 h accumulated rainfall simulated using LIN + YSU + KF combination for all the TCs along with the 48 h accumulated precipitation in TRMM satellite. The simulated rainfall is compared well with the TRMM observed rainfall. However, predicted rainfall is higher compared to TRMM observed rainfall for the TCs VIYARU, PHAILIN and MADI. Whereas, for TC LEHAR the TRMM shows higher rainfall compared to simulated rainfall. But, for the TC HELEN, the simulated precipitation is in good agreement with the TRMM observed precipitation.

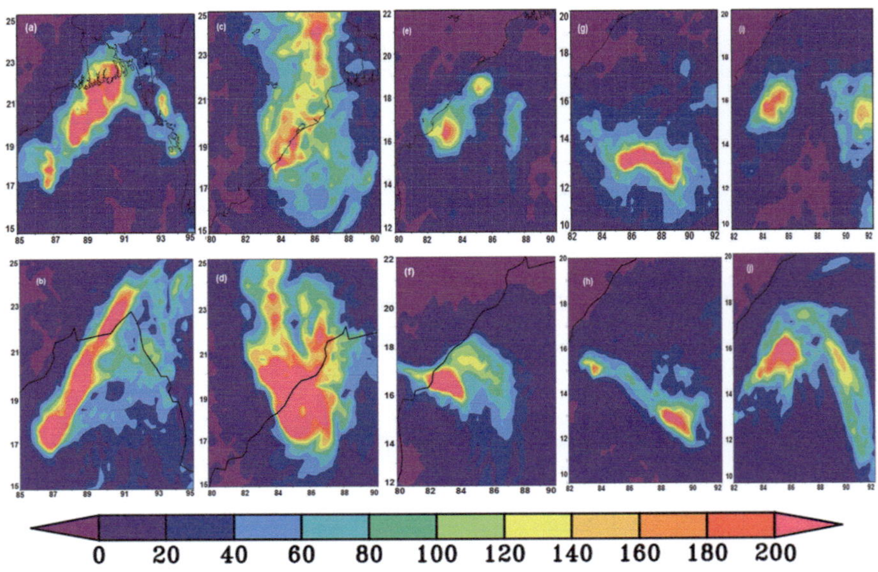

Fig. 12 TRMM observed and Simulated rainfall (mm) using LIN + YSU + KF in (**a**) & (**b**) VIYARU, (**c**) & (**d**) PHAILIN (**e**) & (**f**) HELEN, (**g**) & (**h**) LEHAR and (**i**) & (**j**) MADI TCs respectively during 48 h

4.6 Discussion

Sensitivity study conducted for the Odisha super TC (Bhaskar Rao and Hari Prasad 2007) using MM5 model revealed that the combination of MYJ scheme for PBL and (Kain 2004) Kain–Fritch 2 scheme for CP (Kain 2004) give the best results for intensity and track prediction. Osuri et al. (2012) showed that YSU-PBL scheme simulated track and intensity with less error for the north Indian Ocean TCs using WRF-ARW model. Pattanayak et al. (2012) concluded that the combination of Simplified Arakawa Schubert (SAS) convection, YSU PBL, NMM land surface, and Ferrier microphysics parameterisation schemes in WRF-NMM give better track and intensity forecast for the TC NARGIS. Similarly, Raju et al. (2011b) simulated the TC NARGIS with less track error using KF + YSU + FERR combination with WRF-ARW model. However, in the present study, MYJ scheme simulated the TCs VIYARU, LEHAR and MADI TCs with less track and intensity error whereas both YSU and MYJ schemes simulated the same track and intensity error for TC PHAILIN at 120 h of simulation.

Several researchers (Pattanayak et al. 2012) studied the sensitivity of track and intensity of TCs on simulation time and concluded that the error in the simulated track and intensity decreases as the simulation time decreases. In the present study, MYJ scheme simulated less error at 144, 120 and 96 h of simulations. In TCs VIYARU and PHAILIN, the track error decreases from 144 h of simulation to 96 h of simulation. Whereas, in TCs LEHAR and MADI the simulated error decreased in some schemes and increased in some other schemes. However at 72 h of simulation, YSU scheme simulated less track error compared to MYJ scheme. Also the track and intensity error is less in 72 h of simulation compared to 144, 120 and 96 h of simulations. Mohanty et al. (2010) reported that the FNL and NCMRWF boundary condition predicted the track with minimum error upto the landfall of the four TCs (NARGIS, RASHMI, KHAIMUK and NISHA). The present study, simulated the TCs with various boundary conditions and reported that the impact of boundary conditions on track and intensity predictions.

Previous studies suggested that the air–sea surface flux exchanges the moist enthalpy via surface latent heat and sensible heat fluxes, which are the primary source for the changes in the intensity of the TC (Emanuel 1986). Sun et al. (2013), suggested that the sensible heat flux depends on the surface wind speed and air–sea temperature difference (ASTD), while latent heat flux is dependent on the surface wind speed and air–sea moisture difference (ASMD). Recently, Benjamin and Zhang examined the impacts of model parameters on simulation of Hurricane Katrina using WRF-ARW model and found that the surface flux had a significant effect on structure and intensity of the Hurricane. In the present study, we found that the intensity of the NIO TC depends on SEF. However, for some TCs, the intensity is maximum in YSU scheme whereas the SEF in MYJ scheme at that time and vice-versa. Further, SEF depends on temperature difference and specific humidity. Mohapatra and Sharma (2015) have demonstrated role of humidity on structure of TCs over the NIO.

Mukhopadhyay et al. (2011) studied the influence of moist processes on track and intensity forecast of TCs *Gonu* and *Sidr* over the north Indian Ocean and found that the intensity of TCs depends mainly on the convective heating. Also the WSM6 scheme produces better intensity due to the incorporation of graupel and rain mixing ratio. In the present study, we found that the specific humidity depends on the total mixing ratio. The total mixing ratio is maximum at different pressure levels. This is because FERR scheme simulates higher cloud ice, THOMP scheme simulates higher snow, LIN scheme shows higher rain water and cloud ice and WSM6 shows higher graupel mixing ratio.

5 Summary and Conclusions

Simulation study of track and intensity of five TCs (VIYARU, PHAILIN, HELEN, LEHAR and MADI) originated in the BOB during 2013 is performed using WRF model with KF CP scheme and different MP and PBL schemes. In the simulation of TCs with different MP schemes, the entire MP scheme over-predicted the track of TC VIYARU. A small deviation of track error is observed for the TCs HELEN and MADI. However, there are no considerable variations for TCs PHAILIN and LEHAR. For simulation of five TCs track and intensity with 2-PBL schemes, MYJ scheme simulated better for TCs VIYARU, LEHAR and MADI and YSU simulated for PHAILIN and HELEN. However, the variations in intensities are comparatively higher for the TCs LEHAR and MADI in all the MP and PBL combinations. For these results it is distinctly evident that PBL schemes had a significant effect on the simulated track and intensity of the TCs. Futher these simulation results showed that THOMP + MYJ produced less track error and intensity for VIYARU, FERR + MYJ for LEHAR, FERR + YSU for PHAILIN, THOMP + YSU for HELEN and WSM6 + MYJ for MADI. From all these simulations on five TCs it is noticed that all combinations of MP and PBL simulated track very well but deviated from the observed intensity. To understand the deviations of intensity, an experiment was performed on the sensitivity of track and intensity of TCs on the initial conditions. At 144, 120 and 96 h of simulation, MYJ scheme simulated less track error. However at 72 h of simulation, LIN + YSU combination simulated less track and intensity error.

Further we studied the effect of sensible heat flux on intensity of the TC. The sensible heat flux and wind speed is maximum at the eye-wall region and are minimum at the eye of the TC. However, for the TC PHAILIN and LEHAR the intensity is maximum in YSU scheme whereas the sensible heat flux is maximum in MYJ scheme. For this we observed the air and sea surface temperature difference and specific humidity and found that the sensible heat flux strongly depends on these atmospheric parameters. In addition to the sensible heat flux, we studied the variations in convergence and relative vorticity field in the lower troposphere. The maximum surface wind depends on divergence field and minimum sea-level pressure depends on the relative vorticity field.

To understand the vertical structure of the TC, we simulated the vertical cross-section of wind speed, temperature, specific humidity and hydrometeor mixing ratio. The specific humidity is maximum at the eye-wall region and it decreases towards the eye of the TC. Minimum wind speed and warmer temperatures are observed at the eye of the TC compared to eye-wall regions for all the TCs. The simulated wind speeds are maximum in YSU scheme for the TCs VIYARU, PHAILIN and LEHAR at the eye-wall region. But for the TC MADI, MYJ scheme simulated the maximum wind at the eye-wall region of the TC. These schemes simulated stronger intensities also for the respective TCs.

The PBL schemes which simulate maximum winds show higher specific humidity and total mixing ratio.

From the aforesaid results it can be concluded that the intensity of the TC depends on vertical variation of temperature, wind speed and specific humidity. The specific humidity further depends on hydrometeor mixing ratio. Variation of hydrometeor mixing ratio in different MP schemes have been studied and found that FERR scheme simulates higher cloud ice between 950 and 350 hPa, THOMP scheme simulates higher snow between 500 and 150 hPa, LIN scheme shows higher rain water and cloud ice between 950 & 550 and 350 & 100 hPa and WSM6 shows higher graupel mixing ratio between 550 and 350 hPa.

Acknowledgement We acknowledge to the India Meteorological Department (IMD), Govt. of India, for providing the TCs track information. We acknowledge to Indian Space Research Organization (ISRO), Govt. of India for providing the financial support to carry out this research work. We acknowledge to National Center for Environmental Prediction, USA, for providing real-time GFS analysis/forecast data.

References

Bhaskar Rao, D.V. and Hari Prasad, D. (2006). Numerical prediction of the Odisha super-TC: Sensitivity to the parameterization of convection, boundary layer and explicitmoisture processes. *Mausam*, **57**, 61–78

Bhaskar Rao, D.V. and Hari Prasad, D. (2007). Sensitivity of tropical TC intensification to boundary layer and convective processes. *Natural Hazards*, **41**, 429–445, doi:10.1007/s11069-006-9052-7

Braun, S.A. and Tao, W.-K. (2000). Sensitivity of high resolution simulations of Hurricane Bob (1991) to planetary boundary layer parameterizations. *Monthly Weather Review*, **128**, 3941–3961, doi:10.1175/1520-0493(2000)129<3941:SOHRSO>2.0.CO_2.

Brown, P.R.A. and Swann, H.A. (1997). Evaluation of key microphysical parameters in three-dimensional cloud-model simulations using aircraft and multiparameter radar data. *Quarterly Journal of the Royal Meteorological Society*, **123**, 2245–2275.

Chandrasekar, R. and Balaji, C. (2012). Sensitivity of tropical TC Jal simulations to physics parameterizations. *Journal of Earth System Science*, **121**, pp. 923–946.

Deshpande, M, Pattnaik, S. and Salvekar, P.S. (2010). Impact of physical parameterization schemes of numerical simulation of super TC Gonu. *Natural Hazards*, **55**, 211–231, doi:10.1007/s11069-010-9521-x

Deshpande, M.S, Pattnaik, S. and Salvekar, P.S. (2012). Impact of cloud parameterization on the numerical simulation of a super TC. Annals of Geophysics, **30**, 775–795, doi:10.5194/angeo-30-775-2012

Dudhia, J. (1989). Numerical study of convection observed during the winter monsoon experiment using a mesoscale two-dimensional model. *Journal of Atmospheric Science*, **46**, 3077–3107, doi:10.1175/1520-0469(1989)046<3077:NSOCOD>2.0.CO_2.

Emanuel, K.A. (1986). An air-sea interaction theory for tropical TCs: Part I: Steady state maintenance. *Journal of Atmospheric Science*, **43**, 585–605.

Emanuel, K. A. (1999). Thermodynamic control of hurricane intensity. *Nature*, **401**, 665–669.

Ferrier, B.S, Jin, Y, Lin, Y, Black, T, Rogers, E. and DiMego, G. (2002). Implementation of a new grid-scale cloud and rainfall scheme in the NCEP Eta Model, Preprints, *15th Conf. on Numerical Weather Prediction*, San Antonio, TX, American Meteorological Society, pp. 280–283.

Fovell, R.G, Corbosiero, K.L, Seifert, A. and Liou, K.N. (2010a). Impact of cloudradiative processes on hurricane track. *Geophysical Research Letters*, **37**, L07808, doi:10.1029/2010GL042691

Fovell, R.G, Corbosiero, K. L. and Kuo, H.-C. (2010b). Influence of cloud-radiative feedback on tropical TC motion: Symmetric contributions. Preprints, 29th Conf. on Hurricanes and Tropical Meteorology,Tucson, AZ, American Meteorological Society, 13C.5.

Gray, W.M. (1968). Global view of the origin of tropical disturbances and storms. *Monthly Weather Review*, **96**, 669–700.

Hong, S.-H. and Pan, H.-L. (1996). Nonlocal boundary layer vertical diffusion in a medium-range forecast model. *Monthly Weather Review*, **124**, 2322–2339.

Hong, S.-Y. and Lim, J.-O.J. (2006). The WRF single-moment 6-class microphysics scheme (WSM6). *Journal of the Korean Meteorological Society*, **42**, 129–151.

Hong, S.-Y, Dudhia, J. and Chen, S.-H. (2004). A revised approach to ice microphysical processes for the parameterization of clouds and precipitation. *Monthly Weather Review*, **132**, 103–120.

Janjić, Z.I. (1994). The step-mountain eta coordinate model: Further developments of the convection, viscous sub layer and turbulence closure schemes. *Monthly Weather Review*, **122**, 927–945, doi:10.1175/1520-0493(1994)122<0927:TSMECM>2.0.CO_2.

Janjic, Z.I, Black, T.L, Rogers, E, Chuang, H. and DiMego, G. (2003a). The NCEP Nonhydrostatic Meso Model (NMM) and First Experiences with Its Applications. EGS/EGU/AGU Joint Assembly, Nice, France, 6–11 April.

Janjic, Z.I, Black, T.L, Rogers, E, Chuang, H. and DiMego, G. (2003b). The NCEP Nonhydrostatic Mesoscale Forecasting Model. 12.1, Extended Abstract, 10th Conference on Mesoscale Processes, Portland, OR, American Meteorological Society (Available Online).

Kain, J.S. (2004). The Kain–Fritsch convective parameterization: An update. *Journal of Applied Meteorology*, **43**, 170–181.

Kubota, M, Iwasaka, N, Kizu, S, Konda, M. and Kutsuwada, K. (2002). Japanese ocean flux data sets with use of remote sensing observations (J-OFURO), *Journal of Oceanography*, **58**, 213–225.

Lin, Y.-L, Farley, R.D. and Orville, H.D. (1983). Bulk parameterization of the snow field in a cloud model, *Journal of Climate and Applied Meteorology*, **22**, 1065–1092, doi:10.1175/1520-0450(1983)022<1065:BPOTSF>2.0.CO;2.

Liu, C. and Moncrieff, M.W. (2007). Sensitivity of cloud-resolving simulations of warm-season convection to cloud microphysics parameterizations. *Monthly Weather Review*, **135**, 2854–2868.

Loh, W, Juneng, L. and Tandang, F. (2010). Sensitivity of Typhoon Vamei (2001) simulation to planetary boundary layer parameterization using PSU/NCAR MM5. *Pure and Applied Geophysics*, **168**, 1799–1811.

Lord, S.J, Willoughby, H.W. and Piotrowicz, J.M. (1984). Role of a parameterized ice-phase microphysics in an axisymmetric tropical TC model. *Journal of Atmospheric Science*, **41**, 2836–2848.

Mellor, G.L. and Yamada, T. (1982). Development of a turbulence closure model for geophysical fluid problems. *Reviews of Geophysics and Space Physics*, **20**, 851–875.

Mlawer, E.J, Taubman, S.J, Brown, P.D, Iacono, M.J. and Clough, S.A. (1997). Radiative transfer for inhomogeneous atmosphere: RRTM, a validated correlated-k model for the long wave. *Journal of Geophysical Research*, **102**, 16,663–16,682, doi:10.1029/97JD00237

Mohanty, U.C. and Gupta, A. (1997). Deterministic methods for prediction of tropical TC tracks, *Mausam*, **48**, 257–272.

Mohanty, U.C, Mandal, M. and Raman, S. (2004). Simulation of Odisha Super TC (1999) using PSU/NCAR Mesoscale Model. *Natural Hazards*, **31**, 373–390, doi:10.1023/B:NHAZ.0000023358.38536.5d

Mohanty, U.C, Osuri, K.K, Routray, A, Mohapatra, M. and Pattanayak, S. (2010). Simulation of Bay of Bengal tropical cyclones with WRF modeling system: Impact of initial value and boundary conditions. *Marine Geodesy*, **33**, 294–314.

Mohapatra, M. and Sharma, M. (2015). Characteristics of surface wind structure of tropical cyclones over the North Indian Ocean. *Journal of Earth System Sciences*, **124**, 1573–1598.

Mohapatra, M, Mandal, G.S, Bandyopadhyay, B.K, Tyagi, A. and Mohanty, U.C. (2012). Classification of cyclone hazard prone districts of India. *Natural Hazards*, **63**, 1601–1620.

Mohapatra, M, Nayak, D.P, Sharma, R.P. and Bandopadhyay, B.K. (2013a). Evaluation of official tropical cyclone track forecast over North Indian Ocean issued by India Meteorological Department. *Journal of Earth System Sciences*, **122**, 589–601.

Mohapatra, M, Bandyopadhyay, B.K. and Nayak, D.P. (2013b). Evaluation of official tropical cyclone intensity forecast over North Indian Ocean Issued by India Meteorological Department. *Natural Hazards*, **68**, 433–451.

Mohapatra, M, Nayak D.P, Sharma, M, Sharma, R.P. and Bandyopadhyay, B.K. (2015). Evaluation of landfall forecast over North Indian Ocean issued by India Meteorological Department. *Journal of Earth System Sciences*, **124**, 861–874.

Mukhopadhyay, P, Taraphdar, S. and Goswami, B.N. (2011). Influence of moist processes on track and intensity forecast of TCs over the north Indian Ocean. *Journal of Geophysical Research*, **116**, D05116, 21 pp. doi:10.1029/2010JD014700

Osuri, K.K, Mohanty, U.C, Routray, A. and Mohapatra, M. (2012). Impact of satellite derived wind data assimilation on track, intensity and structure of tropical cyclones over North Indian Ocean. *International Journal of Remote Sensing*, **33**, 1627–1652.

Osuri, K.K, Mohanty, U.C, Routray, A, Mohapatra, M. and Dev, N. (2013). Real-time track prediction of tropical cyclones over the North Indian Ocean Using the ARW Model. *Journal of the Applied Meteorology and Climatology*, **52**, 2476–2492. doi: http://dx.doi.org/10.1175/JAMC-D-12-0313.1

Osuri, K.K, Mohanty, U.C, Routray, A, Mohapatra, M. and Niyogi, D. (2014). Real-time track prediction of tropical cyclones over the North Indian Ocean using the ARW Model. *Journal of Applied Meteorology and Climatology*, **52**, 2476–2492.

Pattnaik, S. and Krishnamurti, T. N. (2007a). Impact of cloud microphysical process on hurricane intensity. Part 1: Control run. *Meteorology and Atmospheric Physics*, **97**, 117–126, doi:10.1007/s00703-006-0247-y

Pattnaik, S. and Krishnamurti, T.N. (2007b). Impact of cloud microphysical process on hurricane intensity. Part 2: Sensitivity experiments. *Meteorology and Atmospheric Physics*, **97**, 127–147, doi:10.1007/s00703-006-0248-x

Pattanaik, S. and Mohanty, U.C. (2008). A comparative study on performance of MM5 andWRF models in simulation of tropical TCs over Indian seas. *Current Science*, **95**, 923–936.

Pattanaik, D.R. and Rama Rao, Y.V. (2009). Track prediction of very severe TC 'Nargis' using high resolution weather research forecasting (WRF) model. *Journal of Earth System Science*, **118**, 309–329.

Pattanayak, S, Mohanty, U.C. and Gopalakrishnan, S.G. (2012). Simulation of very severe TC Mala over Bay of Bengal with HWRF modeling system. *Natural Hazards*, **63**, 1413–1437.

Paul, B. (2010). Human injuries caused by Bangladesh's TC Sidr: An empirical study. *Natural Hazards*, **54**, 483–495.

Prater, B. and Evans, J.L. (2002). Sensitivity of modeled tropical TC track and structure of Hurricane Irene (1999) to the convection parameterization scheme. *Meteorology and Atmospheric Physics*, **80**, 103–115.

Qiang Sun, Y, Yuxin Jiang and Benkui Tan. and Fuqing Zhang (2013). The Governing Dynamics of the Secondary Eyewall Formation of Typhoon Sinlaku (2008). *Journal of Atmospheric Science*, **70**, doi:10.1175/JAS-D-13-044.1

Rahaman, H. and Ravichandran, M. (2013). Evaluation of near-surface air temperature and specific humidity from hybrid global products and their impact on latent heat flux in the North Indian Ocean. *Journal of Geophysical Research: Oceans*, **118**, 1034–1047, doi:10.1002/jgrc.20085

Raju, P.V.S, Jayaraman, P. and Mohanty, U.C. (2011a). Prediction of severe tropical TCs over the Bay of Bengal during 2007–2010 using highresolution mesoscale model. *Natural Hazards*, doi: 10.1007/s11069-011-9918-1

Raju, P.V.S, Jayaraman, P. and Mohanty, U.C. (2011b). Sensitivity of physical parameterizations on prediction of tropical TC Nargis over the Bay of Bengal using WRF model. *Meteorology and Atmospheric Physics*, **113**, 125–137, doi: 10.1007/s00703-011-0151-y

Rozoff, C.M, Nolan, D.S, Kossin, J.P, Zhang, F. and Fang, J. (2012). The roles of an expanding wind field and inertial stability in tropical cyclone secondary eyewall formation. *Journal of Atmospheric Science*, **69**, 2621–2643.

RSMC, New Delhi (2014) Report on Cyclonic Disturbances over north Indian Ocean during, 2013.

Sathi Devi, K, Hari Prasad, D. and Bhaskar Rao, D.V. (2006). The evaluation of Kain–Fritsch scheme in tropical cyclone simulation. *Mausam*, **57**, 395–410.

Sikka, D.R. and Suryanarayana, R. (1972). Forecasting the movement of tropical storms/depressions in the Indian region by computer oriented technique wing climatology and persistence. *Indian Journal of Meteorology and Geophysics*, **23**, 35–40.

Skamarock, W.C. (2008). A Description of the Advanced Research WRF Version 3. – NCAR Tech. Note, June 2008/W.C. Skamarock. Available at: http://www.mmm.ucar.edu/wrf/users/docs (accessed at: 11.01.2013).

Srinivas, C.V, Bhaskar Rao, D.V, Yesubabu, V, Baskaran, R. and Venkatraman, B. (2012). Tropical TC predictions over the Bay of Bengal using the high-resolution advanced research weather research and forecasting model. *Quarterly Journal of the Royal Meteorological Society*, doi:10.1002/qj.2064

Srinivas, C.V, Venkatesan, R, Bhaskar Rao, D.V. and Hari Prasad, D. (2007). Numerical simulation of Andhra severe TC (2003): Model sensitivity to the boundary layer and convective parameterization. *Pure and Applied Geophysics*, **164**, 1465–1487.

Thompson, G, Rasmussen, R.M. and Manning, K. (2004). Explicit forecast of winter precipitation using an improved bulk microphysical scheme. Part I: Description and sensitivity analysis. *Monthly Weather Review*, **132**, 519–542, doi:10.1175/1520-0493(2004)132<0519:EFOWPU>2.0.CO;2.

Thompson, G, Field, P.R, Rasmussen, R.M. and Hall, W.D. (2008). Explicit forecasts of winter precipitation using an improved bulk microphysics scheme. Part II: Implementation of a new snow parameterization. *Monthly Weather Review*, **136**, 5095–5115.

Trivedi, D.K, Sanjay, J. and Singh, S.S. (2002). Numerical simulation of a super CS, Odisha (1999), Impact of initial conditions. *Meteorological Applications*, **9**, 367–376, doi:10.1017/S1350482702003109

Venkatarami Reddy, M, Surendra Prasad, S.B, Murali Krishna, U.V. and Krishna Reddy, K. (2014). Effect of cumulus and microphysical parameteri-zations on the *JAL* TC prediction. *Indian Journal of Radio and Space Physics*, **43**, 103–123.

Wang, Y. (2002). An explicit simulation of tropical TCs with a triply nested movable mesh primitive equation model: TCM3. Part II: Model refinements and sensitivity to cloud micro physics parameterization. *Monthly Weather Review*, **130**, 3022–3036.

Wang, Y. (2009). How do outer spiral rain bands affect tropical cyclone structure and intensity? *Journal of Atmospheric Science*, **66**, 1250–1273, doi:10.1175/2008JAS2737.1

Wu, L. and Wang, B. (2001). Effects of convective heating on movement and vertical coupling of tropical TCs: A numerical study. *Journal of Atmospheric Science*, **58**, 3639–3649.

Zhu, T. and Zhang, D.-L. (2006). Numerical simulation of Hurricane Bonnie (1998). Part II: sensitivity to varying cloud microphysical processes. *Journal of Atmospheric Science*, **63**, 109–126.

Standard Operation Procedure for Tropical Cyclone Vital Parameters over North Indian Ocean

M. Sharma and M. Mohapatra

1 Introduction

There is a growing need for improvement in tropical cyclone (TC) vital parameters (Knaff 2011) in view of the requirements of numerical weather prediction (NWP) models and various stake holders. As the damage due to a TC is directly proportional to the square of the maximum sustained wind (MSW) and loss due to a TC is proportional to cube of MSW, the surface wind structure associated with a TC serves insurance agencies to assess the damage due to a TC. The disaster managers who prepare for the impact of a landfalling TC may use the wind field information as guidance as to where the most severe wind or surge damage may occur. The TC Vital parameters also serve as input to NWP models and storm surge models that are run prior to landfalling events to create synthetic vortex (Chourasia et al. 2013), as most of the NWP models fail to simulate accurately the location and intensity of the TC. The creation of synthetic vortex helps in improving the track and intensity forecast of the model. In the parametric storm surge prediction models, the surface wind structure in the quadrant base form alongwith the radius of maximum wind (RMW) and pressure drop (ΔP) at the centre are utilised to create the wind stress and hence predict the storm surge (Dube et al. 2013). In post-event cases, these wind structure data are utilised for diagnosis of TC and to better plan for future TC forecasts. Engineers and planners rely on historical TC information to determine long-term risks to facilities and infrastructure and to ensure the resilience of communities to potential disasters. Another most important use of this product is the determination of ship avoidance area over the sea due to a TC.

India Meteorological Department (IMD) started issuing TC vital parameters from very severe cyclonic storm (VSCS) Giri in October 2010. Currently, in

M. Sharma • M. Mohapatra (✉)
India Meteorological Department, Mausam Bhavan, Lodi Road, New Delhi 110003, India
e-mail: mohapatraimd@gmail.com

© Capital Publishing Company 2017

M. Mohapatra et al. (eds.), *Tropical Cyclone Activity over the North Indian Ocean*,
DOI 10.1007/978-3-319-40576-6_24

addition to routinely analysing position (and uncertainty), MSW (3-min mean) and mean sea level pressure (MSLP), maximum horizontal extent from the circulation center of a particular sustained wind speed viz., 34, 50 and 64 knots (referred as *R*34, *R*50 and *R*64) in each of four quadrants, viz., northeast (NE), southeast (SE), northwest (NW) and southwest (SW), radius of outermost closed isobar (ROCI), pressure of outermost closed isobar (P_O), pressure at centre (P_C), ΔP defined as $P_O - P_C$, RMW and the vertical depth/extent of the TC (deep/medium/shallow) are estimated every 6 h based on 00, 06, 12 and 18 UTC from the stage of deep depression (MSW of 28 knots or more) onwards, if the system is expected to intensify into a TC. This product is provided in coded format to the NWP and storm surge modelling groups. In addition, the forecast R28, R34, R50 and R64 knots threshold are generated alongwith the MSW and location every 6 h based on 00, 06, 12 and 18 UTC valid upto 120 h in the interval of 6 h for first 24 h forecast period and in the interval of 12 h for the subsequent forecast periods. This product is uploaded on RSMC website (www.rsmcnewdelhi.imd.gov.in) in text and graphic forms and also provided to various users by e-mail.

One of the reasons for limited study on structural aspects of TCs over the NIO is the non-availability of best estimates of structure parameters like RMW, ROCI, R34, R50 and R64. Hence, an attempt has been made to develop a standard operation procedure (SOP) for preparation of best estimates of the TC Vital parameters as mentioned above. With reference to the SOP used by IMD for preparation of the best estimates of R34, R50 and R64, the TC vital parameters for VSCS Phailin and Hudhud have been discussed in a separate study by Mohapatra and Sharma (2015). Detailed review of the past work done over the North Indian Ocean (NIO) region is also presented by Mohapatra and Sharma (2015).

2 Data Utilised for Derivation of TC Vitals

The TC vital parameters are basically the surface parameters including location, movement (speed and direction) of storm during past 6 h, P_c, P_o, ROCI, MSW, RMW, R34, R50 and R64 in four geographical quadrants (NE, SE, SW and NW) and depth of convection. The data for these parameters is collected from various sources including:

- Surface observations from coastal observatories, island stations, ships and buoys and coastal observations from World Meteorological Organisation/Economic and Social Commission for Asia and Pacific (WMO/ESCAP) Panel member countries available through GTS.
- Satellite observations including various geostationary and polar orbiting satellites in visible, infra-red and microwave bands. The satellites include Kalpana, INSAT-3A, INSAT-3D, Meteosat and the products available from US Navy National Research Laboratory (NRL), Co-operative Institute for Meteorological

Satellite Studies (CIMSS) and Co-operative Institute for Research in the Atmosphere (CIRA), USA website.
- Dvorak T. No. estimates available from Satellite Division of IMD, Satellite Services Division of NOAA, Sat Fix bulletin of Joint Typhoon Warning Centre (JTWC), Automated Dvorak Technique estimates from CIRA.
- Multi-platform satellite derived winds from CIRA.
- Radar imageries and products from IMD and Bangladesh.
- Climatology of structure of TCs over NIO (Mohapatra and Sharma 2015).

The best estimates of the TC Vital parameters is a consensus derived mainly from the above mentioned observations following the SOP as mentioned in the following sections.

3 Standard Operation Procedure for Derivation of TC Vitals

The methodology followed by IMD to derive various TC Vital parameters is presented in Sects. 3.1, 3.2, 3.3, 3.4, 3.5, 3.6 and 3.7.

3.1 Location

The location of the centre of the TC is determined based on (a) synoptic, (b) satellite (geostationary and polar orbiting satellites) and (c) Radar observations. When the TC is far away from the coast and not within the radar range, the satellite estimates get more weightage and necessary corrections are carried out based on available ships, buoys and island observations. When the TC is within the Radar range, radar estimates get maximum preference followed by satellite and coastal observations. When TC is very close to coast or over the land surface, coastal observations get the highest preference followed by radar and satellite observations. When the TC is over land, only surface observations will be the determining factor (IMD 2013). In the pressure field, the location of lowest pressure is considered as centre and in the wind field, the centre determined by the streamline analysis of 10-m wind is considered as the centre of TC. In the radar imagery, the centre is determined with the help of a logarithmic spiral, in case of spiral band structure and the centre of the eye in the radar imagery is considered as the centre of the TC with the development of eye (IMD 1976). In satellite method, the centre of eye and the centre of central dense overcast (CDO) are considered as the centre of TC in case of eye and CDO pattern respectively. The centre estimated with a logarithmic spiral is the centre of TC in case of curved band/spiral pattern. In case of shear pattern, the centre of low-level circulation as observed in visible imagery is considered as centre of TC. As during night the visible imagery is not available, the low cloud lines cannot be detected. As a result, the location of the centre during night time cannot be determined correctly. It is more so when it is a low intensity system like depression/cyclonic storm. When the system intensifies into a severe cyclone with appearance of CDO/eye feature, the location of the centre becomes most accurate and it is the centre of the CDO/eye. The eye can be detected in both visible and infra red (IR) imageries. Hence, in case of curved band pattern or CDO pattern, centre can be more accurately determined compared to shear pattern. Thus the error in determining the location of the centre at night is maximum in case of shear pattern. To overcome the above problems, the microwave imageries can be utilised. Microwave imageries at night can detect the cloud features and hence the centre of cyclone as the centre of the TC. As regards availability, the products from geostationary satellites provide half hourly images and capture radiations in visible and IR bands. These radiations are reflected from the top of cloud and hence IR and visible provide centre at the top of cloud and not from the surface level. The resolution for IR is 4/8 km and that for visible imagery is 1 km. The polar orbiting satellites provide microwave imageries which can capture radiations in oxygen band and thus provide the centre from the lower level. However, microwave imageries are not available all the time. So, we take the imagery of nearest available time and interpolate the centre for current time. Detailed procedure is available in Dvorak (1984). The entire process of determining the centre of the TC is shown in Fig. 1. The accuracy of estimation of location and the availability of data over the region are shown in Fig. 2.

3.2 *Intensity*

The intensity of the system is measured in terms of 3-min average MSW at surface level (10 m above ground level) (IMD 2003, 2013). Various steps involved in determination of intensity are shown in Fig. 3.

When TC is over deep sea, Dvorak's technique (Dvorak 1984) is used for estimation of intensity. For this purpose, the intensity of the tropical system is indicated by a code figure called *T* number based on pattern recognition technique by Dvorak (1984). This technique relies on four distinct geophysical properties that relate organised cloud pattern to TC intensity. Two are kinematic namely vorticity and vertical wind shear and the other two are thermodynamical viz. convection and core temperature. The strength and distribution of circular winds (by implication vorticity) in a TC organises the cloud into the patterns that Dvorak relates to MSW. External/environmental shear is a kinematic force that works to distort the vorticity and hence the cloud pattern. Dvorak found that degree of distortion was also related to MSW. Dvorak Technique uses a scale from T1 to T8 in the interval of 0.5. The relation between *T* number, MSW and Pressure drop is shown in Table 1.

This classification of intensity is based on two parameters arrived on an analysis of the cloud features as seen in the satellite pictures of the disturbance. These are (1) the central features (CF) which define the cloud system centre and its relation to

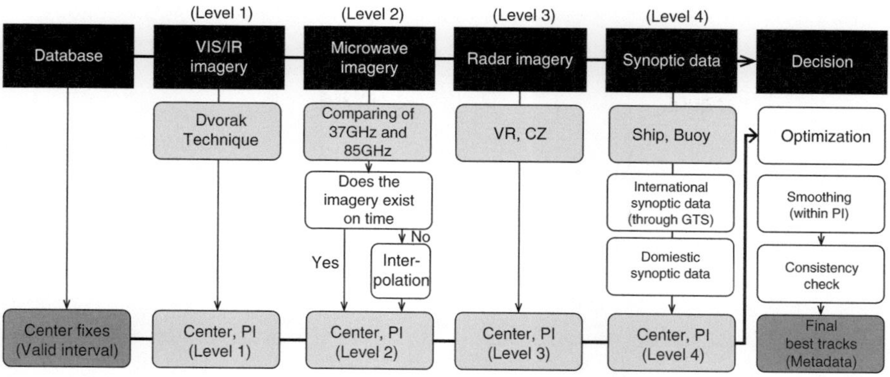

Fig. 1 Steps involved in determination of location of centre of TC

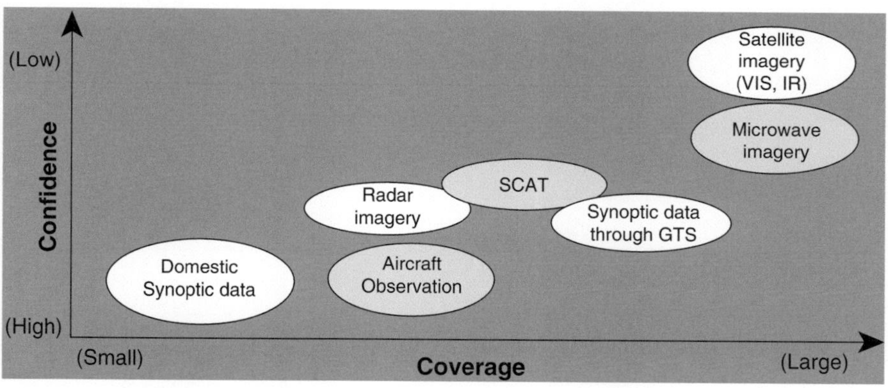

Fig. 2 Relative availability and confidence of forecaster

dense overcast clouds and (2) the outer banding features (BF). The intensity of the tropical system is indicated by a code figure called *T* number which is the sum of the CF and BF.

In central dense overcast pattern, size and temperature of the CDO determines the intensity. In shear pattern, the distance between the low level circulation centre (LLCC) and the sharp boundary of the convective cloud determines the intensity of the system. In the curved band/spiral band pattern, the logarithmic spiral is used to estimate number of parts of the logarithmic spiral covered with convective clouds and accordingly *T* number is determined.

In case of TCs with eyes the technique determines the temperature of the eye and the surrounding eye wall clouds using IR data and relates to the intensity with warmer/cooler eye/wall cloud temperatures respectively indicating greater intensity.

Another feature of the technique is the Current Intensity (CI) number which relates directly to the intensity (in term of MSW) of the TC. The CI number may differ from the *T* number on some occasions to account for certain factors which are

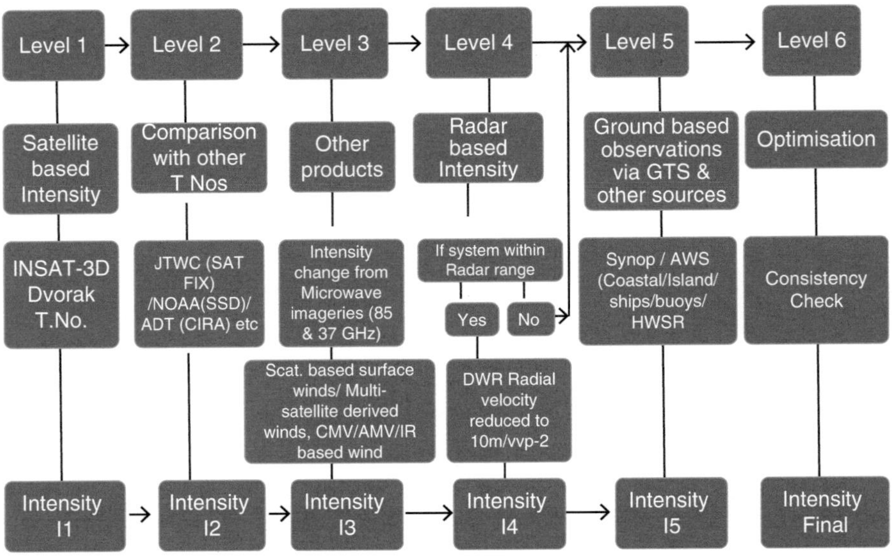

Fig. 3 Steps involved in determination of intensity of a TC

Table 1 Dvorak's classification of cyclonic disturbances

CI Number	Maximum wind speed (knots)	Pressure drop (hPa)
1.0	25	3.1
1.5	25	3.1
2 0	30	4.5
2.5	35	6.1
3.0	45	10.0
3.5	55	15.0
4.0	65	20.9
4.5	77	29.4
5.0	90	40.2
5.5	102	51.6
6.0	115	65.6
6.5	127	80.0
7.0	140	97.2
7.5	155	119.1
8.0	170	143.3

not directly related to cloud features. The empirical relationship between C.I . number and the maximum wind speeds (according to Dvorak) are given in Table 1. Col. 3 of the Table 1 gives the pressure drop as applicable for Indian Sea area using the relation $V_{\max} = 14.2 \times \sqrt{\Delta P}$.

When the system is in the Radar range, we consider the radial velocity observed by Radar for intensity estimation (IMD 1976). The radial velocity observed from

Radar is converted to 10 m wind using appropriate conversion factor (Raghavan 1997). When the system is over the coast, the observations from High Wind Speed Recorders (HWSRs) and conventional anemometers are considered for intensity estimation.

3.3 Pressure Drop, Estimated Central Pressure and Pressure of OCI

Pressure drop (ΔP) at the centre is defined as the difference between P_o and P_c, where P_o is the pressure of outermost closed isobar (OCI) and P_c is the pressure at the centre of TC. It is is obtained using the relation $V_{max} = 14.2 \times \sqrt{\Delta P}$ (Mishra and Gupta 1976). The outermost closed isobar (OCI) is defined as the isobar surrounding the centre of the system which is circular or elliptical in nature and beyond which the isobars are either not closed or deformed. As the Bay of Bengal and Arabian Sea are small Ocean basins and there are observations from coast, island, ships and buoys, it is possible on most of the occasions to find out the OCI. Using P_o and ΔP, P_c can be determined. However, further corrections are applied based on available mean sea level pressure (MSLP) values from nearby ships, buoys and island observations.

3.4 Radius of Outermost Closed Isobar (ROCI)

The ROCI is defined as the radius of OCI, if it happens to be circular. If OCI is elliptical, the average of semi-major and semi-minor axes of the ellipse will be the ROCI.

3.5 Radius of 28, 34, 50 and 64 Knots Wind

The structure of TC is described in terms of maximum radial extent of the winds in four geographical quadrants, viz., NW, NE, SW and SE for thresholds of 28, 34, 50 and 64 knots, referred to as R28, R34, R50 and R64. The primary methods for TC wind field estimation by IMD involves satellite based scatterometer estimates, cloud motion vectors, water vapour based wind vectors, wind estimates from brightness temperatures, multiplatform satellite based wind developed by CIRA, estimates from RADAR products and NWP model analyses products (IMD 2013; Mohapatra and Sharma 2015). Consensus analysis that gathers all the available observation and uses synoptic and climatological guidance are utilised to issue best

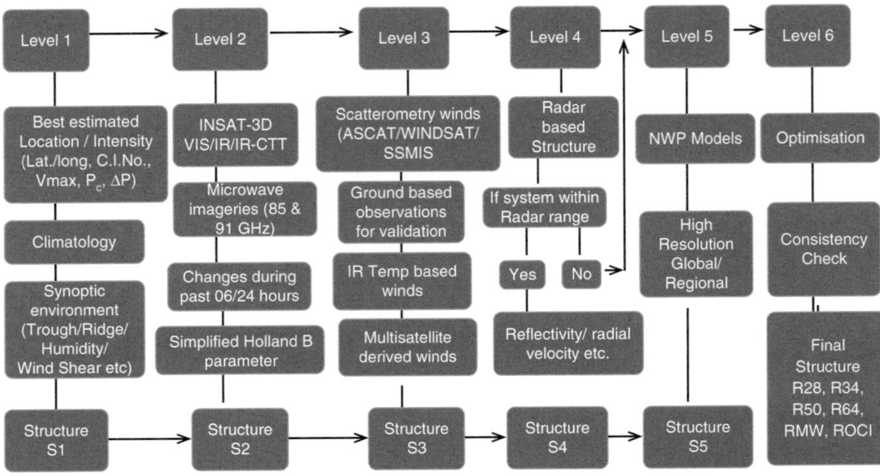

Fig. 4 Steps involved in determination of structure of a TC

estimates of surface wind radii in four geographical quadrants. Various steps involved in determination of structure are shown in Fig. 4.

The climatological guidance is based on the study carried out by Mohapatra and Sharma (2015). Climatologically, it is seen that during cyclonic storm (CS) stage, for an increase in MSW by 5 knots there is an increase in R34 by 15 km. When the system intensifies to a severe cyclonic storm (SCS), for an increase of 5 knots in intensity, the wind radii increases by 25 km and upon intensification into VSCS, there is an increase of 5 km when MSW increases by 5 knots. The detailed classification of TCs over the NIO into CS, SCS and VSCS etc. are given in IMD (2013).

3.6 *Radius of Maximum Wind (RMW)*

RMW is defined as the average distance from the centre of the TC to the location of occurrence of maximum wind in the wall cloud region. RMW can be estimated by the satellite and Radar observations as well as the derived winds from the satellites. Based on the satellite IR imagery with cloud top temperature (CTT) distributions, the zone of maximum reflectivity in the wall cloud region can be determined (Kalsi 2002). The distance from the centre to the location of coldest temperature surrounding the centre determines the radius of maximum reflectivity (RMR). In this process, we assume that the RMW is same as the RMR. As the IR imageries have resolution of 4/8 km, there can be an error of 4/8 km in this process. Further, there can be error in estimation of location leading to error in RMW. As the centre determination is most accurate for the eye pattern followed by CDO, curved band and shear pattern, the accuracy in determination of RMW will also be maximum in eye pattern and minimum in shear pattern of the TC. With respect to the intensity of TC, the confidence in RMW estimate will be lower in case of low intensity storms and will be higher in case of high intensity storms like VSCS Phailin and Hudhud. RMW can be better estimated from Radar, when the system is in the Radar range. In Radar, RMW is defined as RMR (Raghavan 1997, 2013). RMW is also available from the winds derived from multiplatform satellite observations developed by CIRA. Detailed study of CIRA data is given by Knaff et al. (2011). However, there is limitation in estimation of RMW through derived winds as the number of observations in the core region may be less and standard deviation in estimation of RMW is high. As a thumb rule, the RMW is assumed as half of the average radial extension of the core wind around the centre of TC e.g. if a TC is a VSCS (MSW $>$ 64 knots), the average radial extent of 64-knot winds will be considered for calculating the RMW. If it is a severe cyclonic storm (SCS), (MSW $>$ 47 knots) or a cyclonic storm (MSW $>$ 34 knots), the average radial extent of 50- or 34-knot winds respectively will be considered for calculating the RMW.

3.7 *Depth of Convection*

The vertical depth of the system is considered in three categories viz. shallow, medium and deep for the NIO. It is considered as shallow if it is a depression (MSW 17–27 knots), medium if it is a deep depression (MSW 28–33 knots) and deep for TCs (MSW $\geq$ 34 knots).

4 Examples of TC Vital Parameters Derived by IMD

The TC vital parameters are available in textual, graphic and coded format as per the requirement of various users. The TC vital parameters in coded form for TC Phailin and Hudhud are given in Tables 2a and 2b, respectively. The explanation of the codes is given in Table 3. A typical graphic bulletin in association with cyclone, Phailin is shown in Fig. 5. Detailed life history and other characteristics of TC Phailin are available in RSMC New Delhi, 2014.

5 Limitations and Future Scope

In the absence of aircraft reconnaissance, and insufficient buoys/ship observations, radar coverage along the coast and HWSR network, the product has various limitations. The limitations and future scope for improvement of the product are discussed herewith.

Table 2a TC Vital parameters for VSCS, Phailin

	TTTT	Latitude (°)	Longitude (°)	Storm motion	Storm speed (mps)	Central P	Outermost P	ROCI	Vmax	RMW	R34 in four geographical quadrants NE	SE	SW	NW	
YYYYMMDD	UTC	(X10)	(X10)	deg	(X10)	hPa	hPa	km	mps	km	km	km	km	Km	Depth
20131009	0000	130 N	0935 E	320	035	1002	1007	0550	015	075	0000	0000	0000	0000	M
20131009	0600	130 N	0930 E	270	025	1000	1006	0550	015	075	0000	0000	0000	0000	M
20131009	1200	135 N	0925 E	320	035	0999	1006	0360	018	050	0120	0100	0090	0120	D
20131009	1800	140 N	0920 E	320	035	0998	1006	0330	020	045	0130	0110	0100	0130	D
20131010	0000	145 N	0915 E	320	035	0996	1006	0390	023	040	0140	0120	0110	0140	D
20131010	0600	150 N	0905 E	290	045	0984	1006	0330	032	035	0150	0130	0120	0150	D
20131010	1200	155 N	0900 E	320	035	0976	1006	0460	038	030	0180	0160	0150	0170	D
20131010	1800	155 N	0895 E	270	025	0960	1006	0460	048	025	0220	0200	0190	0210	D
20131011	0000	160 N	0885 E	290	045	0946	1006	0460	055	020	0280	0250	0240	0260	D
20131011	0600	162 N	0883 E	320	044	0940	1006	0440	058	020	0300	0260	0250	0270	D
20131011	1200	168 N	0877 E	290	046	0940	1006	0550	058	018	0300	0260	0250	0270	D
20131011	1800	170 N	0870 E	280	012	0940	1006	0500	058	018	0300	0260	0250	0270	D
20131012	0000	175 N	0865 E	290	045	0940	1006	0550	058	018	0300	0260	0250	0270	D
20131012	0600	181 N	0857 E	280	055	0940	1006	0600	058	015	0330	0260	0250	0270	D
20131012	1200	187 N	0852 E	300	043	0940	1006	0700	058	015	0330	0300	0250	0270	D
20131012	1800	195 N	0848 E	330	050	0956	1006	0660	055	025	0330	0320	0250	0260	D
20131013	0000	205 N	0845 E	350	057	0976	1006	0550	038	045	0280	0270	0200	0210	D
20131013	0600	215 N	0840 E	340	061	0996	1006	0470	020	065	0210	0200	0130	0140	D
20131013	1200	225 N	0838 E	350	056	0998	1006	0380	018	080	0200	0190	0120	0130	D
20131013	1800	230 N	0835 E	330	032	1002	1008	0380	015	100					M

Table 2b TC Vital parameters for VSCS, Hudhud

	TTTT	Latitude (°)	Longitude (°)	Storm motion	Storm speed (mps)	Central P	Outermost P	ROCI	Vmax	RMW	NE	SE	SW	NW	
YYYYMMDD	UTC	(X 10)	(X10)	deg	(X10)	hPa	hPa	km	mps	km	km	km	km	km	Depth
20141008	0600	125 N	0925 E	280	032	0996	1004	0280	020	110	120	110	100	120	D
20141008	1200	128 N	0910 E	280	084	0996	1004	0310	020	110	130	120	110	130	D
20141008	1800	132 N	0902 E	290	049	0994	1003	0280	023	124	140	120	110	150	D
20141009	0000	137 N	0892 E	290	061	0990	1004	0400	023	100	170	150	120	170	D
20141009	0600	139 N	0888 E	290	025	0988	1004	0280	028	090	185	160	120	185	D
20141009	1200	141 N	0884 E	290	025	0988	1004	0390	028	080	185	170	110	175	D
20141009	1800	141 N	0879 E	270	028	0988	1004	0310	028	080	185	170	110	185	D
20141010	0000	144 N	0876 E	315	023	0988	1004	0290	031	080	205	170	110	195	D
20141010	0600	148 N	0870 E	315	040	0986	1004	0280	031	070	205	170	120	195	D
20141010	1200	152 N	0867 E	330	027	0982	1002	0390	036	070	205	180	130	185	D
20141010	1800	155 N	0864 E	315	023	0978	1002	0360	038	060	220	180	140	185	D
20141011	0000	159 N	0857 E	280	044	0970	1002	0420	038	050	220	180	140	185	D
20141011	0600	161 N	0851 E	275	035	0966	1002	0390	046	040	230	215	140	200	D
20141011	1200	162 N	0848 E	280	017	0962	1002	0390	048	035	240	215	160	200	D
20141011	1800	164 N	0847 E	330	012	0954	1002	0360	051	030	230	220	160	190	D
20141012	0000	172 N	0842 E	350	052	0950	1002	0410	051	025	230	220	160	180	D
20141012	0600	176 N	0834 E	290	049	0950	1002	0360	051	025	215	205	140	180	D
20141012	1200	180 N	0827 E	280	044	0982	1002	0400	031	050	215	195	125	165	D
20141012	1800	187 N	0823 E	340	044	0987	1002	0330	020	065	170	125	90	155	D
20141013	0000	195 N	0815 E	315	62	0994	1002	0310	015	080					M
10141013	0600	207 N	0815 E	360	66	0998	1003	0280	015	100					M

Table 3 Explanation of the coded form of TC vital

Description of characters			Example
character*4	tcv_center	! Hurricane Center Acronym	IMD
character*3	tcv_storm_id	! Storm Identifier (02B, etc)	01B
character*9	tcv_storm_name	! Storm name	VIYARU
integer	tcv_century	! 2-digit century id (19 or 20)	20
integer	tcv_yymmdd	! Date of observation	130515
integer	tcv_hhmm	! Time of observation (UTC)	1800
integer	tcv_lat	! Storm latitude (*10), always > 0	190 (for 19.0°)
character*1	tcv_latns	! 'N' or 'S'	N
integer	tcv_lon	! Storm longitude (*10), always > 0	0885 (for 88.5°)
character*1	tcv_lonew	! 'E' or 'W'	E
integer	tcv_stdir	! Storm motion vector (in degree)	030 (past 6 h)
integer	tcv_stspd	! Speed of storm movement (m/s × 10)	050 (past 6 h reported in 3 digits)
integer	tcv_pcen	! Min central pressure (mb)	0990 (reported 4 digits)
integer	tcv_penv	! val outermost closed isobar (mb)	1000 (reported 4 digits)
integer	tcv_penvrad	! rad outermost closed isobar (km)	0250 (reported 4 digits)
integer	tcv_vmax	! max sfc wind speed (m/s)	022 (reported 3 digits)
integer	tcv_vmaxrad	! rad of max sfc wind speed (km)	060 (reported 4 digits)
integer	tcv_r15ne	! NE rad of 34 knots winds (km)	0170 (reported 4 digits)
integer	tcv_r15se	! SE rad of 34 knots winds (km)	0130 (−do-)
integer	tcv_r15sw	! SW rad of 34 knots winds (km)	0120 (−do-)
integer	tcv_r15nw	! NW rad of 34 knots winds (km)	0170 (−do-)
character*1	tcv_depth	! Storm depth (S,M,D) X = missing	M

S, shallow (for Dep), M, Medium (for DD), D, Deep (for CS and above) and X, missing
Thus the TC Vital message for the above example is given below
IMD 01B VIYARU 20130515 1800 190 N 0885E 030 050 0990 1000 0250 022 060 0170 0130 0120 0170 M
N.B.:
1. Century and yymmdd are given in one column (column number 4)
2. Latitude of centre of TC is given in 3 digits alongwith N/S (column 6)
3. Longitude of the centre of TC is given in 4 digits alongwith E/W (column 7)
4. Starts from DD stage and in place of storm name give 9 blank spaces
5. Storm identifier is system No. in particular basin and reported as NNB or NNA (reported in 3 digits as 01B)

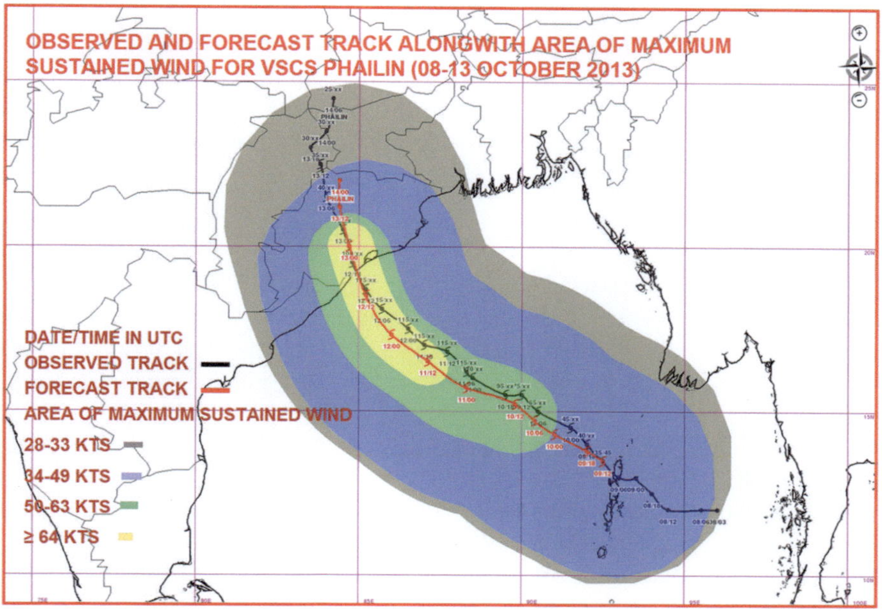

Fig. 5 Typical example of quadrant wind based on 1200 UTC of 9 October 2013 in association with VSCS Phailin

The product is mainly dependent on satellite derived products and satellite observations are subject to various types of errors (Mohapatra et al. 2012) including poor resolution, base map errors, errors due to shifting of orbit of revolution of satellite over the years and parallax error due to viewing angle. Since the cyclone develops over the sea, error is maximum during genesis and over the deep sea. Dvorak technique itself has limitation in estimation of genesis of a TC and low intensity storms (Dvorak 1984).

As regards Radar, we do not have Doppler Weather Radar (DWR) network covering the entire coastal belt of NIO (RSMC New Delhi 2015). Therefore, when the TC comes near the coast, we do not have structure fully dependent on Radar observations. Dvorak technique is also not applicable to the TCs over land and TCs interacting with land surface. There is also subjectivity in the analysis, as the Radar observations are taken in the form of radial velocity and the reflectivity at different heights of the atmosphere (IMD 1976). The same are then converted to find out the wind distribution at 10 m level. The conversion process is based on the mean vertical wind profile as observed over the Atlantic using aircraft (Franklin et al. 2000). It has not been validated over NIO for want of aircraft reconnaissance. Hence there can be error in Radar observations.

The coastal observations are in general taken using ordinary anemometer for maximum sustained wind. However, at the time of landfall high wind speed recorders (HWSR) are required to measure the winds. Only 14 HWSRs are installed along the coast (RSMC New Delhi 2015). Thus at the time of landfall, the HWSR network is not sufficient enough to assess the cyclonic winds correctly.

The improvement in TC Vital parameters will help in better understanding of dynamics of a TC and will reduce error in intensity predictions (Chourasia et al. 2013). To address the above limitations, following observations and analysis tools are to be developed:

1. Modification of Dvorak technique according to the environmental conditions over NIO.
2. Validation of pressure–wind relationship in TCs over the NIO.
3. Development/validation of wind conversion factor for converting 3-min average wind to 1-min average wind (used in Dvorak's technique) and 10-min average wind (as required for preparation of standardised international best tracks archives).
4. Augmentation of Radar network over entire NIO region including WMO/ESCAP member countries.
5. Surface wind observation network over the island stations and along the coast regions may be equipped with HWSRs to assess cyclonic winds correctly.
6. Deployment of scatterometry based satellites.
7. Augmentation of buoys network.

Acknowledgement We thank the Director General of Meteorology, IMD for providing us an opportunity to carry out this work. We also acknowledge the support provided by staff members of Cyclone Warning Division in collection of data.

References

Chourasia, M, Ashrit, R.G. and George, J.P. (2013). Impact of cyclone bogusing and regional assimilation on tropical cyclone, track and intensity predictions, *Mausam*, **64**, 135–148.

Dube, S.K, Poulose, J. and Rao, A.D. (2013). Numerical simulation of storm surge associated with severe cyclonic storm in the Bay of Bengal during 2008–11. *Monsoon*, **64**, 193–202.

Dvorak, V.F. (1984). Tropical cyclone intensity analysis using satellite data. Technical Report (NOAA TR NESDIS 11), National Oceanic and Atmospheric Administration, National Environmental Satellite, Data and Information Service, 47 pp.

Franklin, J.L, Black, M.L. and Valde, K. (2000). Eyewall wind profiles in hurricanes determined by GPS dropwindsondes. NOAA/NWS Report.

IMD (1976). Weather Radar Observations Manual. IMD.

IMD (2003). Cyclone Manual. IMD, New Delhi, India.

IMD (2013). Cyclone Warning in India: Standard Operation Procedure. IMD, New Delhi, India.

Kalsi, S.R. (2002a). Use of satellite imagery in tropical cyclone intensity analysis and forecasting. India Meteorological Department, New Delhi, p 44.

Knaff, J.A, DeMaria, M, Molenar, D.A, Sampson, C.R. and Seybold, M.G. (2011). An automated, objective, multi-satellite platform tropical cyclone surface wind analysis. *Journal of Applied Meteorology and Climatology*, **54**, 624–642.

Mishra, D.K. and Gupta, G.R. (1976). Estimates of maximum wind speed in tropical cyclones occurring in the Indian Seas. *Indian Journal Of Meteorology and Geophysics*, **27**, 285–290.

Mohapatra, M. and Sharma, M. (2015). Characteristics of Surface Wind Structure of Tropical Cyclones over the north Indian Ocean, JESS, accepted for publication.

Mohapatra M, Bandyopadhyay, B.K. and Tyagi, A. (2012). Best track parameters of tropical cyclones over the north Indian Ocean: A review. *Natural Hazards*, **63**, 1285–1317.

Raghavan, S. (1997). Radar observations of tropical cyclone. *Mausam*, **48**, 169–188.

Raghavan, S. (2013). Observational aspects including weather radar for tropical cyclone monitoring. *Mausam*, **64**, 89–96.

RSMC, New Delhi (2014). Report on cyclonic disturbances over the north Indian Ocean during 2013. IMD, New Delhi, India.

RSMC, New Delhi (2015). Report on cyclonic disturbances over the north Indian Ocean during 2014. IMD, New Delhi, India.

Index

© Capital Publishing Company 2017

M. Mohapatra et al. (eds.), *Tropical Cyclone Activity over the North Indian Ocean*,
DOI 10.1007/978-3-319-40576-6

D

E

F

G

H

I

J

K

L

U

V

MIX
Papier aus verantwortungsvollen Quellen
Paper from responsible sources
FSC® C105338

If you have any concerns about our products,
you can contact us on
ProductSafety@springernature.com

In case Publisher is established outside the EU,
the EU authorized representative is:
Springer Nature Customer Service Center GmbH
Europaplatz 3, 69115 Heidelberg, Germany

Printed by Libri Plureos GmbH
in Hamburg, Germany